Environmental Pollution and Management

Environmental Pollution and Management

Edited by **Alfred Muller**

SYRAWOOD
PUBLISHING HOUSE

New York

Published by Syrawood Publishing House,
750 Third Avenue, 9th Floor,
New York, NY 10017, USA
www.syrawoodpublishinghouse.com

Environmental Pollution and Management
Edited by Alfred Muller

International Standard Book Number: 978-1-68286-152-3 (Hardback)

Printed in the United States of America.

Contents

Preface

Pollution is a major threat to our environment. Researchers across the globe are working extensively to devise techniques and methods to mitigate its negative impact. The topics included in this book, such as sustainable use of environmental resources, waste management, effects of pollutants in air, water and soil, environmental quality management, etc., are of utmost significance and bound to provide incredible insights to readers. The extensive content of this text provides the readers with a thorough understanding of the subject. In this book, using case studies and examples, constant effort has been made to aid the sound understanding of the rapid progress made in this significant field.

All of the data presented henceforth, was collaborated in the wake of recent advancements in the field. The aim of this book is to present the diversified developments from across the globe in a comprehensible manner. The opinions expressed in each chapter belong solely to the contributing authors. Their interpretations of the topics are the integral part of this book, which I have carefully compiled for a better understanding of the readers.

At the end, I would like to thank all those who dedicated their time and efforts for the successful completion of this book. I also wish to convey my gratitude towards my friends and family who supported me at every step.

Editor

Impacts of Cement Dust Emissions on Soils within 10km Radius in Ashaka Area, Gombe State, Nigeria

B. B. Amos[1], I. Musa[1], M. Abashiya[1] & I. B. Abaje[2]

[1] Department of Geography, Gombe State University, Gombe, Nigeria

[2] Department of Geography and Regional Planning, Federal University Dutsin-Ma, Katsina State, Nigeria

Correspondence: I. B. Abaje, Department of Geography and Regional Planning, Federal University Dutsin-Ma, P.M.B. 5001 Dutsin-Ma, Katsina State, Nigeria.

Abstract

This study examines the impacts of cement dust emissions on physicochemical properties of soils within 10km radius from the plant (factory) in Ashaka area, Gombe State. Standard procedures used in soil sampling involves collection of soil samples along a transect aligned in the north-east to south-east direction in order to reflect the dominant two opposing air masses direction; the north-easterly (November-April) and south-westerly (May-October) air masses. Physicochemical analysis of the soils was carried out using standard laboratory procedures in the laboratory. Results of the analysis has revealed that the cement dust which contains high calcium has impacted the soils by increased soil pH, calcium (Ca) content, total bases, base saturation and pH dependent cation exchange capacity (CEC). The impact is observed most within the radius of 0-5km from the plant with a diminishing impact from the 5km towards the 10km radius. Similarly, results of the analysis has shown also that there are signs of slight impaction of bicarbonate (HCO_3) and electrical conductivity (EC) arising from the cement dust on the soils. Recommendations were offered to monitor the dust falling on the soils through trapping and utilizing the dust emissions to cement.

Keywords: atmosphere, cement dust, impact, plant, soils

1. Introduction

The impact of the atmospheric pollution on the ecosystems was demonstrated at several times (Bliefert & Perraud, 2001). Otherwise, this form of pollution is caused by industrial activities including the cement industry. Zerrouqi *et al* (2008) reported that the main impacts of the cement activity on the environment are the broadcasts of dusts and gases. These particles or dusts are very numerous and varied. There are basically two types of particles thus: primary particles that are cleared directly in the atmosphere and secondary particles that are formed in the atmosphere following chemical transformation. The particles can enter into soil as dry and humid deposits and can undermine its physiochemical properties (Hosker & Lindberg, 1982). Similar studies have revealed that atmospheric particles can lead to the reduction of biodiversity and the quality of goods and services offered by the ecosystems. Indeed, the dusts can be emitted at every stage of the manufacturing process of the cement production.

Studies revealed that changes in soil properties have been associated with environmental alteration resulting from human activity (Ibanga, 2008). Cement contacts the soil surface and its constituents usually alters the physical and chemical constituents of the soil. Cement has high carbonate content, the dust tends to be highly alkaline. Therefore, it is revealed that soil contaminated by cement will have high pH. The biological, physical and chemical properties of soil, such as water content, electrical conductivity, and pH, were all found to be affected when treated by raw materials of cement (Khan, 1996).

Agriculture remains the central economy activity of the people of Ashaka area with more than 70% of the population engaged in farming activities (the cultivation of cereal crops) such as millet, sorghum, maize and beans. Farmers of the area are also involved in the rising of livestock like sheep, goats and cattle. Soil is the medium through which these activities are made possible. Therefore sustainable soil management is requisite to sustain agriculture (Julio & Carlos, 1999).

Previous studies have reported that chemical and physical degradation affect most of the present agricultural land in Africa in general and of the area in particular. The soils have poor nutrient retention capacity, and many are heavily leached and eroded (Stoorvogel & Smaling, 1990; Smaling, 1993).

Before the establishment of Ashaka cement factory in 1979 there was no environmental impact assessment in order to establish baseline data for the purpose of monitoring the effects of the industrial activities on the environment. The impact of the activities is glaring on the ecosystem components: air, soils, water and vegetation (Adefila, Malgwi, & Balarabe, 2004).

The impact of cement dust was reported to have effects on the microbial population and other soil properties. In comparison to gaseous pollutants, relatively little is known about the effects of particulate pollutants on vegetation, soil microbial population and other soil properties. The determination of soil physical and chemical properties are very important parameters in monitoring environmental pollution. Adefila *et al* (2004) in their study of some elements within the vicinity of the study area reported that the degree of impaction of the cement dust on the soils has not reached the critical stage of soil degradation. However, the soil pH is at the border in the highly and moderately impacted areas. Further emissions of the cement dust will bring about soil degradation. Table 1 to 2 provide details of the dust emissions monitored over a period of time.

Table 1. Dust Emissions from stationery sources

Source	FEPA Limit	Dec '03	Jul '03	Dec'02	Jul '02	Jan '02	Oct '01
No. 1 Kiln	600	980**	320	10,523	NIO	1570	1,700
No. 2 Kiln	600	740**	782	18,000	1,100	750	NIO
Cement Mill 1	500	425	-	-	1,290	535	420
Cement Mill 2	500	465	-	-	NIO	410	425
No. 1 Packer	500	510	-	530	NM	530	510
No. 2 Packer	500	580	-	470	NM	470	580

Source: Adefila *et al* (2004)

Table 2. Atmospheric dust deposition

Sampling Stations	Approx. Distance from Source (km)	Atmospheric Burden (mg/m3)			
		Jun '04	Dec '03	Jul '03	Dec '02
Clinic	0.5	0.110	0.142	0.139	0.0314
CAB	0.3	0.147	0.155	0.154	-
Workers' Village	1.5	0.122	0.148	0.142	0.025
Management Estate	4.0	0.06	0.013	0.008	0.003
Power House	0.1	0.137	0.023	0.016	0.006
Quarry Pit Office	2.0	0.067	0.072	0.065	0.078
Jalingo	2.0	0.146	0.145	0.121	-
Bajoga	10.0	0.08	0.120	0.098	0.0017
Badabdi	5.0	0.07	0.151	0.178	-

Source: Adefila *et al* (2004)

Long-term food productivity is threatened by soil degradation, which is now enough to reduce yields on approximately 16% of the agricultural land, especially cropland in the area. It is estimated that losses in productivity of cropping land if the limestone mining activity continuous unchecked are in the order of 0.5-1% annually. Therefore there is a need for monitoring and evaluation of the activity of limestone mining in order to mitigate the possible negative effects of the activity on the environment (Adefila *et al*, 2004). The main objective of this research is to assess the impact of the dusts given out by a cement factory on the physicochemical characteristics of the soils within 10km radius from the plant in the study area.

2. Study Area

The study area (Figure 1) is located approximately within Latitude $10^0 50'N$ to $10^0 60'N$ and Longitude $11^0 25^0E$

to 10^0 35'E which is some 112km North of Gombe. This makes the area within the northern part of the present Gombe State and close to its border with Yobe State. Figure 1 shows the location of the study area. The population of the area is about 236,087 (Federal Republic of Nigeria, 2010).

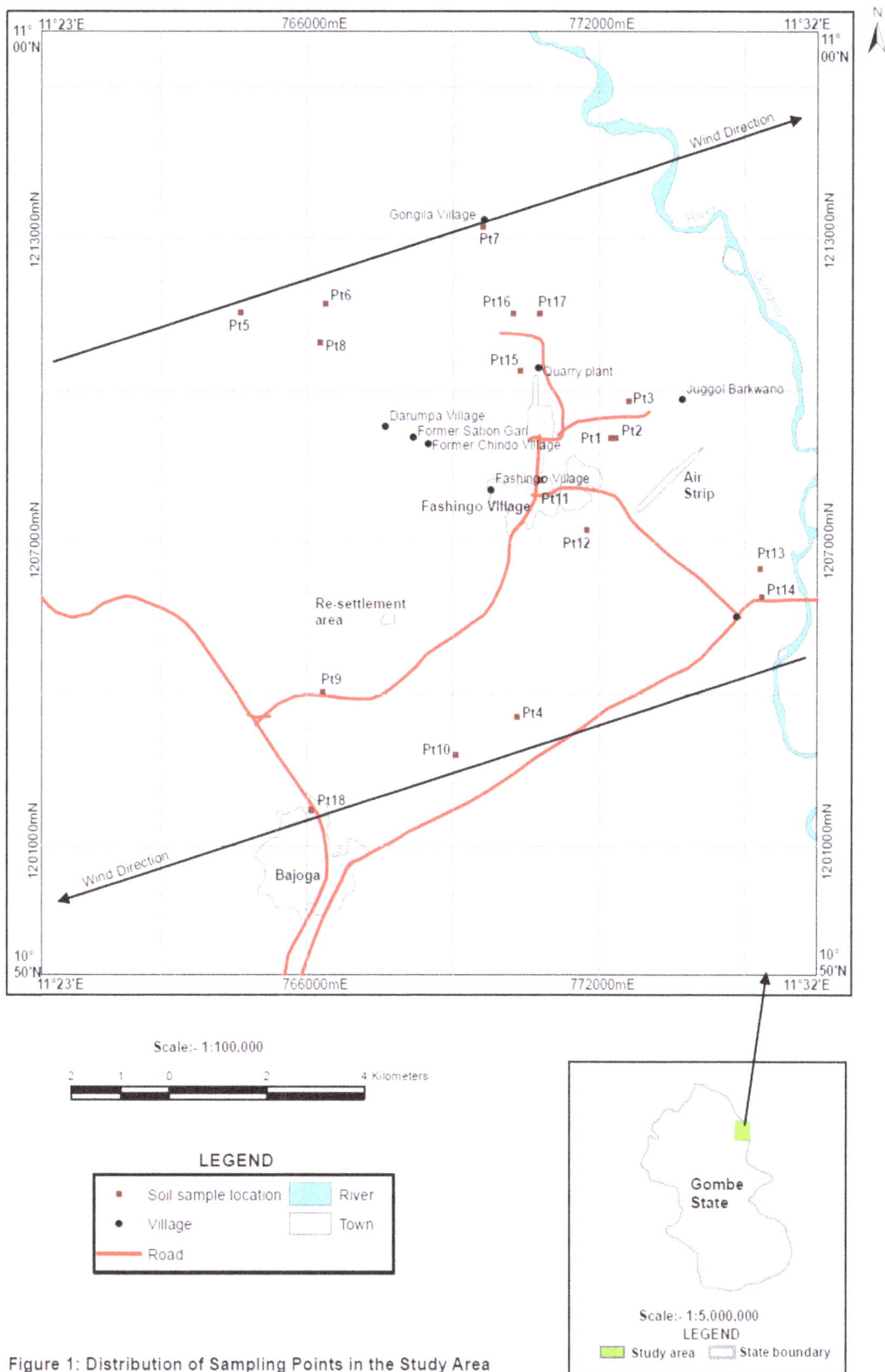

Figure 1: Distribution of Sampling Points in the Study Area

The climate of the area is the seasonally wet and dry type, classified by Koppen as Aw climate (Abaje, Ati, Iguisi & Jidauna, 2013). Like the rest of Nigeria, it is dominated by two air masses. These bring with them the rain bearing south-westerly winds and the cold, dry and dusty North-easterly wind, locally known as the ''Harmattan''. At different times of the year, one or the other of the winds prevail and the area experiences either rainfall or the dry harmattan depending on the advance or retreat of the other (Nyong & Kanaroglou, 1999). The types of soils in the study area are closely related to the parent rock. Where the parent rock is homogenous relatively simple soil association are recognized. The shale give rise to grey heavy loams and clay derived from homogenous mudstones give rise to olive brown clay loam or loamy sand. The dominant soil type is grey mottled, sand and loams with some grey clay. The surface texture of the soils in the study area range from loamy sand, loam, sandy loam, sandy clay loam and sandy clay. The whole area is covered by Sudan vegetation with the density of trees and other plants decreasing as one move northwards (Abaje, 2007; Abaje, Ati & Iguisi, 2012).

3. Material and Methods

The Physicochemical properties of the soils were studied within the radius of 10km from the plant to assess the impact of the cement dust on the soils. Soil samples were mainly collected along a transect aligned in the north east to south east direction. The Global Positioning (GPS) system Garmin 76 with 15m accuracy was used to determine the sampling points. Soil samples were collected at two depths using auger at 0-30 and 30-50cm depths respectively. A total of 46 soil samples were collected at 23 locations in the study area. Soil samples collected from the field were air-dried, grinded and sieved to remove materials larger than 2mm diameter. The less than 2mm fraction was analyzed for the following parameters: Particle size distribution by hydrometer method; pH in water and 0.01 $CaCl_2$ solution at 1:25 soil/liquid ratio using pH meter; Exchangeable bases (Ca, Mg, K, Na) using Ammonium Acetate (1N NH_4OAc) solution; Potassium and Sodium were read on a flame photometer; Calcium and magnesium were read on the atomic absorption spectrophotometer (AAS); exchangeable acidity($H^+ + Al^{3+}$) by leaching of soil samples by 1MKCL followed by titration with NAOH; Effective cation exchange capacity (ECE) by summation of total exchangeable bases and extractable acidity; Cation exchange capacity by the neutral 1N NH_4OAc solution saturation method; Base saturation equal Total exchangeable bases/CEC multiply 100; Electrical conductivity (EC) was determine on a 1:2 soil water saturation extract using conductivity meter; organic carbon by the dichromate wet combustion method; total nitrogen (N) by the Microkjeldah technique; available phosphorus (Ap) by Bray No.1 method and chloride(Cl) was determined from soil saturation extract titrated with silver nitrate.

4. Results and Discussion

The composition of the dust generated from the plant has been analyzed and presented in Table 3. The most important components of the dust that can affect soil properties are oxides of Ca, Mg, Fe, Na and Al in order of significance. The impact of Ca has been significant on the soil properties such as pH and base saturation.

Table 3. Composition of dust around the factory (Weight %)

S/No	Component	Weight %
1	SiO_2	14.24
2	NaO	0.06
3	Al_2O_3	4.11
4	Fe_2O_3	1.89
5	SO_3	0.54
6	MgO	0.54
7	CaO	41.71

Source: Field Analysis, 2013

The results of the physical and chemical analysis are presented in table 4. The dominant texture of the soils are sandy clay loam, sandy loam and pockets of loam sand and clay loam. The direct impact of the dust on the soil texture cannot be assessed without the pre-plant evaluation of the soil texture. However, the different textural class vary in their accumulation and retention of the dust from the plant. The amount of Ca and its impact on the soil pH are clearly evident in the sandy clay loam soils which are fine as compared to the loam sand soils which are coarse.

Table 4. Results of physical and chemical analysis

Sample Depth (cm)	pH H$_2$O	CaCL$_2$	Ca	Mg	K	Na	Total Bases	Al + H	ECEC	CEC
1	Lat. 10^0 55 637N Long. 11^0 29 433'E									
0-30	7.0	6.2	6.8	3.2	0.59	0.32	10.91	0.10	11.01	12.30
30-50	6.8	6.2	7.1	3.4	0.21	0.12	10.83	0.05	10.88	12.00
2	Lat. 10^0 55 637N Long. 11^0 29 492'E									
0-30	7.1	6.4	5.8	3.0	0.79	0.08	9.67	0.05	9.72	9.90
30-50	7.2	6.4	6.6	2.4	0.46	0.08	9.54	0.05	9.59	9.00
3	Lat. 10^0 56 025N Long. 11^0 29 643'E									
0-30	7.2	6.5	5.8	2.6	0.44	0.07	8.91	0.10	9.01	11.50
30-50	7.1	6.4	4.0	2.6	0.16	0.08	6.84	0.10	6.94	12.50
4	Very shallow Soil									
0-20	7.5	6.5	4.0	1.2	0.47	0.04	5.71	0.05	5.76	6.10
5	Within the plant 500m South of the factory									
0-30	7.5	6.6	5.6	2.4	0.79	0.07	8.86	0.05	8.91	7.00
30-50	7.6	6.7	4.2	1.6	0.79	0.05	6.64	0.05	6.69	7.40
6	Gongila river bank by the bridge									
0-45	7.4	6.3	5.6	3.4	0.17	0.12	9.28	0.05	9.33	15.30
45-85	7.1	5.7	4.0	1.0	0.15	0.05	5.20	0.05	5.25	10.00
85-105	6.7	5.7	10.2	3.4	0.46	0.13	14.19	0.05	14.24	18.40
7	Lat 10^0 52'658N Long 11^{0}28'350E									
0-30	7.2	6.5	5.6	3.2	0.38	0.01	9.19	0.10	9.29	12.80
30-50S	7.3	6.4	8.8	3.0	0.44	0.01	12.25	0.01	12.35	13.40
8	Quarrying pit (profile)									
0-35	6.6	5.4	4.4	1.6	0.22	0.05	6.27	0.10	6.37	11.80
35-75	7.2	6.4	9.0	4.3	0.23	0.04	13.57	0.10	13.67	22.30
75-200	6.9	5.8	6.4	2.6	0.21	0.10	9.31	0.10	9.41	22.50
200-280	6.2	5.4	9.0	3.8	0.16	0.05	13.08	0.10	13.19	17.00
9	Lat. 10^0 57' 013N Long. 11^0 25' 289'E									
0-30	6.8	6.0	3.2	1.2	0.28	0.05	4.73	0.10	4.83	4.40
30-50	6.6	5.8	2.4	1.0	0.21	0.05	3.66	0.10	3.76	3.60
10	Lat. 10^0 57' 101N Long. 11^0 26' 246'E									
0-30	6.4	5.7	2.5	1.0	0.25	0.07	3.82	0.10	3.92	8.70
30-50	5.5	5.0	2.8	1.0	0.14	0.05	3.19	0.10	3.92	7.80
11	Lat. 10 57' 90 7N Long. 11^0 28' 009'E									
0-30	6.5	5.8	3.2	0.8	0.39	0.10	4.49	0.10	4.59	7.10
30-50	6.5	5.8	3.69	0.8	0.16	0.14	4.79	0.10	4.89	5.70
12	Lat. 10 56' 681N Long. 11^0 26' 179'E									
0-30	5.8	4.9	4.2	1.2	0.20	0.22	5.82	0.10	5.92	15.70
30-50	6.0	4.8	4.4	1.2	0.14	0.13	5.87	0.10	5.97	19.00

Depth										
13	Lat. 10 52' 940N Long. 11° 26' 179'E									
0-30	6.0	5.1	1.6	1.0	0.140	0.08	2.85	0.05	2.87	3.00
30-50	5.7	5.2	3.8	1.6	0.16	0.10	5.66	0.05	5.71	3.70
14	Lat. 10 52' 263N Long. 11° 27' 664'E									
0-30	6.1	5.5	6.60	2.8	0.12	0.17	9.69	0.05	9.74	10.30
30-50	5.2	4.0	3.4	0.80	0.11	0.44	4.75	0.40	5.15	13.00
15	Lat 10° 55' 195N Long 11° 28' 634E									
0-30	7.1	6.1	4.0	1.0	0.17	0.13	5.30	0.10	5.40	3.50
30-50	7.0	6.2	3.6	1.2	0.14	0.18	5.12	0.05	5.17	20.90
16	Lat 10° 54' 659N Long 11° 29' 152E									
0-30	7.1	6.3	3.4	0.80	0.64	0.12	4.96	0.05	5.01	6.30
30-50	7.0	6.2	3.6	1.2	0.14	0.18	5.12	0.05	5.17	20.90
17	2km from workers village to Ashaka Gari									
0-30	6.6	5.1	2.80	1.20	0.09	0.31	4.40	0.10	4.50	11.60
30-50	6.5	5.6	4.6	1.4	0.1	0.41	6.51	0.10	6.61	10.70
18	Lat. 10° 54'225N Long. 11° 31'130'E									
0-30	5.8	4.8	4.0	1.0	0.1	0.19	5.29	0.10	5.39	24.90
30-50	5.5	4.2	2.3	1.0	0.17	0.23	3.70	0.10	3.80	4.80
19	Lat.10° 53' 924N Long. 11°31'145E									
0-30	6.3	5.2	2.4	1.0	0.17	0.12	3.69	0.05	3.74	4.81
30-50	6.0	4.8	1.6	0.8	0.15	0.13	2.68	0.05	2.73	5.40
20	Lake Northwest of Plant (Shallow soil)									
0-30	6.7	0.5	9.8	3.0	0.88	0.38	14.06	0.05	14.11	16.30
21	Lat. 10° 56' 980N Long. 11° 28' 345E									
0-30	7.3	6.8	7.2	3.6	1.52	0.16	12.48	0.05	12.53	14.10
22	Lat. 10° 56' 982N Long. 11° 28' 643E									
0-30	7.5	6.8	6.2	2.8	0.38	0.14	9.52	0.05	9.57	12.90
30-50	7.4	6.7	3.4	1.8	0.23	0.14	5.57	0.05	5.62	14.10
23	Lat. 10° 51' 681'N Long. 10° 51' 681N									
0-30	6.1	5.5	2.4	1.0	0.13	0.05	3.58	0.05	3.63	6.20
30-50	5.7	4.7	2.2	1.2	0.22	0.13	3.75	0.10	3.85	6.10

Source: Field Analysis, 2013

The results of the analysis revealed that the highly impacted areas fall within 500m radius from the plant and moderately impacted fall between 500m to 1km radius from the plant. Slightly impacted areas found within 1km to 5km radius from the plant, whereas very slightly to non-impacted areas fall between 5km to 10km radius from the plant.

The pH of the soils range from 6.8 to 7.6, 6.7 to 7.5, 6.7 to 7.2, and 5.5 to 6.8 in Zones 1 to 4 respectively. From the plant pH decreases radially from zone 1 to 4. The high pH in zones 1 and 2 has a direct bearing to the high dust received from the plant. The cement dust has high Ca content and has made the soils in zones 1 and 2 slightly alkaline. Further accumulation will lead to strongly alkaline unless the Ca is leached out of the plant root zone. Under strong alkaline condition, the availability of certain plant nutrients such as phosphorus and boron can be reduced to deficiency level. It can also result in low levels of micronutrients (Fe, Mn, Zn, Cu and Co) that can affect plant growth and overall performance is constrained under high pH. The activities of soil micro-organism will also be curtailed and soil structural deterioration increased.

The principal saturating cation in the cement dust and all the soils was Ca in the soils. In the soil, it ranged from zones 1 to 4 from 3.4 to 7.2, 3.0 to 9.8, 1.6 to 9.0 and 1.6 to 4.4 cmol$(+)$kg^{-1}. The calcium content is highest around the plant in zones 1 and 2. There is however high calcium content in soils of zones 3 and 4 which are far away from the plant. This is partly because of high Ca content of the soils developed from shale. There is clear incidence of Ca addition from the dust generated from the plant, which is reflected on the pH of the soils of the different zones. The next saturating cations are Mg and K respectively. The Mg and K contents of the soils follow the same trend to that of Ca. Zones 1 and 2 are highly impacted and decreases in zones 3 and 4 respectively. The exchangeable Na contents of the soils do not clearly indicate the impact arising from the dust.

Exchangeable acidity of all the soils of the four zones are low ranging between 0.5 to 0.10 cmol$(+)$kg^{-1}. This is expected as the pH of the soils do not indicate high levels of exchange acidity. High Ca content leads to surface crusting which increase bulk density, reduced infiltration at the expense of runoff resulting in increased erosion and loss of topsoil.

The total exchangeable bases in zones 1 and 2 are generally high. The values also represent the effective cation exchange capacity (ECEC) which seems to be influenced by the cement dust from the plant. In zones 1 and 2, they ranged from 0.57 to 12.48 and 4.20 to 14.06 cmol$(+)$kg^{-1}. In zones 3 and 4 they ranged from 2.68 to 13.57 and 2.82 to 5.82 cmol$(+)$kg^{-1} respectively. There is a clear decreasing trend from the plant outwards, which can be inferred to reflect the diminishing distribution of the cement dust with distance.

Soils of the highly and moderately impacted zones have higher base saturation. In zones 3 and 4, they have lower base saturation. The high base saturation of zones 1 and 2 are related to the higher pH and Ca content received from the cement dust. There is no clear impact of the availability of phosphorus due to the cement dust. Organic carbon, total, carbonate and bicarbonate, chloride and electrical conductivity do not show apparent relationship with the cement dust. There is a slight relationship of the sites close to the plant with HCO$_3$ and electrical conductivity.

5. Summary and Conclusion

The impact of the cement dust falling on the soils within and around Ashaka area has been evaluated. The impact is glaring on the soils within the radius of 0-5km. This has affected the physical and chemical properties of the soil. Among the physical properties is the surface soil structural deterioration leading to surface crust. The surface crust leads to increased bulk density which can limit seedling emergence and early root development. The chemical impacts on the soils are evidenced in increased soil pH, and exchangeable cations especially Ca and Mg. The high pH has not reached the critical level. However, it is approaching the zone.

There is need to put into practice mitigation measures to check the influx of the cement dust in the soils. This calls for reduction in the amount of cement dust into the air. There is also need for the company to embark on extension programme to the farmers within the vicinity of the plant. This will go a long way in reducing the impact of cement dust on their farms for increase crop yield through better land management practices.

References

Abaje, I. B. (2007). *Introduction to soils and vegetation*. Kafanchan: Personal Touch Productions.

Abaje, I. B., Ati, O. F., & Iguisi, E. O. (2012). Recent trends and fluctuations of annual rainfall in the sudano-sahelian ecological zone of Nigeria: risks and opportunities. *Journal of Sustainable Society, 1*(2), 44-51.

Abaje, I. B., Ati, O. F., Iguisi, E. O., & Jidauna, G. G. (2013). Droughts in the sudano-sahelian ecological zone of Nigeria: implications for agriculture and water resources development. *Global Journal of Human Social Science (B): Geography, Geo-Sciences & Environmental, 13*(2), 1-10. http://dx.doi.org/10.5539/jgg.v6n2p103

Adefila, S. S., Malgwi, W. B., & Balarabe, M. L. (2004). Ashaka environmental auditing report. pp 46-54.

Bliefert, C., & Perraud, R. (2001). Chimie de l'environnement air, eau, sols, déchets. *De Boeck Diffusion*, Paris, 359-365.

Federal Republic of Nigeria. (2010, April). Federal Republic of Nigeria 2006 population and housing census. Priority table Vol.III. Abuja: National Population Commission

Hosker, J. R., & Lindberg, S. E. (1982). Review: atmospheric deposition and plant assimilation of gases and particles. *Atmos Environ., 16*, 889-910. http://dx.doi.org/10.1016/0004-6981(82)90175-5

Ibanga, I. J., Umoh, N. B., & Iren, O. B. (2008). Effects of Cement Dust on soil Chemical Properties in the Calabar Environment, South Eastern Nigeria. *Communication in Soil Science and Plant Analysis, 39*(3-4),

551-558. http://dx.doi.org/10.1080/00103620701826829

Julio, H., & Carlos, B. (1999). Estimating rates of nutrient depletion in soils of agricultural lands of Africa. *Tech. Bulletin-T.48.* Alabama: U.S.A.

Khan, M. R., & Khan, M. W. (1996). The Effect of Fly Ash on plant Growth and Yield of Tomatoes. *Environmental Pollution, 92,* 105-111. http://dx.doi.org/10.1016/0269-7491(95)00098-4

Nyong, & Kanaroglou. (1999). The influence of water resources and their location on rural population distribution in north eastern Nigeria. *Journal of Environmental Sciences, 3*(1), 46-54.

Smaling, E. M. A. (1993). An agro-ecological framework for integrating nutrient management, with special reference to Kenya. Ph.D. Thesis, Wageningen Agricultural University, Wageningen, Netherlands.

Stoorvogel, J. J., & Smaling, E. M. A. (1990). Assessment of soil nutrient depletion in Sub-Saharan Africa. Report 28. The Winand Staring Center, Wageningen, Netherlands.

Zerrouqi, Z., Sbaa, M., Oujidi, M., Elkharmouz, M., Bengamra, S., & Zerrouqi, A. (2008). Assessment of cement's dust impact on the soil using principal component analysis and GIS. *Int. J. Environ. Sci. Tec., 5,* 125-134. http://dx.doi.org/10.1007/BF03326005

Chemical Speciation: A Strategic Pathway for Insightful Risk Assessment and Decision Making for Remediation of Toxic Metal Contamination

Haruna Adamu[1,3], Leke Luter[1,4], Mohammed Musa Lawan[1,5] & Bappah Adamu Umar[2,6]

[1] Department of Chemistry, University of Aberdeen, UK

[2] Department of Geology and Petroleum Geology, University of Aberdeen, UK

[3] Department of Environmental Management Technology, Abubakar Tafawa Balewa University, Nigeria

[4] Department of Chemistry, Benue State University, Nigeria

[5] Department of Chemistry, Yobe State University, Damaturu, Nigeria

[6] National Centre for Petroleum Research and Development, Energy Commission of Nigeria, Abubakar Tafawa Balewa University Research Centre, Nigeria

Correspondence: Haruna Adamu, Department of Chemistry, University of Aberdeen, Scotland, UK; Department of Environmental Management Technology, Abubakar Tafawa Balewa University, Bauchi State, Nigeria. E-mail: aisonhardo2003@yahoo.com

Abstract

Industrial and anthropogenic activities have resulted in high levels of metallic contaminants in the environment, thus creating imbalance in the biotic and abiotic regimes of the ecosystem. This has remained as a stabbing problem in the mind of environmental and agricultural scientists, since metallic contaminants unlike biodegradable contaminants are persistent in affecting the biophysical population of our environment. This problem requires an insightful assessment before the best remediation option can be selected appropriately. In this paper, we present an overview of chemical speciation and its adaptations in environmental cleanup for achieving result oriented remediation technique.

Keywords: metallic contaminants, speciation, remediation

1. Introduction

Contamination of the surface of the earth by metallic contaminants from human activities has been significant (Markham, 1994) and thus, regimes of ecosystem of the aquatic and terrestrial environment of the earth's surface are burdened with a range of metallic pollutants/contaminants. For example, many materials were processed in significant quantities to support the developing technologies underpinning the growth of the developed countries (Markham, 1994; Macklin, 1992). The extraction and utilization of metals as major components of infrastructure and in high value goods became the cornerstone of the development of human civilization (Wedepol, 1991).

With the advent of the industrial revolution in Western Europe, e.g. the availability of steam powered devices, the utilization of fossil fuels increased the capacity of society to manipulate the Earth's resources and process materials (Wedepol, 1991). The net impact was to advance civilization and enhance socio-economic well-being of society and as a direct result of the incomplete efficiency of these processes, release of residual materials to the atmosphere, aquatic and terrestrial environment occurred (Macklin, 1992).

Through human activities, contamination of the Earth's surface by metallic contaminants developed from localized problems associated with mining and initial ore processing, e.g. like the Zamfara Pb poisoning in northern Nigeria (WHO Lead NGR, 2010), through to large scale manipulation and refining, construction, manufacture and finally to waste disposal. This, coupled with the focusing of population center into industrialized cities, resulted in an increased burden on the environment (Lowe & Bowlby, 1992). The consequence of this has increased awareness of human and industrial impacts on environmental systems at the local and global scale. This has resulted in efforts to manage and improve the degraded quality of aquatic and terrestrial systems (Ozdes et al., 2011; Thornton, 1996; Cairney, 1995; Bora, 1998).

The increase in human stress on sensitive surface environments requires the development of reliable management options for soil and water systems, which are often contaminated with a mix group of physical and chemical components. Contamination from processing residues, direct deposition and accidental releases of organic and metallic species create a wide spectrum of environmental hazards (Hagelstein, 2009; Cairney, 1995; Ferguson et al., 1998). An in-depth understanding and appropriate remediation of these situations would rely on an intimate mix of science, technology and socio-economic factors.

It is within this content that the speciation of metallic contaminants provides both the potential hazard (mobility, reactivity, toxicity) (Centeno, 2003) and thus, the pathway to a technological acceptable solution (remediation option). The factors influencing the speciation and changes with both time and environmental conditions provide the greatest challenges and opportunity for environmental science and technology. While the contamination of the surface of the earth by human intervention is a relatively old phenomenon, the evaluation, assessment and remediation of negative impact are relatively recent (Morgan, 2008; Khan et al., 2004; Cairney, 1995; Johnson, 1993; Bernhard et al., 1996). As such, this paper provides an overview of the content associated with metallic contamination and highlights the significant role of metal speciation in risk insight of contamination scenario and achieving result-oriented remediation (if contamination necessitates).

2. Definition of Chemical Speciation

The determination of distinct chemical species often referred to as speciation analysis is now widely acknowledged to be of vital importance in environmental chemistry (Hill, 1997). The term chemical speciation may encompass both functionally defined speciation, i.e. the determination of species that are, for example, available to plants or present as exchangeable forms and operationally defined speciation which refers to the determination of extractable forms of an element (Hill, 1997). While it is often possible to define a particular compound or oxidation state when dealing with solutions, for example, natural waters, it is far more difficult to characterize the actual chemical form of an element in solids such as soils and sediments (Hill, 1997). Thus, speciation tends to be defined somewhat differently by research workers to reflect their field of study. However, one of the most comprehensive formal definitions of speciation is the one recommended by the International Union of Pure and Applied Chemistry (IUPAC), which states that speciation is the process yielding evidence of atomic or molecular form of an analyte (Lobinski & Szpunar, 1999). As such, this definition was subsequently elaborated conspicuously as the specific form of an element defined as to electronic or oxidation state, complex or molecular structure and isotopic composition (Templeton et al., 2000). In addition, International Union of Pure and Applied Chemistry (IUPAC) has recently evaluated and provided a definition of speciation as an analytical activity of identifying and/or measuring the quantities of one or more individual chemical species in a sample; the chemical species are specific forms of an element defined as to isotopic composition, electronic or oxidation state, and/or complex or molecular structure; the speciation of an element is the distribution of an element amongst defined chemical species in a system (Clough et al., 2012). With this elaboration, chemical speciation has now clearly demarcated itself from chemical fractionations, which were previously interchangeable and confusing.

The determination of such specific chemical entities is of course not new to analytical chemists. For example, the determination of NO_3^-, NO_2^-, NH_4^+, and NH_3 where the nitrogen is characterized into its most environmentally important forms is long established (Hill, 1997). However, the characterization of metallic contaminants does not have the same history, although with the increased awareness of the importance of chemical speciation in terms of toxicity, mobility, bioavailability, reactivity and persistence in the environment, a range of sensitive yet specific analytical techniques have now been developed to address a wide range of environmental complex problems.

3. Analytical Approaches to Speciation Studies

Many analytical techniques have been developed for chemical speciation of elements in the environmental samples: typically, sequential leaching methods (Koschinky & Hein, 2003), hyphenated techniques such as GC-ICP-MS (analysis of organometallic elements) (Hirata et al., 2006), X-ray spectroscopic techniques (Ohnuki et al., 2005). These methods of analysis provide significant information on the chemical forms of elements in the environment (Hirose, 2006). In terms of analysis, it is possible to identify and quantify species in environmental samples. Speciation studies on environmental samples were initially focused mainly on separation of specific elemental species is groups of species (Ebdon et al., 1987; Harrison & Rapsomaniskis, 1989). Species separation is achieved mainly by one of the following well known techniques: liquid chromatography (LC), gas chromatography (GC), capillary electrophoresis (CE), and gel electrophoresis (GE) (Hill, 1997). The choice is determined by the chemical properties of the species, the available skills and infrastructure in the laboratory, and last but not the least, the available resources.

However, during the last two decades substantial progress was made in incorporating separation methods with powerful detection methods for reliable quantitative measurements. Majority of work performed coupling GC with atomic spectrometry has so far been achieved using flame spectrometry as a detector (Webster & Karmahan, 1992). This approach has been used most commonly for analytes that are present in relatively high concentration e.g. determining Organo-Lead species in fuel or roadside dust (Hill, 1997). A potential limitation of using GC is that often the sample must be derivatised to make it volatile enough for analysis. This can greatly increase sample preparation time, which may cause loss of analyte and uncertainty about the identity of the original species in the sample. To deal with such problems, high performance chromatography (HPLC) is one of the most common methods used for separation of non-volatile analytes and has been extensively coupled to atomic spectroscopy for quantification. There are a large number of resin supports that are being in use as the stationary phase with anion exchange, cation exchange, size exclusion, chelating and reversed phase, e.g. octadecylsilane which is often used in conjunction with an ion pairing reagent such as tetrabutylammonium phosphate, diethyl dithiocarbamate or 8–hydroxyquinoline to enhance separation of analytes. In this way, analytes with different oxidation states e.g. Cr^{3+} and Cr^{6+} are separated and quantified (Peter, 2001). In the same vein, arsenic has been speciated into As^{3+}, As^{5+}, monomethyl arsonic acid, dimethyl arsinic acid and arsenobetaine using an anion exchange resin with an ammoniacal potassium sulphate mobile phase (Hill, 1997; Ebdon et al., 1987). In addition, the HPLC separation of arsenic species has been coupled with online microwave digestion and hydride generation to facilitate the direct determination of reducible and non–reducible forms of arsenic (Ure & Davidson, 1995). The other convenient method for on-line coupling are inductively coupled plasma atomic emission spectrometry (ICP-AES), which sometimes referred to as ICP–optical emission spectrometry (ICP-OES) and inductively coupled plasma mass spectrometry (ICP-MS). ICP-MS has enabled multi-elemental detection, isotopic discrimination with extremely high sensitivity. Moreover, HPLC and GC were successfully coupled to ICP-MS and now referred to as hyphenated technique for speciation analysis. This technique is now the most frequent used analytical technique for speciation purposes. For instance, Fan et al. (2012) determined the concentration of Cu and its chemical forms in soil extracts using inductively coupled plasma optical emission spectrometry (ICP-OES).

Similarly, eletroanalytical method, for example, anodic stripping voltammetry (ASV) and other electronalyical techniques such as amperometry potentiometry have also been applied for quantification of the various oxidation states of an element. (Fe^{2+}/Fe^{3+}, Cr^{6+}/Cr^{3+}, Tl^{3+}/Tl^{+}, Sn^{4+}/Sn^{2+}, Mn^{4+}/Mn^{2+}, Sb^{5+}/Sb^{3+}, As^{5+}/As^{3+}, and Se^{6+}/Se^{4+}) its organometallic species, or metal complexes (Harrison & Rapsomaniskis, 1989; Caroli, 1996; Quevanviller & Maier, 1994).

Of recent, various mathematical modeling and computer aided programs have been developed for speciation analysis. For example, in many studies in which metal speciation in soil solution is connected to metal uptake by target organism, the Free Ionic Activity Model (FIAM) was used (Campbell, 1995). This model was later extended into the Biotic Ligand Model (BLM) in order to include the influence of competing cations like Ca^{2+} and H^{+} (Di-Toro et al., 2001). With more advances in this aspect, Donnan Membrane Technique (DMT) Diffusive Gradient in Thin films (DGTs) have been developed and successfully applied in speciation analysis. Similarly, Windermere Humic Aqueous Model (WHAM VII) is the latest version of the mathematical model and computer aided program, which is developed and thus world widely embraced (Tipping, 1998). The WHAM VII model contains database for cation binding to humic and fulvic acids, including all rare earth elements and silver. It simulates the precipitation of iron (III) and aluminium hydroxides, and binding of ions to precipitate. The binding activity of dissolved organic matter is used to generate distributions of predicted free ions concentrations (Lofts & Tipping, 2011).

4. Role of Metal Speciation in Risk Insight and Remediation

The identification and quantification of specific metal species in environmental samples is no longer an academic curiosity but also a substantial deal of particular interest to researchers in industries, as well as government agencies and legislative bodies for environmental pollution/contamination monitoring and management. However, the capacity and importance of metal speciation have not yet considered and incorporated in risk insight and remediation strategies, because it has not been appraised to be amongst the prerequisite approaches to determining the most technically and cost effective method for remediating a particular contaminated site(s) (if need there be). Although, remediation procedures and techniques are becoming increasingly more costly, however, many of today's remediation costs can be minimized without jeopardizing effectiveness by first gaining a better understanding of contamination scenario by metal speciation. This in turn, will allow remediation experts to put in appropriate remediation option to certain contaminants, which will undoubtedly respond efficiently and effectively upon the metallic contaminants. More explicitly, metal speciation research is needed to improve the ability of decision makers of environmental pollution/contamination to evaluate the risks of metallic contamination and develop effect and economic remediation strategies. The speciation or chemical form of metals governs its fate, toxicity, mobility and bioavailability in contaminated soils and water. To assess these chemical properties and to accurately gauge their impact on human health and the environment, metallic contaminants need to be characterized into different species. The importance of metal speciation research to the development of pollution law and control technologies (remediation options) is increasingly evident (USEPA, 2008a). Research on the chemical, biological, toxicological, and ecological effects of chemical species in the environment and methodology for the determination of chemical species and their bioavailability are providing new insight on previously hard to understand problems. For instance, studies conducted by the Land Research Program in the United State of America (USA) have enabled EPA to better predict the mobility, bioavailability and fate of metallic contaminant in environmental systems and developed cost-effective remediation strategies in response (USEPA, 2008b).

However, before remediation exercise to be undertaken or not, a reliable and proper analysis and identification of the contamination problem must be carried out, both its exact nature and extent. The more thorough the analysis the less likely the cost surplus will surface at a later stage. This involves site investigation by chemical speciation (Van Loon, 1975) and therefore, metal speciation is at best a difficult task but worth doing for appropriate remedy option selection and risk insights assessment.

Remediation of sites contaminated with metallic contaminants is a complex problem as the metallic contaminants exist in different oxidation states, inorganic and organic complexes and the degree of toxicity depends on which of those chemical species or forms of the contaminants (Hill, 1997). In turn, therefore, the metallic contaminants should be characterized first into their different chemical species if a reliable management option is to be best selected for achieving a result–oriented remediation technique. It is within this context that the chemical speciation of metallic components provides both the potential hazard (mobility and toxicity) and the pathway to a technological acceptable solution to the contamination scenario (Todd & Raina, 2003). Thus, involvement of chemical speciation in the remediation of contaminated site(s) is of great importance since effectiveness and efficiency of any remediation technique rely on reaction specificity. Because, this is found and applied in medical sciences for the treatment/medication of a particular illness/disease by medical doctors, where a patient must first be diagnosed, after which prescribed medicine(s) or drug(s) would specifically act upon the diagnosed illness/disease and systematically remove the illness/disease shortly afterwards. With this analogy, generation of data from chemical speciation will definitely provide important supporting information for treatment proposal (remediation) and risk assessments.

Remediation of metallic contaminants can only be brought about by their removal from the site or by establishing conditions which favor their retention in the solid phase. This can be achieved by ensuring that the described conditions for the initial removal or immobilization process are met and maintained over a suitable period of time (Morgan, 2008; Pratt, 1993). In the contest of soil and water contamination and remediation, we need to consider all aspect of transfer of metallic contaminants between phases (absorption, solubility, coagulation, coprecipitation, volatilization), transformation of species by electron transfer, ligand exchange availability and bioaccumulation and physical transport processes affecting a particular component of the metal species or group of interest (Tangahu et al., 2011; Morgan & Stumm, 1991; Sparks, 1995; Hering, 1995). This will impact on the influence of the contaminant status, reaction within the deposited environment, suitability and effectiveness of remediation process and long-term behavior of remediation results essentially a true four dimensional context. To adequately address this, data are required from chemical speciation which will allow appropriate assessment to be made (Tangahu et al., 2011; Mach et al., 1996). Therefore, when chemical speciation is applied in an appropriate manner, information is always gained and its role in subsequent interpretation should be the driving force for considering any remediation technique. For example, chromium exists as hexa and trivalent oxidation state which are the key features of its beneficial and detrimental contribution, and also its geochemical behavior in surface environment. Simply, toxic chromium (VI) is generally found as a mobile component in surface environments, while chromium (III) is an essential nutrient and relatively immobile in these systems (Miller, 1991). With this information from chemical speciation of chromium, the most common approach to treat Cr(VI) contamination is to reduce Cr(VI) to Cr(III) following by precipitation of Cr(III) (Ozer et al., 1997).

Furthermore, characterization of metallic contaminants through chemical speciation research into free ions of different oxidation states, inorganic and organic complexes will enable remediation experts to better predict the mobility, bioavailability and fate of metallic contaminants in surface environmental systems and developed effective and economic remediation strategies in response like the earlier one presented in the case of chromium (VI) contamination. It is generally accepted that the bioavailability of metallic contaminants and their physiological and toxicological effect depend on the actual species present and not on the total concentration (Brummer et al., 1986; Hendershot et al., 2000). More precisely, for example, Mn(III) species are more toxic than +2, +4, +6 and +7 oxidation states. Organometallic of Hg, Pb and Sn are more toxic than inorganic forms. In contrast, inorganic species of Cu, As and Al are more toxic than organic ones (Chutia et al., 2009; Tangahu et al., 2011; Hill, 1997). In addition, distribution of a metallic contaminant amongst different species profoundly affects its transport by determining such properties as–oxidation state (gauge for toxicity, solubility) and diffusion coefficient (organic complexes, key determinant for macro-molecular size) (Sager, 1992). In addition, as well as in the practical term, the elevated levels of Pb found in the water quality test during the investigation of the environmental impact of Rhandirmwyn Mine in South Wales, UK suggests that anglesite (one of the chemical forms of Pb) which was the most common occurring form observed during the SEM analysis and found to be more soluble than all the forms observed, may be responsible for the elevated concentration of Pb in the water of the major river and the major source of water for public utilization in the area. Therefore, the situation at which Pb becomes most toxic ecotoxicologically depends on its chemical form which exists in the environment, which is the centripetal force that was used to direct decision towards arriving at an appropriate conclusion that the river water and the Rhandirmwyn Mine sites require remediation treatments (Umar, 2010). Upon all these information, it is quite clear that chemical speciation drastically models contaminants behavior and thus, the approach can be used in decision making in choosing and developing reliable remediation option.

Much more, without a full understanding of the local interaction of the contaminants and potential ecological impacts the derivation of total concentration levels for contaminants is meaningless (Wuana & Okieimen, 2011; Martins & Bardos, 1995; Peter & Shem, 1995). In recognition of this issue, regulatory regimes worldwide are now moving towards advocating a risk–based assessment process for contaminated sites (Lai et al., 2010; Fergusen et al., 1998). This is a crucial point for the speciation community as it will derive the development of appropriate testing systems, and stimulate the collection of data linking speciation to environmental impact assessment of metallic contaminants. By applying these criteria to the investigation process and incorporating them in the evaluation of remediation targets, the risk-based approach will allow a more effective remediation strategy to be defined and within the context of limitation of the site investigation (sampling intensity versus cost), will provide a more realistic evaluation of the volumes of material/site required to be remediated.

The identification of contamination problem is only a small part of the contamination scenario. The decision to remediate must come from the information gained on the chemical speciation research of metallic contaminants, as it provides predictive insights on toxicity, mobility, bioavailability and fate of metallic contaminants. Besides, the influence of speciation on the behavior of metal species in contaminated soils/water is fundamental to the response of contaminants to remediation technique(s) (Wuana & Okieimen, 2011; Newson, 1992).

5. Conclusion

With the good understanding of the capacity of chemical speciation of environmental pollutants/contaminants, from the laboratory to field scale evaluation of contaminants behavior and the approach to be taken to assess and recommend the reliability of remediation option will no longer be a hard to understand problem, since chemical (oxidation/valence state) forms determine toxicity; chemical forms determine the mobility of an element in the environment; speciation of metals can affect the bioavailability of metals and therefore their associated risks. By implication this indicates that total concentrations of toxic metals in the environment should not be generalized as threshold for remedial action, but rather the chemical forms of the toxic metals. Hence, the speciation community has a sustained contribution to the management of metal contamination of the natural environment. In a nutshell, chemical speciation has a pivoted role to play in hazard and risk assessment, remediation strategies and in the long term management of remediated sites.

References

Bernhard, M., Brinckman, F. E., & Sadler, P. J. (1996). *The importance of speciation in environmental processes.* Berlin: Springer-Verlag.

Bora, A. (1998). *Green to contaminated and derelict land* (pp. 10-18). London: Thomas Telford.

Brummer, G. W. (1986). In M. Bernhard, F. E. Brickman, & P. J. Sadler (Eds), *The importance of chemical speciation in the environmental process* (p. 170). Berlin, Heidel berg, New York: Springer.

Cairney, T. (1995). *There use of contaminated land.* Chichester: John Wiley.

Campbell, P. G. C. (1995). Interactions between trace metals and aquatic organisms: a critique of the free ion activity model. In A. Tessier, & D. R. Turner (Eds.), *Metal speciation and bioavailability in aquatic systems, 3,* 45-102.

Caroli, S. (1996). *Element speciation in bioinorganic chemistry.* Wiley.

Centeno, J. A. (2003). Speciation of trace elements and toxic metal ions species. Retrieved from www.cprm.gov.br/publique/media/jose-speciation_lecture.pdf

Chutia, P., Kato, S., Kojima, T., & Satokawa, S. (2009). Arsenic adsorption from aqueous solution on synthetic zeolite. *J. Hazard. Mat., 162*(1), 440-447. http://dx.doi.org/10.1016/j.jhazmat.2008.05.061

Clough, R., Lindsay, R., Drennan, H., Harrington, C. F., Hill, S. J., & Tyson, J. F. (2012). Atomic spectrometry Update: Elemental speciation. *J. Anal. Atomic Spectr., 27,* 1185-1224. http://dx.doi.org/10.1039/c2ja90037h

Di-Toro, D. M., Allen, H. E., Bergman, H. L., Meyer, J. S., Paqiun, P. R., & Santore, R. C. (2001). Biotic Ligand model of the acute toxicity of metal. 1. Technical basis. *Environ Toxicol Chem., 20,* 2383-2396. http://dx.doi.org/10.1002/etc.5620201034

Ebdon, L., Hill, S. J., & Ward, R. (1987). Chemical speciation of metal species in surface environment. *Analyst, 112,* 1. http://dx.doi.org/10.1039/an9871200001

Fan, J., He, Z., Ma, L. Q., Nogueira, T. A. R., Wang, Y., Liang, Z., & Stoffela, P. J. (2012). Calcium water treatment residue reduces copper phytotoxicity in contaminated sandy soils. *J. Hazard. Mat., 199-200,* 375-382. http://dx.doi.org/10.1016/j.jhazmat.2011.11.030

Feng, D. X., Dang, Z., Hung, W. L., & Yang, C. (2009). Chemical speciation of fine particle bound trace metals. *Int. J. Environ. Sci. Technol., 6*(3), 337-346.

Ferguson, C., Dermendraid, K., Freier, K., Jensen, B. K., Jensen, J., Kasamas, H. ... Vegter, J. (1998). *The risk assessment for contaminated sites in Europe, Volume1, scientific basic.* Nottingham: LQM Press.

Hagelstein, K. (2009). Globally sustainable manganese metal production and use. *J. Environ. Manag., 90*(12), 3736-3740. http://dx.doi.org/10.1016/j.jenvman.2008.05.025

Harrison, R. M., & Rapsomaniskis, S. (1989). *Environmental analysis using chromatography interfaced with atomic spectroscopy.* Ellis Harwood.

Hendershot, W. H., Murray, P., & Ge, Y. (2000). Trace metal speciation and bioavailability in urban soils. *Environ. Pollut., 107,*137-144. http://dx.doi.org/10.1016/S0269-7491(99)00119-0

Hering, J. G. (1995). In H. E. Allen, C. P. Huang, G. W. Bailey, & A. R. Bower (Eds.), *Metal speciation and contaminated soil* (pp. 59 -86). Boca Raton, FL: CRC Press.

Hill, S. J. (1997). Speciation of trace metals in the environment. *Chemical Society Review, 26,* 291-298. http://dx.doi.org/10.1039/cs9972600291

Hirata, S., Toshimitsu, H., & Aihara, M. (2006). Determination of arsenic species in marine samples by HPLC-ICP-MS. *Anal. Sci., 22,* 39-43. http://dx.doi.org/10.2116/analsci.22.39

Hirose, K. (2006). Chemical speciation of trace metals in seawater: A review. *Anal. Sci., 22,* 1055-1062. http://dx.doi.org/10.2116/analsci.22.1055

Johnson, S. (1993). *Remedial processes for contaminated land* (pp.1-17). Institute of chemical engineers, Rugby, Warwickshire.

Khan, F. I., Husain, T., & Hejazi, R. (2004). An overview and analysis of site remediation technologies. *J. Environ. Manag., 71,* 95-122. http://dx.doi.org/10.1016/j.jenvman.2004.02.003

Koschinsky, A., & Hein, J. R. (2003). Uptake of elements from seawater by ferromanganese crusts: Solid phase association and seawater speciation. *Mar. Geol., 198,* 331-351. http://dx.doi.org/10.1016/S0025-3227(03)00122-1

Lai, H-Y., Hseu, Z-Y., Chen, T-C., Chen, B-C., Guo, H-Y., & Chen, Z-S. (2010). Health risk-based assessment and management of heavy metals-contaminated soil sites in Taiwan. *Int. J. Environ. Res. Pub. Heal., 7*(10), 3595-3614. http://dx.doi.org/10.3390/ijerph7103596

Lobinski, R., & Szpunar, J. (1999). Biochemical speciation analysis by hyphenated techniques. *Anal. Chim. Acta., 400,* 321-332. http://dx.doi.org/10.1016/S0003-2670(99)00628-5

Lofts, S., & Tipping, E. (2011). Assessing WHAM/Model VII against field measurement of free metal ion concentrations: Model performance and the role of uncertainty in parameters and inputs. *Environ. Chem.,* *8*(5), 501-516. http://dx.doi.org/10.1071/EN11049

Lowe, M. S., & Bowlby, S. R. (1992). *Environmental issues in the 1990s* (pp. 117-130). Chichester: John Wiley.

Mach, M. H., Nott, B., Scott, J. W., Maddalone, R. F., & Widdon, M. (1996). Metal speciation and contaminated soils. Amended with sewage sludge. *Water, Air, Soil pollution,* *90,* 269. http://dx.doi.org/10.1007/BF00619287

Macklin, M. (1992). *Managing the human impact on the natural environment: Pattern and processes.* London: Belhaven press.

Markham, A. (1994). *A brief history of pollution.* London: Earths scan publications.

Martins, I., & Bardos, P. (1995). *A review of full scale treatment technologies for the remediation of contaminated soil.* EPP publications, Richmond surrey.

Miller, B. S. (1991). *Heavy metals in the environment* (pp. 199-203). Edinburg: CEP Consultants..

Morgan, J. J., & Stumm, W. (1991). *Metals and their compounds in the environment* (pp. 67-103). Weinhein: VCH.

Morgan, P. (2008). Contaminated land remediation technology. *Land remediation yearbook* (pp.39-42). Retrieved from www. eic-yearbook.co.uk

Newson, M. (1992). *Managing the human impact on the natural environment: Pattern and processes* (pp. 14-36). London: Belhaven press.

Ohnuki, T., Yoshida, T., Nankawa, T., Ozaki, T., Kozai, N., Sakamono, F., ... Francis, A. J. (2005). A continuous flow system for in-situ XANES measurements of change in oxidation state of Ce(III) to Ce(IV). *J. Nucl. Radiochem. Sci., 6,* 65-67.

Ozdes, D., Diran, C., & Senturk, B. H. (2011). Adsorptive removal of Cd (II) and Pb (II) ions from aqueous solutions by using Turkish illitic clay. *J. Environ. Manag., 92*(12), 3082-3090. http://dx.doi.org/10.1016/j.jenvman.2011.07.022

Ozer, A., Altundogan, H. S.,Erdem, M., & Tumen, F. (1997). Chemical speciation and remediation of metallic contaminants. *Environ. pollut., 97,*107

Peter R.W., & Shem, L. (1995). In H. E. Allen, C. P. Huang, G. W. Bailey, & A. R. Bowers (Eds), *Metal speciation and contamination of soil* (pp. 255-274). Boca Raton, FL: CRC press.

Peter, V. D. (2001). *Trace element speciation for environment, food and health* (pp. 233-235). Thomas Graham house, Cambridge, UK: Royal society of chemistry.

Pratt, M. (1993). *Remedial process for contaminated soils* (p. 14). Rugby Warwickshire.

Quevanviller, P., & Maier, E. A. (1994). *Research trends in the field of environmental analysis.* European commission, Brussels.

Reeder, R. J., & Schoonan, M. A. A. (2006). Metal speciation and its role in bioaccessibility and bioavailability. *Reviews in Mineralogy and Geochem., 64,* 312-319. http://dx.doi.org/10.2138/rmg.2006.64.3

Sager, M. (1992). *Hazardous metals in the environment: Techniques and instruments in analytical chemistry* (pp. 133-175). Amsterdam: Elsevier. http://dx.doi.org/10.1016/S0167-9244(08)70106-9

Sparks, D. L. (1995). In H. E. Allen, C. P. Huang, G. W. Bailey, & A. R. Bower (Eds.), *Metal speciation and contaminated soil* (pp. 35-58). Boca Raton, FL: CRC Press.

Tangahu, B. V., Abdullah, S. R. S., Basri, H., Idris, M., Anuar, N., & Mukhlisin, M. (2011). A review on heavy metals (As, Pb and Hg) uptake by plants through phytoremediation. *Int. J. Chem. Eng., 2011,* 1-31. http://dx.doi.org/10.1155/2011/939161

Templeton, D. M., Ariese, F., Cornelis, R., Danielson, L. G., Muntan, H., Van Leuwen, H. P., & Labinski, R. (2000). Chemical speciation and Fractionation. *Pure Appl. Chem., 72,* 1453-1470. http://dx.doi.org/10.1351/pac200072081453

Thornton, I. (1996). Remediation of metallic contaminants in natural environment. *Appl. Geochem, 11,* s335.

Tipping, E. (1998). Humic ion–bioding model VI: An improved description of the interaction of proton and metal ions with humic substances. *Aquat Geochem., 4,* 3-48. http://dx.doi.org/10.1023/A:1009627214459

Todd, R. S., & Raina, M. M. (2003). Impact of metals on the biodegradation of organic pollutants. *Environ. Health Perspectives, 111*(8), 1093-1101. http://dx.doi.org/10.1289/ehp.5840s

Umar, B. A. (2010). *Environmental Impact and Regeneration of Rhandirmwyn Mine* (MSc thesis, Applied Environmental Geology, Department of Geology, Cardiff University, UK).

Ure, A. M., & Davidson, C. M. (1995). *Chemical speciation in the environment.* Black Academic and professional.

USEPA. (2008a). *Building a scientific foundation for sound environmental decision.* Retrieved from www.Epa.gov/ord/lrp

USEPA. (2008b). *Science in action: Land research programme.* Retrieved from www.Epa.gov

Van Loon, J. C. (1975). Heavy metals and environment. *Proceedings of International conference on heavy metal in the environment* (p. 349).

Webster, G. K., & Karmahan, J. W. (1992). Element specific: Chromatographic detection by atomic emission spectroscopy. *American Chemical Society Synposium, 479*, 218.

Wedepol, K. H. (1991). *Metals and their compounds in the environment* (pp. 3-17). Weinheim: VCH.

WHO Lead NGR. (2010). Retrieved from http://www.who.int/scr/don/2010_07_07/en/index.html

Wuana, R. A., & Okieimen, F. E. (2011). Heavy metals in contaminated soils: A review of sources, chemistry, risks and best available strategies for remediation. *ISRN Ecology, 2011*, 402647, 20.

3

Indoor Air Quality of Typical Malaysian Open-air Restaurants

Yusri Yusup[1], Mardiana Idayu Ahmad[1] & Norli Ismail[1]

[1] Environmental Technology, School of Industrial Technology, Universiti Sains Malaysia, Penang, Malaysia

Correspondence: Yusri Yusup, Environmental Technology, School of Industrial Technology, Universiti Sains Malaysia, 11800 Penang, Malaysia. E-mail: yusriy@usm.my

Abstract

This paper reports the indoor air quality state of typical open-air restaurants in Malaysia. The measured air pollutant parameters include respirable coarse particulate matter (PM10), carbon monoxide (CO) and microorganisms (bacteria and fungi). We determined the effects of occupancy, number of vehicles on nearby roads, temperature, wind speed and relative humidity on the indoor concentrations of PM10, CO and microorganisms. The indoor air quality of the restaurants tested was moderate, only in the 75th percentile, and the CO concentrations were slightly elevated indoors. Among the ambient parameters measured, only wind speed and temperature affect the PM10 concentrations. The indoor and outdoor values of wind speed and temperature were similar. We observed a strong positive correlation between the PM10 concentrations and concentration of airborne microorganisms. Further microbiological analyses showed that Gram-positive bacteria were abundant compared to Gram-negative bacteria. Gram-positive cocci (micrococci, streptococci, staphylococci and diplococci) were the dominant microbial morphologies, followed by pathogenic Gram-negative enterobacteriaceae and Gram-positive bacilli.

Keywords: indoor air quality, PM10, CO, bacteria, fungi

1. Introduction

Indoor air quality in working, dining and residential locations concerns a considerable number of people because they spend increasing fractions of their lives indoors (Lee & Chang, 2000; Zhao & Wu, 2007). Both scientific and public interest in indoor air pollutants have recently increased due to the negative impact of poor air quality on environmental and occupational health (Rajasekar & Balasubramanian, 2011).

Criteria air pollutants, such as particulate matter (PM) and carbon monoxide (CO), are important indicators of both indoor and outdoor air quality due to their harmful effects on human health. PM originates from a variety of sources and may contain particles with different shapes, sizes and physicochemical compositions (El-Fadel & Massoud, 2000). PM has been linked to cardiovascular and respiratory diseases (Wan Mahiyuddin et al., 2013). Coarse particulate matter (PM10) consists of particles with aerodynamic diameters between 2.5 and 10 μm. Indoors, this class of PM, also known as respirable particulate matter (RSP), originates from cooking and smoking activities (Wallace, 1996). Table 1 lists studies reporting particulate matter concentrations (including PM10 and PM2.5, which have particle aerodynamic sizes of < 2.5 μm) in different building types. CO is a colourless, tasteless and odourless gas generated from the incomplete combustion of fuel and biomass, e.g., in gas stoves, and it causes asphyxiation at high concentrations. Although not a criteria pollutant, airborne microorganisms, or bio-aerosols, are also an important air quality parameter (Reanprayoon & Yoonaiwong, 2012), especially in restaurants, where they can increase the risk of food contamination and may originate from occupants, microbial growth and organic waste (Fabian et al., 2005).

Table 1. Respirable particulate matter (PM10 and PM2.5) concentrations in different building types from published literature (n is number of data points)

Study	Building type	Season	Sampling duration	Pollutant	n	Mean	Median	*S.D.	Range
(Baek, 1997)	Restaurant	Summer & Winter	2 hr	[a]RSP	24	171	159	101	33-475
(Fromme, 2007)	Classroom	Winter	8 hr	PM2.5	79	23.0	19.8	-	-
			8 hr	PM10	79	105	91.5	-	-
		Summer	8 hr	PM2.5	74	13.5	12.7	-	-
			8 hr	PM10	74	71.7	64.9	-	-
(Lee, 2001)	Restaurant (Chinese hot pot)	-	1 hr	PM2.5	-	81.1	-	10.0	49.1-136.5
		-	1 hr	PM10	-	105.3	-	19.9	39.3-129.5
(Bohac, 2010)	Restaurants (before ban)	-	10 min	PM2.5	62	77.1	52.1	-	-
	Restaurants (after ban)	-	10 min	PM2.5	62	2.9	1.9	-	-
(Asadi, 2011)	Hotel	-	-	PM10	-	-	-	-	59-94
(Li, 2001)	Shopping malls	-	1 hr	PM10	-	-	-	-	35-380
(Guo, 2004)	Wet markets	-	2 hr	PM10	24	-	-	-	49-167

* Standard Deviation

- Data not available

a Respirable suspended particulates (or respirable particulate matter)

In Malaysia and many other Asian countries, numerous restaurants and cafes are open or semi-open air buildings and often situated beside roads, factories and construction sites. The served food is commonly exposed to air, which could be contaminated due to poor indoor and outdoor air quality. Studies of indoor air quality in these dining establishments are scarce, and thus, no guidelines or best practices can be developed. This study attempts to investigate the indoor air quality state of three typical open-air restaurants by conducting an assessment of respirable coarse particulate matter (PM10), carbon monoxide (CO) and PM-laden microorganisms together with ambient temperature, wind speed, relative humidity, number of occupants and number of vehicles on nearby roads.

The objectives of this study are (1) to assess PM10 and CO concentrations at three open-air restaurants, (2) to quantify and identify airborne microorganisms and (3) to determine the relationships between the concentrations of PM, CO and airborne microorganisms and ambient (or surrounding) wind speed, temperature and relative humidity.

2. Materials and Methods

2.1 Site Description

Restaurant #1 (RF): Open-air Restaurant Far from a road

This open-air canteen had 28 tables with four chairs each and seated a maximum of 112 occupants at a time. The restaurant consisted of a 26 m^2 kitchen and a 164 m^2 dining area, for a total area of 190 m^2. The floor of the cafe was made of tiles; the layout of the cafe is illustrated in Figure 1 (a). There were six ceiling fans. Most of the walls enclosing the cafe were not solid, and the windows were mostly open. The cafe had four full-time workers. The cafe opened from 08:00 LST to 16:30 LST (local standard time) during work days only (normally from Monday to Friday). In the kitchen, there were four gas stoves, and the styles of cooking were open-wok frying and deep-frying. Food was served on countertops and tables, mostly with no covers. This canteen-type establishment is common in Malaysia.

(a)

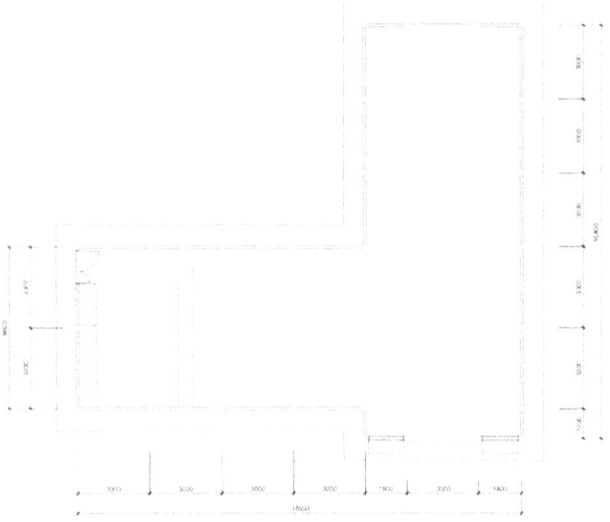

(b)

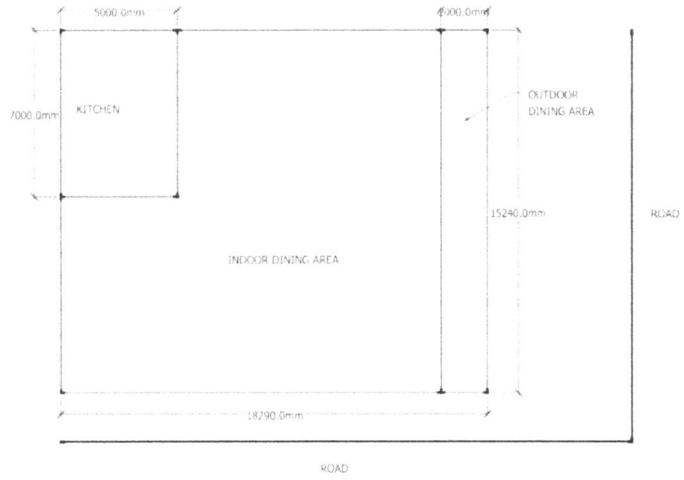

(c)

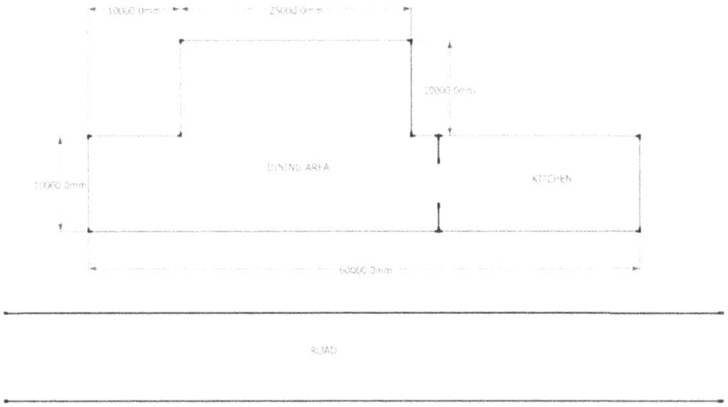

Figure 1. (a) Floor plan of restaurant RF (open-air Restaurant Far from a road) (b) floor plan of restaurant RBR (open-air Restaurant Beside a Road) (c) floor plan of restaurant RNR (open-air Restaurant Near a Road) and the location of roads for (b) and (c); dimensions are in mm

Restaurant #2 (RBR): Open-air Restaurant beside a Road

This 30-table Thai food restaurant was situated beside a busy four-lane highway. The restaurant had an area of 240 m^2 (60 m × 40 m) and was located 5 m from the road (Figure 1 (b)). This restaurant was also located near a car workshop. Restaurant 2 was open for 9 hr a day from 17:00 LST to 02:00 LST. For ventilation purposes, the restaurant employed five standing fans. The floors were made of cement. The restaurant employed ten workers, including waiters and cooks, and the kitchen contained four gas stoves for cooking. The kitchen was openly connected to the dining area. This restaurant sporadically used charcoal burners for seafood barbecues, but the main styles of cooking were open-wok frying and deep-frying.

Restaurant #3 (RNR): Open-air Restaurant Near a Road

This restaurant had 70 tables with four chairs each, able to cater to a maximum of 280 occupants. The total area of the restaurant was 279 m^2 (Figure 1 (c)). The floor was made of tile, and for ventilation, there were 11 ceiling fans. The restaurant had ten full-time workers and remained open 24 hr daily. In the kitchen, there were five gas stoves, and the style of cooking was deep-frying. The kitchen was separated from the dining area by a wall. Food was served on a countertop with covers. "Sheesha" (or water-pipes) were available only between 18:00 to 00:00 LST daily.

2.2 Measurements of Coarse Particulate Matter (PM10) and Carbon Monoxide (CO)

Dustmate (Turnkey Instruments, UK) was used to measure the concentrations of PM10 in the indoor air. Dustmate operates by continuously pulling in air samples through a nephelometer, a light-scattering based sensor. The nephelometer analyses the particles passing through its laser beam to determine particle concentrations and relative size fractions. The particle size range of Dustmate is 0.5 - 20 μm, and the detection limit of PM10 is 0.01 μg/m^3. The Dustmate sampled air at a flow rate of 0.6 L/min. Readings were taken at an interval of 5 min over a total sampling period of 30 min for RF and 2 hr for RBR and RNR.

A Crowcon Tetra gas detector (Crowcon Detection Instruments, UK) was used to measure carbon monoxide (CO) concentrations in parts per million (ppm). The Crowcon Tetra can measure CO in the range 0 - 500 ppm with a resolution of 0.1 ppm.

The Dustmate was stationed at a height of 1 m above the ground. This height was chosen to ensure the samples collected were representative of the air breathed by a seated person (Baek et al. 1997). At site RF, this was performed thrice a day at 08:00 LST (morning; occupants are present), 11:00 LST (late morning; occupants are present) and 18:00 LST (evening; cafe closed, occupants not present). At site RBR, measurements were taken from 21:00 LST to 23:00 LST and at site RNR, from 17:30 LST to 19:30 LST. All measurements were collected within four weeks from the initial sampling time for each location. Both indoor and outdoor measurements were taken, where the "indoor" location was a point in the centre of the restaurant, while the "outdoor" location was a point 2 m from the edge of the restaurant's boundaries.

A gravimetric measurement of PM10 using MiniVol Portable Air Sampler (Airmetrics, USA) was also employed to calibrate the Dustmate measurements. In this study, we used a size-selector impactor of < 10 μm with a sampling flow rate of 5 L/min for 2 hr. A 5-decimal analytical balance (OHAUS Discovery, USA) was used to measure the initial and final weight of Pallflex® fibre glass filter papers (Pall, USA).

Gravimetric-based PM measurements are more accurate than light-scattering based measurements (Cambra-López et al. 2012). We found that the relation between PM10 concentration using Dustmate (light-scattering) and MiniVol (gravimetric) is linear (Figure 2) and can be expressed by Equation (1).

$$C_{PM10,corrected} = 0.54(C_{PM10,Dustmate}) + 40 \qquad (1)$$

where C is the PM10 concentration. Subsequent PM10 Dustmate results were corrected using Equation (1).

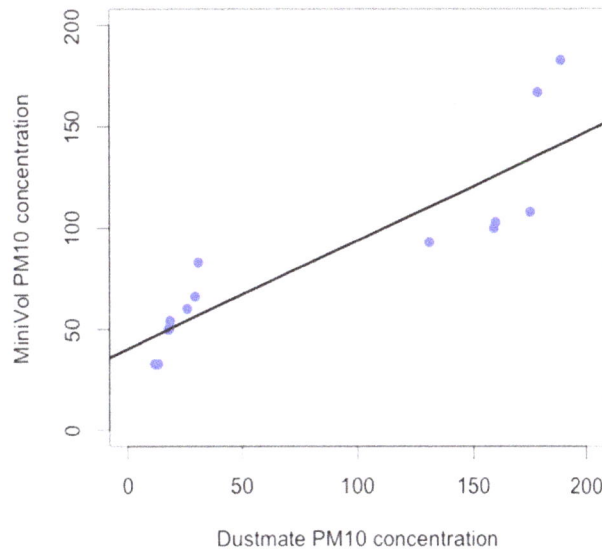

Figure 2. Calibration/corrective line (the solid line, Eq. (1): $C_{PM10,corrected} = 0.54 \, (C_{PM10,Dustmate}) + 40$) of light-scattering based measurement (Dustmate) against gravimetric-based measurement (MiniVol); $R^2 = 0.76$ and highest p-value = 3.157E-5

2.3 Ambient Parameter Measurements

For restaurant RF, the temperature (°C) was collected at 10-min intervals for an hour from a MKIII-LR weather station (Rainwise, USA) located nearby. The measurement range of the temperature sensor was -54 - 74°C, and its accuracy was ±0.5°C.

For restaurants RBR and RNR, wind speed and relative humidity were measured at intervals of 15 min for 2 hr in addition to temperature. Wind speed was measured using a hot-wire thermo-anemometer (Testo 405; Testo, USA). The detectable wind speed range was 0 - 10 m/s, with a resolution of 0.01 m/s. A humidity-temperature sensor (model 44550; Extech Instruments, USA) was used to measure relative humidity and temperature. For relative humidity, the instrument's resolution was 0.01% with a detectable range of 20 - 90%. For temperature, the resolution was 0.01°C with a detectable range of -10 - 50°C.

For sites RBR and RNR, vehicles on the nearby road were counted every 15 min for 2 hr and averaged. The number of occupants for all sites was counted for the entire sampling duration: 30 min at restaurant RF and every 15 min over the 2 hr measurement period at restaurants RBR and RNR. For RBR and RNR, the occupancy measurements were averaged.

For restaurant RF, the activities of a nearby construction site were also qualitatively observed. The types of activities at the construction site included loading and unloading of material, hammering, drilling, welding and transportation of materials.

2.4 Sampling and Analysis of Biological Indicators

Samples of airborne microorganisms in restaurant RF were collected using the exposed petri plate method (Benson 2001). Petri plates filled with nutrient agar were exposed to air for 5 min at a height of 1 m. Samples were collected three times a day.

Bacteria were cultured, and their morphology was determined by Gram staining. Total bacteria and fungi were enumerated and reported as colony forming units per plate (CFU per plate). Different bacteria were isolated and cultured individually on separate petri dishes using the streaking method according to the standard laboratory operating procedures detailed in Microbiological Applications (Benson, 2001) and incubated for two days at 35°C (Lee et al., 2002). This incubation temperature was chosen based on a preliminary study that identified the optimal incubation temperature of the bacteria present in the samples. After two days, the petri dishes were removed from the incubator, and the growth patterns and morphologies of the cultured bacteria were studied to identify potentially pathogenic species.

Gram staining was carried out on the cultured bacteria according to the method described in Microbiological Applications (Benson, 2001) to determine bacteria morphology and Gram characteristics. Staining was repeated for each separate bacterial colony that was cultured. The glass slides with the bacterial smears were then examined under an immersion oil microscope (Nikon Eclipse E200, Japan) and reported based on their genus level and Gram characteristics.

2.5 Data and Statistical Analyses

Exploratory and statistical analyses were completed using RStudio© (R Development Core Team 2012). Sample populations were compared using t tests according to Welch's t test methodology. RStudio© was also used to conduct linear regression analyses and correlation analyses, and the Pearson correlation coefficients ($1 > R > -1$) between all parameters measured were calculated. A correlation coefficient of ($R \approx 1$) indicates a strong positive relationship, whereas a correlation coefficient of ($R \approx -1$) indicates a strong negative relationship. A value of ($R \approx 0$) indicates a weak relationship.

3. Results

We found that the indoor median and mean PM10 (96 $\mu g/m^3$ and 120 $\mu g/m^3$, respectively) and CO (2.7 ppm and 3.1 ppm, respectively) concentrations in the open-air restaurants studied were lower than the standards of Malaysia's Department of Occupational Safety and Health (DOSH) (8 hr average PM10 = 150 $\mu g/m^3$) and Department of Environment (24 hr average PM10 = 150 $\mu g/m^3$; 8 hr average CO = 10 ppm). For PM10, the 75th percentile (3rd quartile) concentration value was slightly above the PM10 limit (see Figure 3 (a) and Table 2).

Table 2. Range of indoor and outdoor air PM10 and CO concentrations collected at three open-air restaurants; n is number of data points

[1]Location	PM10 concentration ($\mu g/m^3$)						
	n	Min	1st quartile	Median	Mean	3rd quartile	Max
RF	24	25	60	77	97	110	280
RBR (overall)	19	18	98	130	210	250	900
Indoor	11	30	100	130	190	200	500
Outdoor	8	18	100	150	250	290	900
RNR (overall)	20	12	30	55	98	160	350
Indoor	10	19	53	130	120	180	240
Outdoor	10	12	20	33	76	62	350
[1]Location	CO concentration (ppm)						
	n	Min	1st quartile	Median	Mean	3rd quartile	Max
RF	-	-	-	-	-	-	-
RBR (overall)	19	<LOD	0.9	2.5	2.8	3.8	11.3
Indoor	11	<LOD	1.9	2.7	2.7	3.8	5
Outdoor	8	<LOD	0	1.9	2.9	3.8	11.3
RNR (overall)	20	<LOD	0.38	1.15	2.15	2.68	9.8
Indoor	10	<LOD	1.1	2.8	3.5	5	9.8
Outdoor	10	<LOD	0.3	0.4	0.8	1.3	2.4

[1] RF - Open-air Restaurant Far from a road (averaging time 30 min)

RBR - Open-air Restaurant beside a Road (averaging time 15 min)

RNR - Open-air Restaurant near a Road (averaging time 15 min)

LOD - Lower than Detection limit (detection limit of 0.1 ppm)

Table 3. Indoor Pearson correlation matrix of air quality parameters measured for two open-air restaurants: RBR (open-air Restaurant beside a Road) and RNR (open-air Restaurant near a Road); bolded values are correlated

Parameters	CO	Wind speed	Temperature	Relative humidity	No. of people	No. of vehicle	PM10
CO	1	-0.1	0.29	-0.18	-0.07	-0.16	-0.13
Wind speed		1	0.45	-0.22	-0.44	-0.48	-0.45*
Temperature			1	-0.71	-0.75	-0.75	-0.45*
Relative humidity				1	0.87	0.73	-0.11
No. of people					1	0.89	0.15
No. of vehicle						1	0.26
PM10							1

* statistically significant, p-value < 0.05

The World Health Organisation's (WHO) 2005 limit is 50 $\mu g/m^3$ for 24 hr for PM10 and 26 ppm for 1 hr for CO. Based on this stricter limit, the PM10 concentration in the open-air restaurants was unhealthy. The indoor CO concentrations were higher than the outdoor concentrations, but this result was not statistically significant (Welch's t test; p-value = 0.1093).

We found no relationship between the number of occupants and the PM10 and CO concentrations for any site or between number of vehicles counted on the road and PM10 and CO concentrations for any site, indoors and outdoors.

Only the indoor ambient parameters, such as wind speed and temperature, affect the PM10 concentrations (refer to Table 3). Wind speed and temperature were inversely correlated with PM10 concentrations for the indoor sites (p-value = 0.0420 and 0.0427, respectively). The CO concentration was not affected by ambient indoor or outdoor parameters.

For open-air restaurants, the indoor and outdoor temperatures, wind speeds and relative humidities can be considered identical (refer Figure 4). Both the indoor and outdoor temperatures were related to the relative humidity and wind speed, but the relative humidity was not related to the wind speed (refer Table 3 and Table 4). The number of occupants within the restaurant was positively correlated with relative humidity, but it was negatively correlated with temperature and wind speed.

Table 4. Outdoor Pearson correlation matrix of air quality parameters measured for two open-air restaurants: RBR (open-air Restaurant beside a Road) and RNR (open-air Restaurant near a Road); bolded values are correlated; note that no correlations are statistically significant

Parameters	CO	Wind speed	Temperature	Relative humidity	No. of people	No. of vehicle	PM10
CO	1	-0.33	-0.33	0.23	0.2	0.23	0.28
Wind speed		1	0.35	-0.22	-0.29	-0.26	-0.38
Temperature			1	-0.83	-0.64	-0.64	-0.22
Relative humidity				1	0.64	0.67	-0.15
No. of people					1	0.93	0.35
No. of vehicle						1	0.29
PM10							1

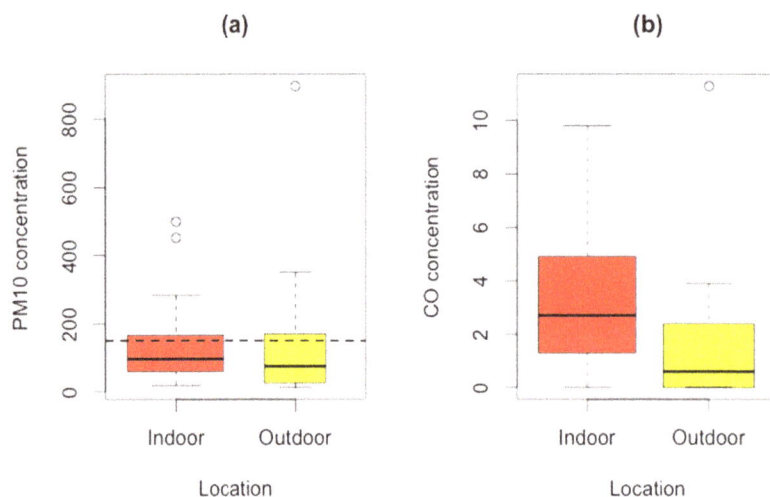

Figure 3. Indoor and outdoor concentrations of (a) PM10 in μg/m³ (indoor: n = 45; outdoor: n = 18) and (b) CO in ppm (indoor: n = 21; outdoor: n = 18) of the three open-air restaurants; in (a), dashed line is 150 μg/m³; width of box-plot is proportional to n

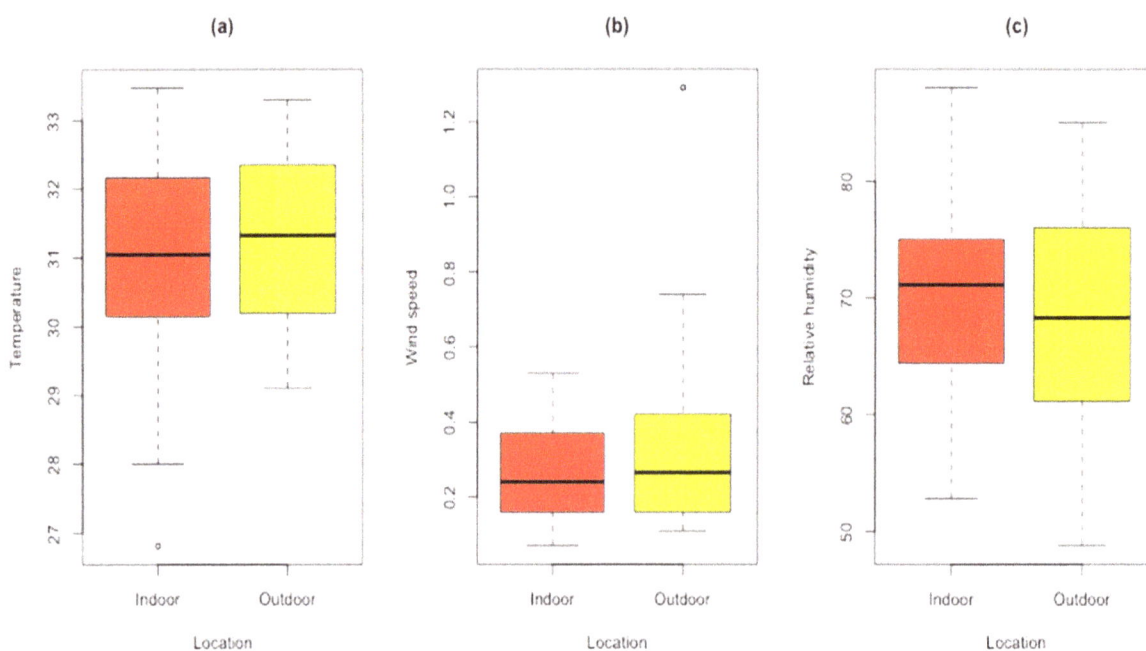

Figure 4. Indoor and outdoor (a) temperature (°C), (b) wind speed (m/s) and (c) relative humidity (%) (indoor: n = 21; outdoor = 18) of two (RBR and RNR) open-air restaurants; width of box plot is proportional to n; temperature, only RBR and RNR are included between because they can be directly compared; wind speed and relative humidity are similar indoors and outdoors

Qualitative observation at RF showed that smoking, ventilation and outside construction activity affect the indoor air quality in open-air restaurants. Relative to other times of day, the PM10 concentrations were high in the morning due to cigarette smoke in this restaurant. Ventilation was low because fans were also turned off at this time. We discovered a correlation between the number of smokers and PM10 (R = 0.60), although its linear regression line is not statistically significant (highest p-value < 0.1). Through qualitative observation of the nearby construction site, we found that loading and unloading and transportation of construction material using heavy vehicles corresponded to higher PM10 concentrations in the restaurant.

At RF, we found a strong positive relationship between PM10 and total colony count (Figure 5; R = 0.90; p-value = 8.56E-13). There were no significant relationships discovered between CFU and temperature or the number of occupants. The majority of bacteria present in open-air restaurants were Gram-positive, particularly cocci, which include diplococci, micrococci, staphylococci and streptococci (refer to Table 5). Gram-positive bacilli comprised a low percentage of the total bacteria count (1.2% to 1.8%). The total percentage of pathogenic gram-negative bacteria found in the restaurant was 0.6% to 1.8% (refer to Table 5 and Figure 6), with the higher range corresponding to greater restaurant occupancy. Only one Gram-negative genera was identified as a member of the family Enterobacteriaceae, which contains E.Coli and Salmonella (Northcutt et al., 2004), associated with food poisoning (Hirsh & Martin, 1983).

Table 5. Percentages (%) of Gram-positive genera, Gram-negative genera and fungi at restaurant RF at different times of day

Bacteria/Fungi	08:00 LST	11:00 LST	18:00 LST
Gram positive genera	29.5	27.7	27.1
Rods	-	1.2	1.8
Bacili	-	1.2	1.8
Cocci	29.5	26.5	25.3
Diplococci	2.4	1.8	-
Micrococci	13.9	12.1	11.5
Staphylococci	4.2	5.4	4.8
Streptococci	9.0	7.2	9.0
Gram negative genera	0.6	1.8	1.8
Enterobacteriaceae	0.6	1.8	1.8
Fungi	5.4	4.8	1.2

- Bacteria not present

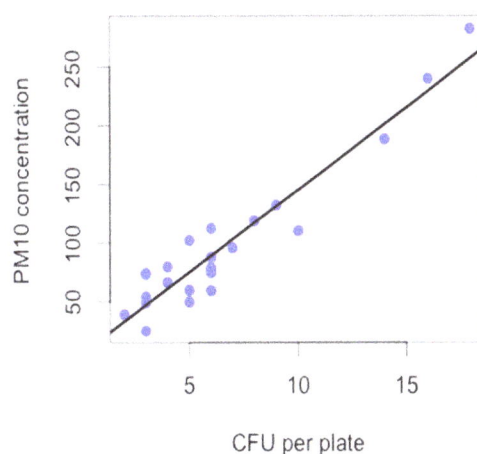

Figure 5. PM10 concentration ($\mu g/m^3$) and CFU per plate; solid black line is regression line (R = 0.90; p-value = 8.56E-13)

(a)   (b)

(c)   (d)

(e)   (f)

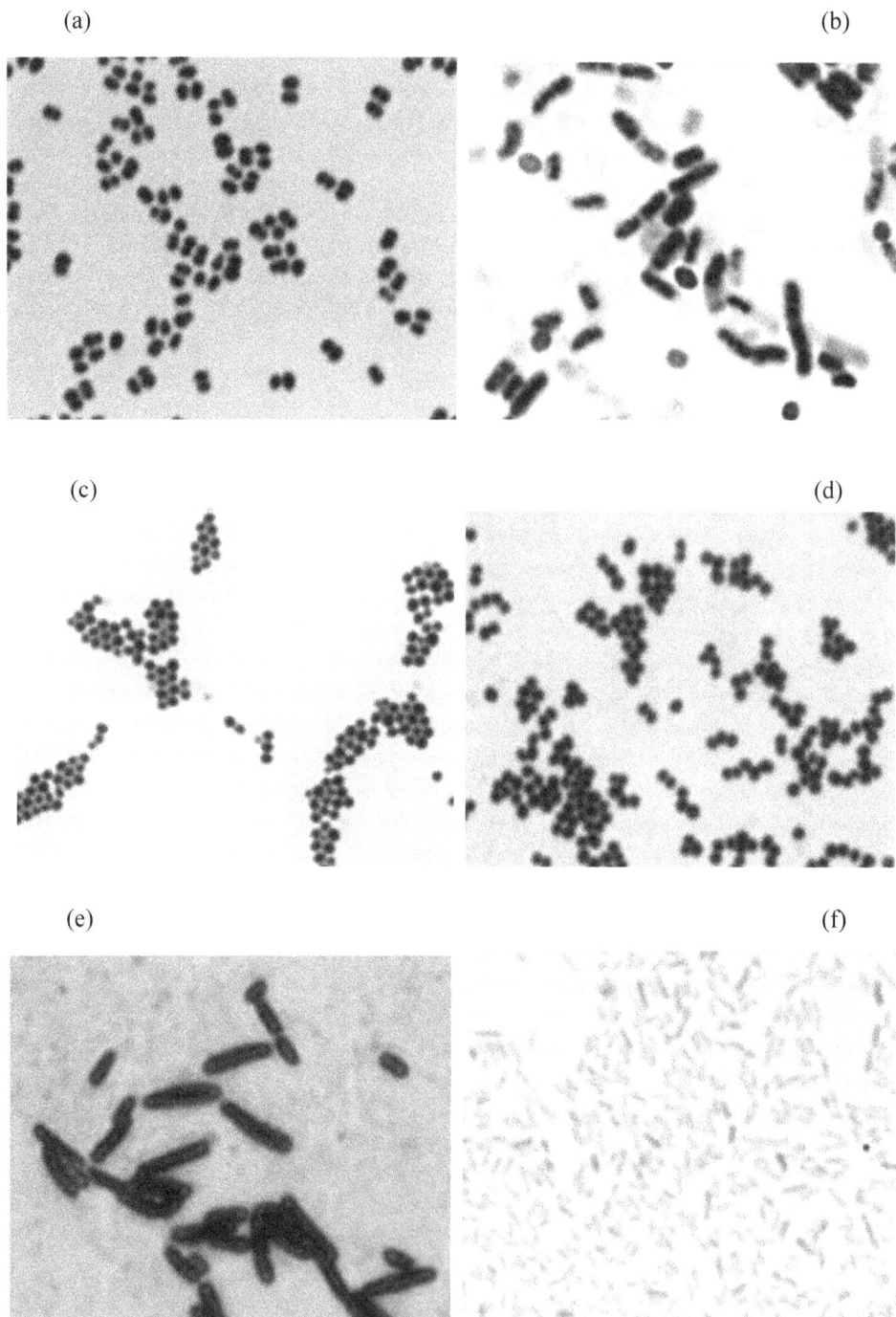

Figure 6. Bacteria found in the open-air restaurant RF: Gram positive genera: (a) Diplococci, (b) Streptococci, (c) Staphylococci, (d) Micrococci and (e) Bacilli; Gram negative genera: (f) Enterobacteriaceae

4. Discussion

We found that the general indoor air quality of open-air restaurants was moderate, exceeding standards established by the Department of Occupational Safety and Health (DOSH) and the Malaysian Department of Environment (DOE) only for the 75th percentile (3rd quartile) of measurements made (Figure 3 (a)). The contaminant concentration ranges in open-air restaurants were found to be similar to those in indoor, closed restaurants (Baek et al., 1997), and air quality was consistent among the surveyed restaurants regardless of their proximity to major streets. Previously reported PM10 concentrations in indoor environments, including wet markets (Guo et al., 2004), hotels and shopping malls, had large ranges (Asadi et al., 2011), as shown in Table 1.

This study used averaging times of 15 min and 30 min; therefore, the results are not directly comparable to national and international standards, which use 8-hr and 24-hr averaging times. However, these are the only available standards that can be used to assess air quality; thus, they are applied here. Numerous studies still adopt this comparison (Baek et al., 1997; Lee et al., 2002; Lee et al., 2001). The exposure time of occupants to PM and CO in the restaurants is less than 8 hr or 24 hr; therefore, the measured values can be considered as peak values within the larger averaging time set.

Restaurants have higher indoor air pollutant concentrations compared with homes or offices. However, the open design of the restaurant promotes air circulation, and thus indoor and outdoor concentrations of PM10 and CO are often similar. However, CO concentrations can be slightly elevated indoors due to local sources, such as gas stoves (Figure 3 (b)). A study on the indoor air quality of restaurants showed that heating and cooking (using kerosene/gas stoves) activities and inadequate ventilation worsen indoor air quality (Baek et al., 1997), particularly PM10 and PM2.5 levels (Lee et al., 2001).

In the surveyed restaurants, kerosene/gas stoves and background sources of PM10 could have overwhelmed any contribution from occupants. CO originates from the incomplete combustion of organic material and accumulates in closed environments; thus, no significant relationship is expected between number of occupants and the CO level. The correlation matrix tables (refer to Table 3 (indoor) and Table 4 (outdoor)) show that occupants were not the main source of PM10 in open-air restaurants. Cooking activities may be the predominant source, though coarse particulate matter can also originate from human activities such as walking (Qian et al. 2008), depending on the floor type (Zhang et al., 2008). Because the restaurants' occupants were mainly seated, human activities were not likely a large contributor to the increased PM10 concentrations.

Although cooking can significantly increase PM concentrations (Chao & Wong, 2002), no cooking occurred at RF in the mornings; thus, RF's morning PM10 concentrations were solely from smokers. Smoking is a major contributor to the PM10 concentrations in Hong Kong shopping malls (Li et al., 2001). Other studies have shown that cigarette smoke is a major source of PM (Wallace 1996), but in confined areas with low ventilation rates (McDonald et al., 2003), its dispersion is limited (Saha et al., 2012). In a study assessing the effects of a ban on smoking in restaurants, researchers discovered that the ban greatly reduced fine PM concentrations (Bohac et al., 2010). However, the dataset on number of smokers and PM concentrations is limited because only one to three smokers were present during measurements.

Lunchtime pollutant concentration measurements were high relative to other times of day, such as breakfast and dinner, due to high occupancy. Given the open-air nature of the restaurant, the ventilation rate is more important to dispersion for the restaurants studied than for a closed building. When construction materials are moved, they are exposed to wind, causing resuspension of coarser PM from the materials (Muleski et al., 2005).

The results obtained showed that wind speed and temperature had negative correlations to indoor PM10 concentrations and did not affect CO concentrations. A study has shown that naturally ventilated buildings have higher pollutant levels than air-conditioned buildings (Kukadia & Palmer, 1998) but do not retain external air pollutants for long periods because of their high air-change rates. Studies have also shown that concentrations of larger particulates, such as PM10, are correlated with ambient temperature (Celik & Kadi, 2007; Zaharim et al., 2007). Ventilation systems, outdoor air pollutant penetration factors, particulate deposition and resuspension rates and background air pollutants affect indoor air quality (Fromme et al., 2007). In most Asian countries, where buildings have no ventilation systems, indoor air pollution is higher than outdoor air pollution, particularly for buildings located near roads and/or construction sites (Montgomery & Kalman, 1989; Chao & Wong, 2002) because outdoor air pollutant sources can affect indoor air quality in addition to indoor sources, such as cooking, cleaning and smoking activities.

Total colony count was likely correlated with PM10 because the large surface area of course particles facilitates microorganism attachment. The bacteria sampling method employed is based on the theory of sedimentation and can therefore be biased towards larger particles. One limitation of this method is that only a fraction of the total bio-aerosols collected may be culture-able. Allergens have been detected on indoor soot particle surfaces (Ormstad et al., 1998). High concentrations of PM10 indicates that microorganisms are transported faster and in larger quantities per unit area. High bacteria counts could be caused by high occupancy, lack of proper sanitation and inadequate ventilation rates (Lee et al., 2002). The effect of temperature on airborne biological contaminants is minimal relative to ventilation and human activities indoors (Balasubraman et al., 2011).

Other studies have shown that Gram-positive bacteria are dominant in both indoor and outdoor environments (Aydogdu et al., 2010). Cocci were found to be prevalent in 100 large U.S. office buildings (Tsai & Macher, 2005). Gram-positive micrococci, staphylococci and streptococci are commonly found on human skin and mucus membranes. These organisms disperse and spread through direct skin-to-skin contact and are expelled through respiration (Chikere et al., 2008). The presence of cocci in the air of the restaurants implies overcrowding combined with inadequate ventilation (Awad, 2007). Gram-positive bacilli are associated with outdoor sources, such as soil emissions, water, dust, air, faeces, vegetation, wounds and abscesses (Aydogdu et al., 2010). They are found indoors as a result of transport by wind-blown particulates (Yassin and Almouqatea 2010). Although the majority of these species are considered harmless to human health, certain species can cause infections, especially in individuals with weak immune systems (Mckernan et al., 2008). Enterobacteriaceae are related to E. coli and Salmonella (Northcutt et al., 2004), which are associated with food poisoning (Hirsh & Martin, 1983). Possible sources of these pathogens include improper handling, serving and preparing of food. The potential presence of these species in restaurants is a risk to human health, which could easily increase with the number of occupants.

5. Conclusions

We surveyed three typical open-air Malaysian restaurants and found that the indoor air quality of these establishments was moderate based on the Department of Environment Malaysia and Department of Occupational Safety and Health standards, only exceeding standards in the 75th percentile of measurements. Indoor PM10 is influenced by wind speed and temperature. The CO concentrations were higher indoors than outdoors, but they were not correlated with temperature or wind speed, suggesting that CO originated from within the restaurants. Indoor ambient parameters, including wind speed, temperature and relative humidity, were similar to outdoor parameters. We found a strong positive relationship between PM10 and CFU per plate. The majority of bacteria detected were Gram-positive (84.3%) and non-pathogenic to humans, with minute percentages of Gram-negative bacteria, enterobacteriaceae (4.2%). The presence of airborne bacteria indicated overcrowding and inadequate ventilation within the open-air restaurants. Wind speed is the sole ambient parameter to consider when trying to improve indoor air quality in open-air restaurants.

Acknowledgments

The authors like to thank Universiti Sains Malaysia (USM) for providing the short-term grant (Grant No.: 1001/PTEKIND/6311006) that assisted them to complete this research. The authors acknowledge the assistance given by Vivienne Sim, Nor Atika Marzan and Ummi Amirah Fadhilah Ayub.

References

Asadi, E., Costa, J. J., & Gameiro da Silva, M. (2011). Indoor air quality audit implementation in a hotel building in Portugal. *Building and Environment, 46*(8), 1617-1623. http://dx.doi.org/10.1016/j.buildenv.2011.01.027

Awad, A. H. A. (2007). Airborne dust, bacteria, actinomycetes and fungi at a flourmill. *Aerobiologia, 23*(1), 59-69. http://dx.doi.org/10.1007/s10453-007-9049-z

Aydogdu, H., Asan, A., & Otkun, M. T. (2010). Indoor and outdoor airborne bacteria in child day-care centers in Edirne City (Turkey), seasonal distribution and influence of meteorological factors. *Environmental Monitoring and Assessment, 164*(1-4), 53-66. http://dx.doi.org/10.1007/s10661-009-0874-0

Baek, S.-O., Kim, Y.-S., & Perry, R. (1997). Indoor air quality in homes, offices and restaurants in Korean urban areas—indoor/outdoor relationships. *Atmospheric Environment, 31*(4), 529-544. http://dx.doi.org/10.1016/s1352-2310(96)00215-4

Balasubraman, R., Nainar, P., & Rajasekar, A. (2011). Airborne bacteria, fungi, and endotoxin levels in residential microenvironments: A case study. *Aerobiologia*, 1-16.

Benson, H. J. (2001). *Microbiological Applications: Laboratory Manual in General Microbiology* (8th ed.). Boston: McGraw-Hill.

Bohac, D. L., Hewett, M. J., Kapphahn, K. I., Grimsrud, D. T., Apte, M. G., & Gundel, L. A. (2010). Change in indoor particle levels after a smoking ban in Minnesota bars and restaurants. *American Journal of Preventive Medicine, 39*(6, Supplement 1), S3-S9. http://dx.doi.org/10.1016/j.amepre.2010.09.012

Cambra-López, M., Winkel, A., Mosquera, J., Ognik, N. W. M., & Aarnik, A. J. A. (2012). *Comparison between light scattering and gravimetric devices for sampling PM10 mass concentration in livestock houses*. Paper presented at the Ninth International Livestock Environment Symposium, Valencia, Spain, July 8–12, 2012.

Celik, M. B., & Kadi, I. (2007). The relation between meteorological factors and pollutants concentrations in Karabuk city. G.U. *Journal of Science, 20*(4), 87-95.

Chao, C. Y., & Wong, K. K. (2002). Residential indoor PM10 and PM2.5 in Hong Kong and the elemental composition. *Atmospheric Environment, 36*(2), 265-277.

Chikere, C. B., Omoni, V. T., & Chikere, B. O. (2008). Distribution of potential nosocomial pathogens in a hospital environment. *African Journal of Biotechnology, 7*(20), 3535-3539.

El-Fadel, M., & Massoud, M. (2000). Particulate matter in urban areas: health-based economic assessment. *Science of the Total Environment, 257*(2-3), 133-146. http://dx.doi.org/10.1016/S0048-9697(00)00503-9

Fabian, M. P., Miller, S. L., Reponen, T., & Hernandez, M. T. (2005). Ambient bioaerosol indices for indoor air quality assessments of flood reclamation. *Journal of Aerosol Science, 36*(5-6), 763-783. http://dx.doi.org/10.1016/j.jaerosci.2004.11.018

Fromme, H., Twardella, D., Dietrich, S., Heitmann, D., Schierl, R., Liebl, B., et al. (2007). Particulate matter in the indoor air of classrooms—exploratory results from Munich and surrounding area. *Atmospheric Environment, 41*(4), 854-866. http://dx.doi.org/10.1016/j.atmosenv.2006.08.053

Guo, H., Lee, S. C., & Chan, L. Y. (2004). Indoor air quality investigation at air-conditioned and non-air-conditioned markets in Hong Kong. *Science of the Total Environment, 323*(1-3), 87-98. http://dx.doi.org/10.1016/j.scitotenv.2003.09.031

Hirsh, D. C., & Martin, L. D. (1983). Detection of Salmonella Spp in milk by using Felix-01 bacteriophage high-pressure liquid-chromatography. *Applied and Environmental Microbiology, 46*(5), 1243-1245.

Kukadia, V., & Palmer, J. (1998). The effect of external atmospheric pollution on indoor air quality: A pilot study. *Energy and Buildings, 27*(3), 223-230. http://dx.doi.org/10.1016/S0378-7788(97)00044-3

Lee, S. C., & Chang, M. (2000). Indoor and outdoor air quality investigation at schools in Hong Kong. *Chemosphere, 41*(1-2), 109-113. http://dx.doi.org/10.1016/S0045-6535(99)00396-3

Lee, S. C., Li, W.-M., & Yin Chan, L. (2001). Indoor air quality at restaurants with different styles of cooking in metropolitan Hong Kong. *Science of the Total Environment, 279*(1-3), 181-193. http://dx.doi.org/10.1016/s0048-9697(01)00765-3

Lee, S.-C., Guo, H., Li, W.-M., & Chan, L.-Y. (2002). Inter-comparison of air pollutant concentrations in different indoor environments in Hong Kong. *Atmospheric Environment, 36*(12), 1929-1940. http://dx.doi.org/10.1016/s1352-2310(02)00176-0

Li, W.-M., Lee, S. C., & Chan, L. Y. (2001). Indoor air quality at nine shopping malls in Hong Kong. *Science of the Total Environment, 273*(1-3), 27-40. http://dx.doi.org/10.1016/s0048-9697(00)00833-0

McDonald, J. D., Zielinska, B., Sagebiel, J. C., McDaniel, M. R., & Mousset-Jones, P. (2003). Source apportionment of airborne fine particulate matter in an underground mine. *Journal of the Air & Waste Management Association, 53*(4), 386-395. http://dx.doi.org/10.1080/10473289.2003.10466178

Mckernan, L. T., Wallingford, K. M., Hein, M. J., Burge, H., Rogers, C. A., & Herrick, R. (2008). Monitoring microbial populations on wide-body commercial passenger aircraft. *Annals of Occupational Hygiene, 52*(2), 139-149. http://dx.doi.org/10.1093/annhyg/mem068

Montgomery, D. D., & Kalman, D. A. (1989). Indoor/Outdoor air quality: Reference pollutant concentrations in complaint-free residences. *Applied Industrial Hygiene, 4*(1), 17-20. http://dx.doi.org/10.1080/08828032.1989.10389885

Muleski, G. E., Cowherd, C., & Kinsey, J. S. (2005). Particulate emissions from construction activities. *Journal of the Air & Waste Management Association, 55*(6), 772-783. http://dx.doi.org/10.1080/10473289.2005.10464669

Northcutt, J. K., Jones, D. R., & Musgrove, M. T. (2004). Airborne microorganisms during the commercial production and processing of Japanese quail. *Poultry Science, 83*(10), 1812-1812.

Ormstad, H., Johansen, B. V., & Gaarder, P. I. (1998). Airborne house dust particles and diesel exhaust particles as allergen carriers. *Clinical and Experimental Allergy, 28*(6), 702-708. http://dx.doi.org/10.1046/j.1365-2222.1998.00302.x

Qian, J., Ferro, A. R., & Fowler, K. R. (2008). Estimating the resuspension rate and residence time of indoor particles. *Journal of the Air & Waste Management Association, 58*(4), 502-516. http://dx.doi.org/10.3155/1047-3289.58.4.502

R Development Core Team. (2012). *R: A Language and Environment for Statistical Computing*. Vienna, Austria: R Foundation for Statistical Computing.

Rajasekar, A., & Balasubramanian, R. (2011). Assessment of airborne bacteria and fungi in food courts. *Building and Environment, 46*(10), 2081-2087. http://dx.doi.org/10.1016/j.buildenv.2011.04.021

Reanprayoon, P., & Yoonaiwong, W. (2012). Airborne concentrations of bacteria and fungi in Thailand border market. *Aerobiologia, 28*(1), 49-60. http://dx.doi.org/10.1007/s10453-011-9210-6

Saha, S., Guha, A., & Roy, S. (2012). Experimental and computational investigation of indoor air quality inside several community kitchens in a large campus. *Building and Environment, 52*, 177-190. http://dx.doi.org/10.1016/j.buildenv.2011.10.015

Tsai, F. C., & Macher, J. M. (2005). Concentrations of airborne culturable bacteria in 100 large US office buildings from the BASE study. *Indoor Air, 15*, 71-81. http;//dx.doi.org/10.1111/j.1600-0668.2005.00346.x

Wallace, L. (1996). Indoor particles: A review. *Journal of the Air & Waste Management Association, 46*(2), 98-126.

Wan Mahiyuddin, W. R., Sahani, M., Aripin, R., Latif, M. T., Thach, T.-Q., & Wong, C.-M. (2013). Short-term effects of daily air pollution on mortality. *Atmospheric Environment, 65*(0), 69-79. http://dx.doi.org/10.1016/j.atmosenv.2012.10.019

Yassin, M. F., & Almouqatea, S. (2010). Assessment of airborne bacteria and fungi in an indoor and outdoor environment. *International Journal of Environmental Science and Technology, 7*(3), 535-544. http://dx.doi.org/10.1007/BF03326162

Zaharim, A., Shaharuddin, M., Nor, M. J. M., Karim, O. A., & Sopian, K. (2007). Relationships between airborne particulate matter and meteorological variables using non-decimated wavelet transform. *European Journal of Scientific Research, 27*(2), 308-312.

Zhang, X. Y., Ahmadi, G., Qian, J., & Ferro, A. (2008). Particle detachment, resuspension and transport due to human walking in indoor environments. *Journal of Adhesion Science and Technology, 22*(5-6), 591-621. http://dx.doi.org/10.1163/156856108x305624

Zhao, B., & Wu, J. (2007). Particle deposition in indoor environments: Analysis of influencing factors. *Journal of Hazardous Materials, 147*(1-2), 439-448. http://dx.doi.org/10.1016/j.jhazmat.2007.01.032

4

Remediation of Dicofol Type Ddts-Contaminated Sediments by Ferrous Activated Sodium Persulfate Oxidation

Chun-You Zhu[1], Peng Bao[2], Yu-Xin Ba[1], Jing Hua[1], Xiao-Ning Liu[1], Guo-Hua Hou[2], Chun-Zao Liu[2] & Zheng-Yi Hu[1]

[1] College of Resources and Environment, University of Chinese Academy of Sciences, Beijing, P. R. China

[2] State Key Lab of Urban and Regional Ecology, Research Center for Eco-Environmental Sciences, Chinese Academy of Sciences, Beijing, P. R. China

Correspondence: Zheng-Yi Hu, College of Resources and Environment, University of Chinese Academy of Sciences, 19A, Yuquan Road, Beijing 100049, China. E-mail: zhyhu@ucas.ac.cn

Abstract

In recent years, contamination by dicofol-type DDTs has attracted immense concern as a new source of DDT pollution. In this study, sediment samples from a dicofol manufacturing factory in Tianjing, China exhibited serious DDT contamination [p,p'-DDE (115.27 mg kg^{-1}) and p,p'-DDT (11.84 mg kg^{-1})]. Results of the batch experiments showed that total DDT degradation rates increase as $S_2O_8^{2-}/Fe^{2+}$ molar ratios increase. The $S_2O_8^{2-}/Fe^{2+}$ molar ratios used in this study were as follows: 60/10 < 10/30 < 20/30 < 60/50 < 60/20 < 40/30 < 60/40 < 60/30 < 80/30. Their corresponding degradation rates were 31, 43, 52, 69, 70, 71, 72, 89, and 91 µg g^{-1}, respectively. The optimal $S_2O_8^{2-}/Fe^{2+}$ molar ratio was 60/30, which resulted in 64% and 96% degradation of p,p'-DDE and p,p'-DDT, respectively. However, when an excessive amount of ferrous ion was used (<$S_2O_8^{2-}/Fe^{2+}$ molar ratio of 60/30), then competition for SO_4^- between ferrous ion and DDTs resulted in decreased DDT degradation efficiency and increased persulfate decomposition (represented by the generated amount of sulfate). Our results implied that a slow and steady production of sulfate free radicals is favorable for DDT degradation, and that Fe^{2+} availability plays an important role in controlling persulfate reactions activated by ferrous ion. Fe^{2+}-activated persulfate oxidation may be significant in developing environment friendly and fast-remediation options for DDT-contaminated sediments and soil. Therefore, this study contributes to current knowledge on remediating DDT contamination.

Keywords: sediments, dicofol-type DDT contamination, persulfate, ferrous ion, oxidative degradation

1. Introduction

DDT [1,1,1-trichoro-2,2-bis(p-chlorophenyl)-ethane] is one of the persistent organic pollutants (POPs) identified by the Stockholm Convention on POPs which has been extensively used for controlling agricultural pests and disease-carrying insects such as malaria vectors (Zitko, 2003; Kamanavalli & Ninnekar, 2004). DDT is more stable than other organochlorine pollutants because of its chlorinated aliphatic and aromatic structures. Exposure to DDTs (DDT and its homologues) may damage the human nervous and reproductive systems (Guo et al., 2009). Although the manufacture and application of DDTs have been restricted since the 1970s because of their negative effects, traces of DDTs can still be detected in air, water, soil, sediments, and organisms (Bettinetti et al., 2008; Yao et al., 2006). Moreover, the 2001 Stockholm Convention on POPs still allows the use of DDTs in several countries, such as South Africa, to control the transmission of malaria. DDE [1-Chloro-2-[2,2-dichloro-1-(4-chlorophenyl)ethenyl]-benzene] is a common metabolite of DDT (Ssebugere et al., 2010; Yang et al., 2010). Environmental DDE originates from the metabolites of DDT resulting from aerobic biotic, abiotic, and photochemical degradation, as well as from technical-grade DDT contaminants (Thomas et al., 2008). DDE has been reported to be more persistent than DDT and can be detected in soil decades after the application of DDT (Thomas et al., 2008). According to the United States Geological Survey (USGS), p,p'-DDE content in America was 60% in urban areas and 48% in rural areas in 1999 (Thomas et al., 2008). As a potent androgen antagonist (Kelce et al., 1995), DDE has also been found to be the most abundant DDT component in sediments (Eganhouse & Pontolillo, 2008), fish, and humans (Kamanavalli & Ninnekar, 2004).

Dicofol [2,2,2-trichloro-1,1-bis(4-chlorophenyl)ethanol] is a non-systemic acaricide extensively used in controlling mites. Dicofol is usually synthesized from technical p,p'-DDT. During synthesis reaction (Scheme 1), p,p'-DDT is first chlorinated into Cl-DDT, and then hydrolyzed into dicofol (Qiu et al., 2005).

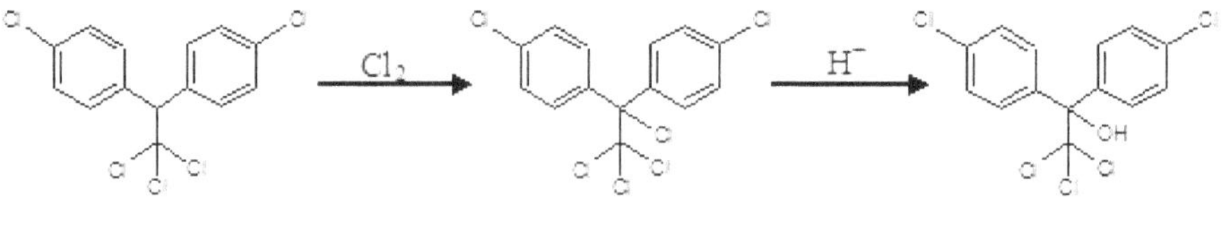

p, p' - DDT   Cl-DDT   Dicofol

Scheme 1. Synthesis reaction of dicofol by p,p'-DDT

To date, China still produces 5000 tons to 6000 tons of DDT per year as raw and processed materials for dicofol production (Huang et al., 2007). Moreover, approximately 8770 tons of DDTs were released into the environment in China by dicofol-type DDT contamination from 1988 to 2002 (Qiu et al., 2005; Turgut et al., 2009).

Although biodegradation and anaerobic reductive dechlorination for remediating DDTs have been well studied (Li et al., 2010; You et al., 1996), investigations on aerobic oxidative degradation for remediating DDTs is limited. Chemical oxidation via persulfate oxidation activated by ferrous ion has been evaluated as an option for treating chlorinated organic contaminants, such as trichloroethylene (TCE; Liang et al., 2004a, 2004b; Liang et al., 2008); tetrachloroethylene, dichloroethylene, and dichloroethane (Abranovic et al., 2006); polychlorinated biphenyls and polycyclic aromatic hydrocarbons (Block et al., 2004); and lindane (γ-HCH; Cao et al., 2008). Sulfate free radicals (SO_4^-) can be formed rapidly through persulfate-ferrous ion reaction at ambient temperature (20 °C) (Liang et al., 2004a). These free radicals can potentially degrade organic contaminants within soil mass by in situ chemical oxidation. The stoichiometric reaction between persulfate and ferrous ion is shown in the following equations (Kolthoff et al., 1951):

$$Fe^{2+} + S_2O_8^{2-} \rightarrow Fe^{3+} + SO_4^{-}\cdot + SO_4^{2-} \tag{1}$$

$$SO_4^{-}\cdot + Fe^{2+} \rightarrow Fe^{3+} + SO_4^{2-} \tag{2}$$

The ratio of reaction between $S_2O_8^{2-}$ and Fe^{2+} is dependent on the concentration of each reactant. When the reaction is near to stall, increasing the concentration of Fe^{2+} will accelerate the reaction shown in Equation (1). However, the target chlorinated organic contaminant and the excess Fe^{2+} will compete for SO_4^-, as shown in Equation (2). Gradual addition of small quantities of Fe^{2+} is necessary to optimize $S_2O_8^{2-}$ oxidative degradation of the target chlorinated organic contaminant and to control the reaction. To our knowledge, no Fe^{2+}-activated persulfate oxidation technique for remediating DDT contamination in sediments has yet been reported.

The primary purpose of this study is to investigate the contamination caused by dicofol-type DDTs in a dicofol manufacturing factory in Tianjin, China and to evaluate the effectiveness of Fe^{2+}-activated $S_2O_8^{2-}$ oxidative degradation for DDTs (p,p'-DDE and p,p'-DDT). In addition, the effects of various initial Fe^{2+} and $S_2O_8^{2-}$ concentrations on DDT degradation in sediments at ambient temperature (20 °C) are also investigated.

2. Materials and Methods

2.1 Chemicals

Standard samples of p,p'-DDT [p,p'-dichlorodiphenyltrichloroethane] (>99.5% purity) and p,p'-DDE [1-Chloro-2-[2,2-dichloro-1-(4-chlorophenyl)ethenyl]-benzene] (>99.5% purity) were obtained from Dr. Ehrenstorfer GmbH (Augsburg, Germany). Sodium persulfate ($Na_2S_2O_8$; >99% purity), ferrous sulfate ($FeSO_4 \cdot 7H_2O$; >99% purity), anhydrous sodium sulfate (Na_2SO_4; >99% purity), and other chemical reagents were of analytical grade, as required. Super pure GC hexane was obtained from J&K Chemical Ltd. (China). Sodium persulfate and ferrous ion solutions were prepared with 18 MΩ deionized water (Milli-Q™ 18 MΩ system, Millipore Corporation, MA, USA) before use. All pieces of glassware were washed twice with hexane prior to use. Serum bottles (20 mL) were used as batch reactors.

2.2 Sediment Collection and Preparation

Sediment samples were collected from Tianjin Renong Pharmaceutical Factory in China (Scheme 2). Sampling sites were highly contaminated by DDTs because of dicofol manufacturing before 2002. The three sampling sites were the factory floor, sewage, and drainage ditch. Sediments were collected from 20 cm to 40 cm sections mainly composed of fine sand with silt. The vessels were filled with sediments and completely sealed. The sediments were air-dried and ground with mortar until they could pass through a 2.0 mm sieve. The DDTs in the samples were detected via gas chromatography with mass selective detection (Table 1). Sample no. 3 was selected as the object for the batch study because it is highly contaminated with p,p'-DDE (115.27 mg kg^{-1}) and p,p'-DDT (11.84 mg kg^{-1}). The soluble sulfate of sample no. 3 was 56.28 mg kg^{-1} and its pH was 7.7. In addition, small amounts of p,p'-DDD and o,p'-DDE were also detected in all samples. However, no attempt was made to quantify p,p'-DDD and o,p'-DDE because their concentrations were very low.

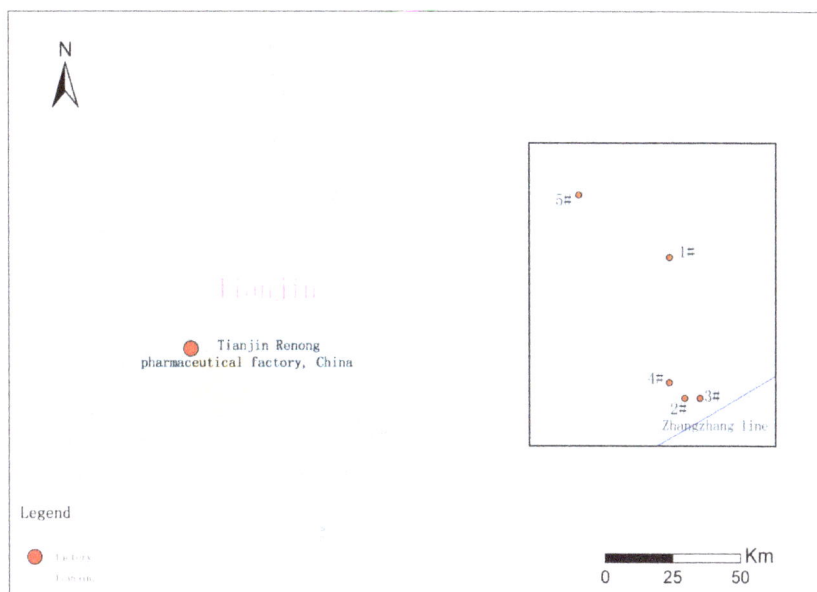

Scheme 2. The study area (Tianjin, China) and the sampling site (Tianjin Renong Pharmaceutical Factory)

2.3 Batch Studies of DDT Oxidation Using Persulfate Activated by Ferrous Ion

Two grams of air-dried sediment sample was added into the serum bottle with 4 mL deionized water and mixed completely. Two sets of experiments were performed to determine the following: (1) effect of $S_2O_8^{2-}$ concentration on DDT degradation and (2) effect of Fe^{2+} concentration on DDT degradation. Six treatments were initiated for set (1): (i) control, (ii) $Na_2S_2O_8$ (10 mmol L^{-1}) + $FeSO_4 \cdot 7H_2O$ (30 mmol L^{-1}), (iii) $Na_2S_2O_8$ (20 mmol L^{-1}) + $FeSO_4 \cdot 7H_2O$ (30 mmol L^{-1}), (iv) $Na_2S_2O_8$ (40 mmol L^{-1}) + $FeSO_4 \cdot 7H_2O$ (30 mmol L^{-1}), (v) $Na_2S_2O_8$ (60 mmol L^{-1}) + $FeSO_4 \cdot 7H_2O$ (30 mmol L^{-1}), and (vi) $Na_2S_2O_8$ (80 mmol L^{-1}) + $FeSO_4 \cdot 7H_2O$ (30 mmol L^{-1}). Five treatments were initiated for set (2): (i) control, (ii) $FeSO_4 \cdot 7H_2O$ (10 mmol L^{-1}) + $Na_2S_2O_8$ (60 mmol L^{-1}), (iii) $FeSO_4 \cdot 7H_2O$ (20 mmol L^{-1}) + $Na_2S_2O_8$ (60 mmol L^{-1}), (iv) $FeSO_4 \cdot 7H_2O$ (40 mmol L^{-1}) + $Na_2S_2O_8$ (60 mmol L^{-1}), and (v) $FeSO_4 \cdot 7H_2O$ (50 mmol L^{-1}) + $Na_2S_2O_8$ (60 mmol L^{-1}). For each set of experiment, 1 mL reagent solution was added in the following order: (1) sodium persulfate and (2) ferrous ion (one-fifth every 5 min) (Liang et al., 2004a). Afterwards, the sediments and solutions were vortex-mixed and incubated at ambient temperature (20 °C). Sampling and detection of the remaining amounts of DDTs were conducted at 0, 1, 2, 3, and 4 h. The concentration of sulfate in the supernatant was detected by ion chromatography at 4 h.

2.4 Extraction and Analysis of DDTs

The DDTs in the sediments were extracted using the ultrasonic extraction method. Hexane (10 mL) was added to the serum bottles containing the sediments. The bottles were then sonicated for 30 min in 6 L water in an ultrasonic bath (Soniclean, Australia) to ensure particle and solvent mixing (Thangavadivel et al., 2011). Furthermore, the contents were vortex-mixed and the solvent with DDTs was separated and passed through a funnel filled with 2.0 g anhydrous sodium sulfate (Na_2SO_4) to eliminate the remaining water in the samples (Hussen et al., 2006). Then, the extraction solvent with DDTs was evaporated to 2.0 mL using nitrogen gas.

The DDTs were analyzed by an Agilent 7890A gas chromatograph (Agilent Technologies, CA, USA) equipped with an Agilent 5975C mass selective detector (Agilent Technologies, CA, USA). The column used was HP-5MS, 30.00 m × 0.25 mm id with a film thickness of 0.25 μm. The injector temperature was 250 °C and the helium gas flow rate was 1.0 mL min^{-1}. The column temperature was initially set to 40 °C for 1 min, and then later increased at a rate of 30 °C min^{-1} to 130 °C min^{-1}. Afterwards, column temperature was switched to a rate of 5 °C min^{-1} to 160 °C min^{-1} for 6 min, and then, to a rate of 10 °C min^{-1} to 190 °C for 3 min. Finally, the temperature was switched to a rate of 20 °C min^{-1} to 300 °C and maintained isothermally for 2 min (Manirakiza et al., 2000). The injection volume was 1.0 µL in a splitless mode.

2.5 Analysis of the Sulfate in the Supernatant

According to Equations (1) and (2), persulfate will decompose into sulfate during reactions activated by ferrous ion. To evaluate the amount of decomposed persulfate, the concentration of supernatant sulfate was determined by ion chromatography. After 4 h reaction, 0.5 mL supernatant was passed through a 0.45 µm filter to eliminate impurities which may interfere with ion chromatography. The ion chromatography system, Dionex Ionpac AS14 column (4.6 mm × 3100.0 mm; Thermo Scientific, CA, USA), comprises a GP50 gradient pump, a column oven LC25, and an electrochemical detector ED50. Elution buffer was made of 3.5 mM sodium bicarbonate and 1.0 mM sodium carbonate. The flow ratio was 1.2 mL min^{-1}.

3. Results and Discussion

3.1 Assessment of DDT Contamination in Sampling Sites

Although the sampling site was a dicofol manufacturing factory which closed in 2002, no dicofol was detected in the samples because this pesticide is highly degradable in natural environments. However, DDT contamination remains serious even after nine years of production cessation. Dicofol impurity was proposed to possibly contribute to DDTs in the environment. This hypothesis was supported by the investigation of air samples collected over Taihu Lake, China during the summer of 2002, where very high concentrations of DDTs were found to be related to dicofol applications (Qiu et al., 2004), thus suggesting that dicofol is a possible source of DDTs which may constantly evaporate from soil to air.

According to the guidelines of the Chinese Environmental Quality Standard for Soil (GB15618-1995), the quality of soil can be classified as: with background pollution (grade I), low pollution (grade II), and high pollution (grade III). All three sampling sites were highly contaminated by DDTs, as shown in Table 1. *p,p'*-DDE and *p,p'*-DDT were the main DDT components and their concentration (1 mg kg^{-1}) was much higher than that of grade III soil. Moreover, *p,p'*-DDE concentration was higher than *p,p'*-DDT concentration in all samples. The results confirmed previous findings that DDE is the most abundant DDT component in soil and sediments (Guo et al., 2009), and is hardly degraded compared with other DDT components (Thomas et al., 2008; de la Cal et al., 2008). Further investigations and an effective remediation option for dicofol-type DDT contamination are recommended.

Table 1. Concentrations of DDTs in samples obtained from the three sampling sites

Station	Orientation	p,p'-DDT (mg kg^{-1}) (Mean±SE)	p,p'-DDE (mg kg^{-1}) (Mean±SE)
No. 1	N: 39°14'26.1" E: 117°06'33.7"	4.38±0.45	22.99±1.21
No. 2	N: 39°14'23", E: 117°06'34.14"	10.27±1.54	14.34±1.61
No. 3	N:39°14'23.3", E:117°06'33.7"	11.84±0.97	115.27±7.52

The samples were collected in 2010. Mean ± SE values (mg kg^{-1}) are shown (n = 3). Means are significantly different (one-way ANOVA: p < 0.05).

3.2 Influence of $S_2O_8^{2-}$ Concentration on DDT Degradation

To investigate the effect of persulfate contents on DDT degradation at Fe^{2+} concentration of 30 mM at ambient temperature (20 °C), different concentrations of persulfate were used. For all $S_2O_8^{2-}$/Fe^{2+} molar ratios, DDT degradation and persulfate decomposition (represented by sulfate formation) were observed (Figures 1 and 2). DDT degradation occurred almost instantaneously and stabilized because of the high reactive activity and unstability of SO$_4^-$ generated from $S_2O_8^{2-}$–Fe^{2+} reaction. For *p,p'*-DDE, the remaining ratios after 4 h decreased from 71% to 40% at $S_2O_8^{2-}/Fe^{2+}$ molar ratios of 10/30 to 80/30, respectively (Table 2). An increase in persulfate concentrations when Fe^{2+} level is 30 mM resulted in increased DDT degradation. However, an insignificant increase in DDT degradation was observed when $S_2O_8^{2-}/Fe^{2+}$ molar ratio was higher than 60/30. The p,p'-DDT remaining ratios after 4 h were 23%, 19%, 25%, 4%, and 5% at $S_2O_8^{2-}/Fe^{2+}$ molar ratios of 10/30, 20/30, 40/30, 60/30, and 80/30, respectively (Table 2). The influence of various persulfate concentrations on the degradation time course of DDTs (p,p'-DDE and p,p'-DDT) (Figure 1) indicated that total degradation ratios (in ascending order) were 10/30 < 20/30 < 40/30 < 60/30 < 80/30. Increasing the amount of persulfate content did result in an increase in sulfate formation and a proportional increase in DDT degradation except for the 80/30 $S_2O_8^{2-}/Fe^{2+}$ treatment (Figure 2). The concentration amounts of sulfate originating from the decomposition of persulfate were 10.04, 10.14, 10.28, 12.64, and 12.83 mg g^{-1} at $S_2O_8^{2-}/Fe^{2+}$ molar ratios of 10/30, 20/30, 40/30, 60/30, and 80/30, respectively. The amount of DDT degradation were 43, 52, 71, 89, and 91 µg g^{-1} at $S_2O_8^{2-}/Fe^{2+}$ molar ratios of 10/30, 20/30, 40/30, 60/30, and 80/30, respectively. The $S_2O_8^{2-}/Fe^{2+}$ molar ratio of 60/30 is the most effective and economical proportion. The observed optimal reaction ratio of $S_2O_8^{2-}/Fe^{2+}$ is greater than the theoretical stoichiometric ratio of 1:1 according to Equation (1). Persulfate anions were depleted by several reactions, such as organic substance oxidation, apart from those activated by Fe^{2+} (Kolthoff et al., 1951). The result was similar to that reported by Liang et al. (2004a) who suggested that the $S_2O_8^{2-}/Fe^{2+}$ molar ratio of 40/30 is the most effective proportion for TCE degradation in aqueous systems, and that beyond this ratio, adding persulfate only resulted in an insignificant increase in TCE degradation. In this study, an excess concentration of persulfate was necessary because persulfate was consumed by sediment materials. As for $S_2O_8^{2-}/Fe^{2+}$ molar ratios of 10/30, 20/30, and 40/30, the initial persulfate concentrations were insufficient when considering the stoichiometric ratio of reactions between persulfate and ferrous ions. Without adding persulfate and ferrous ion, no DDT degradation was observed for 4 h.

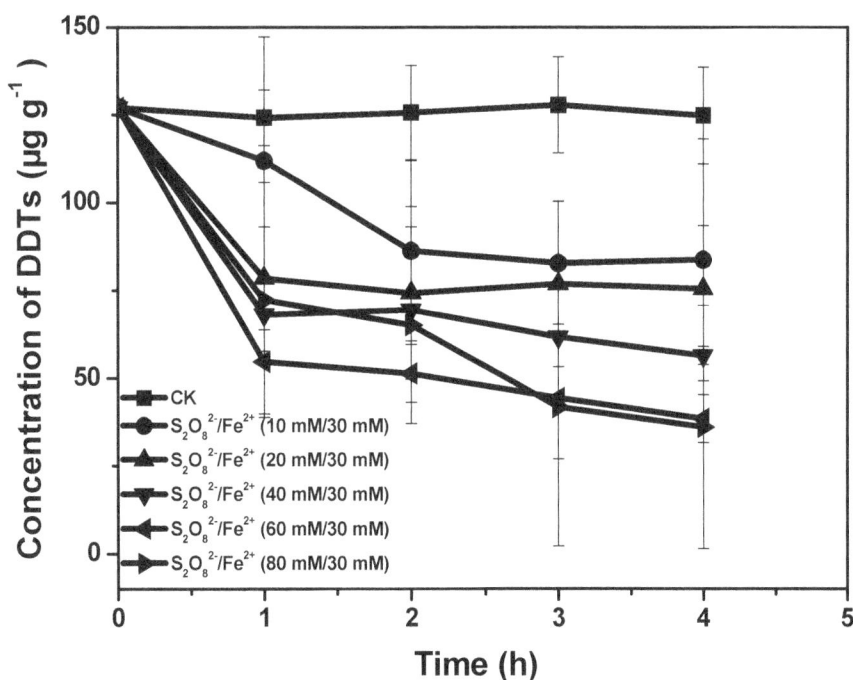

Figure 1. Dynamics of DDT degradation with different persulfate concentrations

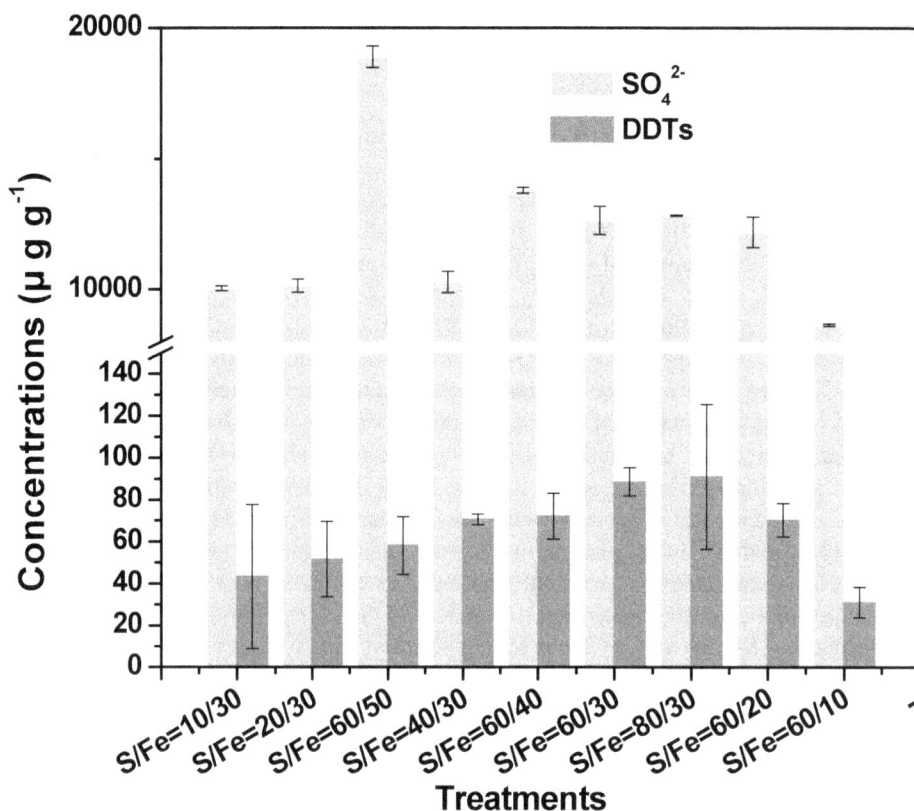

Figure 2. DDT degradation and sulfate formed by persulfate decomposition after 4 h

Table 2. The influence of different persulfate concentrations on the degradation of (A) DDE and (B) DDT

$S_2O_8^{2-}/Fe^{2+}$ molar ratio	Degradation ratio (C_t/C_0 %)	
	DDE (Mean±SD)	DDT (Mean±SD)
10/30	29.06±11.02	76.95±29.30
20/30	36.21±9.20	80.84±23.00
40/30	50.67±2.55	74.93±3.17
60/30	64.09±15.64	95.77±20.35
80/30	39.94±35.14	94.66±86.32

3.3 The Influence of Fe^{2+} Concentration on DDT Degradation

To further elucidate the effect of available ferrous ion at ambient temperature (20 °C) on the degradation of DDTs, ferrous ion was added in one-fifth increments to the reaction bottles containing $S_2O_8^{2-}$ concentration of 60 mM at 5 min intervals. Figure 3 shows the effect of different Fe^{2+} concentrations on DDT degradation and persulfate decomposition. DDT degradation and persulfate decomposition (indicated by sulfate formation) were observed in all $S_2O_8^{2-}/Fe^{2+}$ molar ratios. After five successive additions of Fe^{2+}, the initial $S_2O_8^{2-}/Fe^{2+}$ molar ratios were 60/10, 60/20, 60/30, 60/40, and 60/50. The remaining p, p'-DDE after 4 h were 100%, 63%, 36%, 60%, and 59%, respectively (Table 3). In the case of the 60/10 $S_2O_8^{2-}/Fe^{2+}$ molar ratio, p, p'-DDE degradation was insignificant because the concentration of Fe^{2+} was insufficient. Increasing Fe^{2+} concentration from a $S_2O_8^{2-}/Fe^{2+}$ molar ratio of 60/10 to 60/30 resulted in approximately 64% increase in p, p'-DDE degradation. Theoretically, increasing Fe^{2+} concentration would produce more SO_4^- which could promote DDT degradation. However, p, p'-DDE degradation decreased when $S_2O_8^{2-}/Fe^{2+}$ molar ratio was higher than 60/30. Similar phenomena have been reported by other researchers (Kislenko et al., 1995; Liang et al., 2004a). The possible consumption of sulfate free radicals may result from reactions with H_2O, $S_2O_8^{2-}$, and excess Fe^{2+} (Kolthoff et al.,

1951; Mcelroy & Waygood, 1990), as shown in Equations (2), (3), and (4).

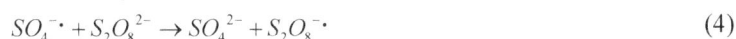

$$SO_4^- \cdot + H_2O \rightarrow OH + HSO_4^- \cdot \tag{3}$$

$$SO_4^- \cdot + S_2O_8^{2-} \rightarrow SO_4^{2-} + S_2O_8^- \cdot \tag{4}$$

The remaining p, p'-DDT after 4 h were 4.92%, 3.29%, 4.23%, 6.07%, and 2.29% at $S_2O_8^{2-}/Fe^{2+}$ molar ratios of 60/10, 60/20, 60/30, 60/40, and 60/50, respectively (Table 3). The rates of p,p'-DDT degradation were all significant whether Fe^{2+} content was sufficient or insufficient. Fe^{2+} has also been reported to result in significant p,p'-DDT reductive degradation (Li et al., 2010). The observed $S_2O_8^{2-}$-Fe^{2+} system in this study showed that the competition for SO_4^- between excess Fe^{2+} and DDTs was significant (Figure 2). Further increases in Fe^{2+} contents resulted in proportional increases in persulfate decomposition but not in DDT degradation (Figure 2). The sulfate obtained from the decomposition of persulfate, accompanied by increasing Fe^{2+} amounts, were 8.64, 12.19, 12.64, 13.79, and 18.89 mg g^{-1} at $S_2O_8^{2-}/Fe^{2+}$ molar ratios of 60/10, 60/20, 60/30, 60/40, and 60/50, respectively. The rates of DDT degradation were 31, 70, 89, 72, and 69 µg g^{-1} at $S_2O_8^{2-}/Fe^{2+}$ molar ratios of 60/10, 60/20, 60/30, 60/40, and 60/50, respectively. In this case, the rates of DDT degradation were not in agreement with the sulfate concentrations, thus indicating that high amounts of Fe^{2+} resulted in a competition for SO_4^- between excess Fe^{2+} and DDTs at $S_2O_8^{2-}/Fe^{2+}$ molar ratios less than 60/30. Moreover, the degradation time course of DDTs (p, p'-DDE and p, p'-DDT) (Figure 3) indicated that total degradation ratios (in ascending order) were 60/10 < 60/50 < 60/20 < 60/40 < 60/30. Therefore, sulfate free radicals formed through ferrous-ion activation are capable of effective DDT oxidative degradation, and that the optimum molar ratio for $S_2O_8^{2-}/Fe^{2+}$ was 60/30.

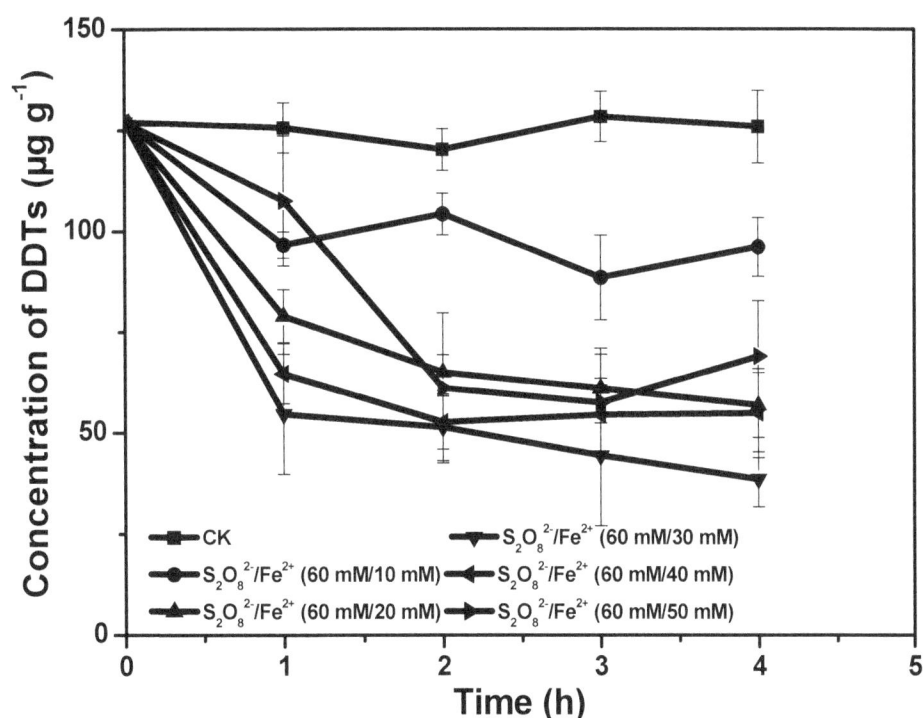

Figure 3. Time course of DDT degradation with different concentrations of ferrous ion

Table 3. Influence of different ferrous ion concentrations on the degradation of (A) DDE and (B) DDT

$S_2O_8^{2-}/Fe^{2+}$ molar ratio	Degradation ratio (C_t/C_0 %)	
	DDE (Mean±SD)	DDT (Mean±SD)
60/10	0	95.08±7.56
60/20	37.37±6.21	96.72±13.20
60/30	64.09±13.90	95.77±21.21
60/40	39.98±8.07	93.93±16.98
60/50	40.14±9.96	97.71±13.32

4. Conclusions

The assessment of the site for a dicofol manufacturing factory showed serious DDT (p, p'-DDE and p, p'-DDT) contamination in sediments. DDT degradation and persulfate decomposition were observed by calculating the amount of ferrous ion and persulfate at ambient temperature (20 °C). This study demonstrated that sulfate free radicals formed by ferrous ion activation are capable of degrading DDTs in sediments. Total DDT degradation ratios at $S_2O_8^{2-}/Fe^{2+}$ molar ratios of 31, 43, 52, 69, 70, 71, 72, 89, and 91 µg g^{-1} were 60/10 < 10/30 < 20/30 < 60/50 < 60/20 < 40/30 < 60/40 < 60/30 < 80/30, respectively. However, ferrous ion would react with SO_4^- and DDT degradation efficiency would decrease if ferrous ion contents became excessive. Further increases in persulfate concentration beyond an $S_2O_8^{2-}/Fe^{2+}$ molar ratio of 60/30 resulted in insignificant increases in DDT degradation. Thus, an $S_2O_8^{2-}/Fe^{2+}$ molar ratio of 60/30 was the most effective and economical proportion. This result indicated that a slow and steady production of free radicals is most desirable, and that Fe^{2+} availability plays an important role in controlling persulfate reactions activated by ferrous ion. Fe^{2+}-activated persulfate may be significant in developing an effective and environment friendly remediation option for DDT-contaminated sediments and soils. However, further work is necessary to determine how practical applications can be accomplished.

Acknowledgments

This research was jointly supported by the National Science Foundation of China (Grant No. 20777092) and the Ministry of Science and Technology of China (2007CB407304).

References

Abranovic, D. J., Brown, D., & Chemburkar, A. (2006). Persulfate stability is limiting factor for ISCO in fine-grained, iron-rich media, In *Proceedings of the Fifth International Conference on Remediation of Chlorinated and Recalcitrant Compounds*, Monterey, CA. 22-25 May. Columbus, Ohio: Battelle Press, D-75.

Bettinetti, R., Quadroni, S., Galassi, S., Bacchetta, R., Bonardi, L., & Vailati, G. (2008). Is meltwater from Alpine glaciers a secondary DDT source for lakes? *Chemosphere, 73*(7), 1027-1031. http://dx.doi.org/10.1016/j.chemosphere.2008.08.017

Block, P. A, Brown, R. A., & Robinson, D. (2004). Novel activation technologies for sodium persulfate in situ chemical oxidation, In *Proceedings of the Fourth International Conference on Remediation of Chlorinated and Recalcitrant Compounds*, Monterey, CA. Gavaskar, A. R., & Chen, A. S. C. (Eds.). Columbus, Ohio: Battelle Press.

Cao, J., Zhang, W.-X., Brown, D. G., & Sethi, D. (2008). Oxidation of lindane with Fe (II)-activated sodium persulfate. *Environmental Engineering Science, 25*(2), 221-228. http://dx.doi.org/10.1089/ees.2006.0244

De la Cal, A., Eljarrat, E., Raldúa, D., Durán, C., & Barceló, D. (2008). Spatial variation of DDT and its metabolites in fish and sediment from Cinca River, a tributary of Ebro River (Spain). *Chemosphere, 70*(7), 1182-1189. http://dx.doi.org/10.1016/j.chemosphere.2007.08.036

Eganhouse, R. P., & Pontolillo, J. (2008). DDE in sediments of the Palos Verdes Shelf, California: In situ transformation rates and geochemical fate. *Environmental Science & Technology, 42*(17), 6392-6398.

http://dx.doi.org/10.1021/es7029619

Guo, Y., Yu, H.-Y., & Zeng, E. Y. (2009). Occurrence, source diagnosis, and biological effect assessment of DDT and its metabolites in various environmental compartments of the Pearl River Delta, South China: A review. *Environmental Pollution, 157*(6), 1753-1763. http://dx.doi.org/10.1016/j.envpol.2008.12.026

Huang, Y., Zhao, X., & Luan, S. (2007). Uptake and biodegradation of DDT by 4 ectomycorrhizal fungi. *Science of the Total Environment, 385*(1), 235-241. http://dx.doi.org/10.1016/j.scitotenv.2007.04.023

Hussen, A., Westbom, R., Megersa, N., Retta, N., Mathiasson, L., & Björklund, E. (2006). Optimisation of pressurised liquid extraction for the determination of p, p′-DDT and p, p′-DDE in aged contaminated Ethiopian soils. *Analytical and bioanalytical chemistry, 386*(5), 1525-1533. http://dx.doi.org/10.1007/s00216-006-0667-z

Kamanavalli, C. M., & Ninnekar, H. Z. (2004). Biodegradation of DDT by a Pseudomonas species. *Current microbiology, 48*(1), 10-13. http://dx.doi.org/10.1007/s00284-003-4053-1

Kelce, W. R., Stone, C. R., Laws, S. C., Gray, L. E., Kemppainen, J. A., & Wilson, E. M. (1995). Persistent DDT metabolite p,p′-DDE is a potent androgen receptor antagonist. *Nature, 375*, 581-585. http://dx.doi.org/10.1038/375581a0

Kislenko, V. N., Berlin, A. A., & Litovchenko, N. V. (1995). Kinetics of Glucose Oxidation with Persulfate Ions, Catalyzed by Iron Salts. *Russian Journal of General Chemistry Part, 657*, 1092-1096.

Kolthoff, I., Medalia, A., & Raaen, H. P. (1951). The Reaction between Ferrous Iron and Peroxides. IV. Reaction with Potassium Persulfate1a. *Journal of the American Chemical Society, 73*(4), 1733-1739. http://dx.doi.org/10.1021/ja01148a089

Li, F., Li, X., Zhou, S., Zhuang, L., Cao, F., Huang, D., ... Feng, C. (2010). Enhanced reductive dechlorination of DDT in an anaerobic system of dissimilatory iron-reducing bacteria and iron oxide. *Environmental Pollution, 158*(5), 1733-1740. http://dx.doi.org/10.1016/j.envpol.2009.11.020

Liang, C., Bruell, C. J., Marley, M. C., & Sperry, K. L. (2004a). Persulfate oxidation for in situ remediation of TCE. I. Activated by ferrous ion with and without a persulfate–thiosulfate redox couple. *Chemosphere, 55*(9), 1213-1223. http://dx.doi.org/10.1016/j.chemosphere.2004.01.029

Liang, C., Bruell, C. J., Marley, M. C., & Sperry, K. L. (2004b). Persulfate oxidation for in situ remediation of TCE. II. Activated by chelated ferrous ion. *Chemosphere, 55*(9), 1225-1233. http://dx.doi.org/10.1016/j.chemosphere.2004.01.030

Liang, C., Lee, I., Hsu, I., Liang, C.-P., & Lin, Y.-L. (2008). Persulfate oxidation of trichloroethylene with and without iron activation in porous media. *Chemosphere, 70*(3), 426-435. http://dx.doi.org/10.1016/j.chemosphere.2007.06.077

Manirakiza, P., Covaci, A., & Schepens, P. (2000). Single step clean-up and GC-MS quantification of organochlorine pesticide residues in spice powder. *Chromatographia, 52*(11), 787-790. http://dx.doi.org/10.1007/BF02491005

Mcelroy, W. J., & Waygood, S. J. (1990). Kinetics of the reactions of the SO_4^- radical with SO_4^-, $S_2O_8^{2-}$, H_2O and Fe^{2+}. *Journal of the Chemical Society, Faraday Transactions, 86*, 2557-2564.

Qiu, X., Zhu, T., Li, J., Pan, H., Li, Q., Miao, G., & Gong, J. (2004). Organochlorine pesticides in the air around the Taihu Lake, China. *Environmental Science & Technology, 38*(5), 1368-1374. http://dx.doi.org/10.1021/es035052d

Qiu, X., Zhu, T., Yao, B., Hu, J., & Hu, S. (2005). Contribution of dicofol to the current DDT pollution in China. *Environmental Science & Technology, 39*(12), 4385-4390. http://dx.doi.org/10.1021/es050342a

Ssebugere, P., Wasswa, J., Mbabazi, J., Nyanzi, S. A., Kiremire, B. T., & Marco, J. A. (2010). Organochlorine pesticides in soils from south-western Uganda. *Chemosphere, 78*(10), 1250-1255. http://dx.doi.org/10.1016/j.chemosphere.2009.12.039

Thangavadivel, K., Megharaj, M., Smart, R. S. C., Lesniewski, P. J., Bates, D., & Naidu, R. (2011). Ultrasonic Enhanced Desorption of DDT from Contaminated Soils. *Water, Air, & Soil Pollution, 217*(1), 115-125. http://dx.doi.org/10.1007/s11270-010-0572-0

Thomas, J. E., Ou, L. T., & Al-Agely, A. (2008). DDE Remediation and Degradation, In D. M. Whitacre. (Ed.), *Reviews of Environmental Contamination and Toxicology* (pp. 55-69). Gainesville, FL 32611, U.S.A.

Turgut, C., Gokbulut, C., & Cutright, T. J. (2009). Contents and sources of DDT impurities in dicofol formulations in Turkey. *Environmental Science and Pollution Research, 16*(2), 214-217. http://dx.doi.org/10.1007/s11356-008-0083-3

Yang, L., Xia, X., Liu, S., & Bu, Q. (2010). Distribution and sources of DDTs in urban soils with six types of land use in Beijing, China. *Journal of Hazardous Materials, 174*(1), 100-107. http://dx.doi.org/10.1016/j.jhazmat.2009.09.022

Yao, F. X., Jiang, X., Yu, G. F., Wang, F., & Bian, Y. R. (2006). Evaluation of accelerated dechlorination of p,p′-DDT in acidic paddy soil. *Chemosphere, 64*(4), 628-633. http://dx.doi.org/10.1016/j.chemosphere.2005.10.066

You, G., Sayles, G. D., Kupferle, M. J., Kim, I. S., & Bishop, P. L. (1996). Anaerobic DDT biotransformation: enhancement by application of surfactants and low oxidation reduction potential. *Chemosphere, 32*(11), 2269-2284. http://dx.doi.org/10.1016/0045-6535(96)00121-X

Zitko, V. (2003). Chlorinated Pesticides: Aldrin, DDT, Endrin, Dieldrin, Mirex. In H. Fiedler. (Eds.), *The handbook of environmental chemistry Vol. 3, Part O Persistent organic pollutants* (pp. 47-84). NB, E5B 1A1, Canada.

5

Evaluation of Status of Heavy Metals Pollution of Sediments in Qua-Iboe River Estuary and Associated Creeks, South-Eastern Nigeria

Ikama E. Uwah[1], Solomon F. Dan[2], Rebecca A. Etiuma[1] & Unyime E. Umoh[1]

[1] Department of Pure & Applied Chemistry, University of Calabar, Calabar, Nigeria

[2] Department of Marine Biology, Faculty of ocean Science and Technology, Akwa Ibom State University Ikot Akpaden, Nigeria

Correspondence: Ikama E. Uwah, Department of Pure & Applied Chemistry, University of Calabar, Calabar, Nigeria. E-mail: ikamauwah@yahoo.co.uk

Abstract

Sixteen bottom sediment samples collected from Qua-Iboe River estuary and associated creeks were analyzed for Cd, Cr, Cu, Fe, Pb, Zn, Ni, pH, Organic carbon (orgC),and grain size in order to assess the current pollution status in sediment of the study area. Concentration data were processed using Pearson correlation analysis. Sediment pollution assessment was carried out using Enrichment factor, Geo-accumulation index and Modified degree of contamination. The calculated enrichment factor showed that the sediment was enriched with Cd, Zn, Cu and Pb. The results of geo-accumulation index (Igoe) indicated that sediments are unpolluted with Fe, moderately polluted with Cr, Cu, Pb, strongly polluted with Cd and extremely polluted with Ni. This was attributed mainly to oil contaminating wastes and metal scraps. The results of the modified degree of contamination (mC_d) revealed that the sediment of Qua Iboe estuary and associated creeks fall between $8 \leq mC_d \leq 16$ indicating very high degree of contamination.

Keywords: heavy metals, enrichment factor, geo-accumulation index, sediment quality

1. Introduction

The pollution of aquatic environment (especially estuaries) by heavy metals has been a source of serious concern to government regulatory agencies, environmentalist and the public at large (Manahan, 1991). This is particularly important because estuaries are rich in nutrient and as a result mothers varieties of fishery resources. Heavy metals play important roles in our society as most of them are vital raw materials in most industries. As trace elements, some heavy metals (e.g. Cu, Se, and Zn etc.) are essential in the maintenance of some metabolic activities in human bodies. However, at certain concentrations they become toxic. They are natural components of the earth's crust with large variations in concentration. They cannot be degraded nor destroyed due to their persistence in the environment. Their distribution in aquatic environment has been evidenced in human health effects and aquatic life disruptions due to long term exposure and bioaccumulation (Dahlia, Apodaca, Emerson, Tui, & Allyn, 2003). Marine sediments are the ultimate sinks of pollutants in the marine environment and it constitutes an important medium for scientific research. Like soils in the terrestrial environment, marine sediments in the aquatic ecosystem are the sources of substrate nutrients and become the basis of support to living aquatic organisms (Abdullah, Sidi, & Aris, 2007). The enrichment of metal in a sink is shown mainly by an increase in their concentrations in the bottom sediment. Their occurrence in the environment results primarily from anthropogenic activities. Also, natural processes, such as weathering of rocks and volcanic activities play a significant role in the enrichment of heavy metals in water bodies (Forstner & Wittmann, 1981; Forstner & Wittmann, 1983; Nriagu, 1989).

Heavy metals accumulate in sediments through complex physical and chemical adsorption mechanisms depending on the nature of the sediment matrix and the properties of the adsorbed compounds (Maher & Aislabie, 1992; Leivouri, 1998; Ankley et al., 1992). Several processes enhance the association of heavy metals with solid phase such as direct adsorption by fine-grained inorganic particles of clays, adsorption of hydrous ferric and manganic oxides which may in turn be associated with clays, adsorption on natural organic substances, which may also be associated with inorganic particles and direct precipitation as new solid phases (Gibbs, 1973). The dissolution and adsorption processes are influenced by several physicochemical parameters such as: pH, dissolve oxygen, salinity, redox potential, organic and inorganic carbon contents, and the presence in water phase of some anions and cations that can bind or co-precipitate the water-dissolved or suspended pollutants (Di Toro et al., 1991; Calmano, Hong, & Forstner, 1993; Wen & Allen, 1999).

2. Study Area

The Qua Iboe River estuary (Figure 1) lies within latitude 4°30′ to 4°45′ N and longitude 7°30′ to 8°00′ E on the south eastern Nigeria coastline. It is a mesotidal estuary having tidal amplitude of 1m and 3m at neap and spring phases respectively. The river originating from Umuahia hills traverses mainly sedimentary terrains of cretaceous to recent ages and develops into extensive meanders before emptying into the Atlantic Ocean. Creeks and channels islands are common throughout the length of the estuary while sand bars occur at the mouth as a result of interplay between the long shore drift which runs approximately in a west-east direction (parallel to the shoreline) and the river current.

Whereas the area has some coastal plain sands which are not older than the quaternary age, the Creeks have younger alluvial covers. Sediments are brought into the estuary by long shore drift, tide flow, waves and river transport. Coarse to medium-grained sand occurs mostly in the mouth of the estuary and middle of the main channels where the tidal currents are strong but most parts of the banks and Creeks, where they are weak are characterized by fine sand, silt and clay. The latter has a high affinity for pollutants such as hydrocarbon and heavy metals.

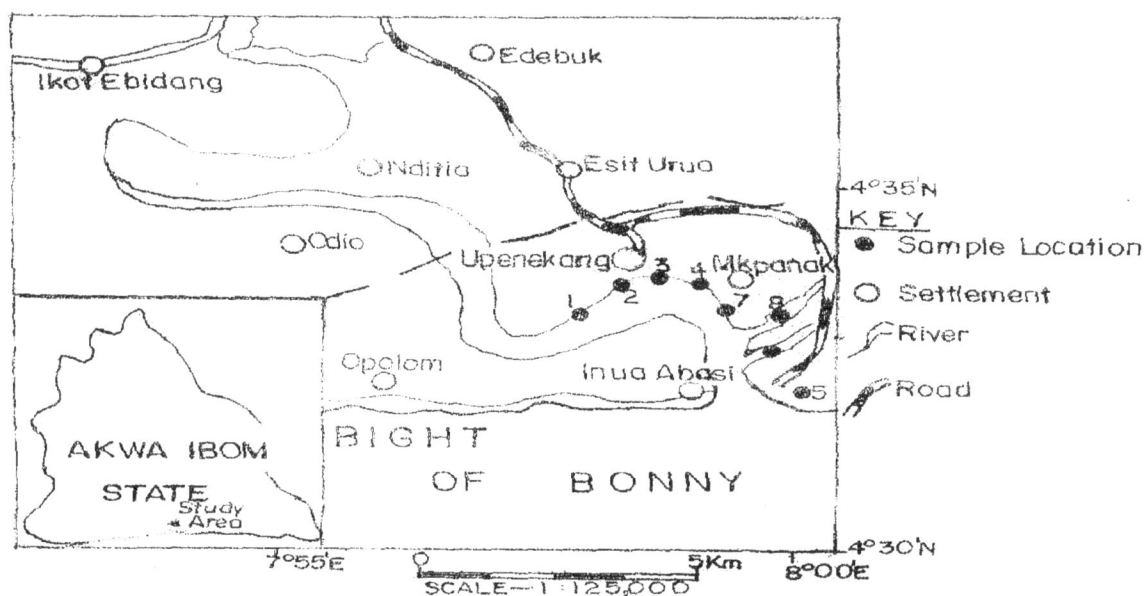

Figure 1. Map of the study area

The climate of the area is characterized by a long wet season usually lasting from April to November and a short spell of dry weather from December to March. The main occupation of the inhabitants include small scale fish farming, boat construction, transportation along the river, oil exploration, sand excavation for commercial purposes as well as timber logging of mangrove vegetation as fuel wood (Ekwere, Akpan, & Ntekim, 1992).

3. Materials and Methods

3.1 Sampling and Grain Size Analysis

A total of sixteen sediment samples were collected from the study area from eight stations using a Van Veen grab sampler. The samples for heavy metals analysis were placed in polyethylene bags while samples for organic carbon determination were placed in aluminum foil. The samples were immediately placed in an ice box as soon as retrieved and then taken to the laboratory. The sediment samples were oven dried to a constant weight at 60 °C for 24 h and were further disaggregated in an agate mortar and sieved to 63 μm sizes. The 63 μm mesh was chosen in order to normalize the result by a chemical factor. Grain size analysis was done using the sieve analysis procedure with the aid of a sieve shaker machine. The different laboratory test sieve ranged from; 2 mm

→ 1mm → 0.50mm → 0.11mm → 0.050 mm → Pan (clay). 100 grams of each sample was placed on the arranged laboratory test sieve, and inserted into the sieve shaker machine for 10 minutes. After this duration, particles that passed through were retained on the standard set of sieves of various sizes and were measured for the weight percentage of particles. The procedure was repeated for each sample and the grain size % was calculated using the formula as given by Wikipedia (2013) and modified as given below:

$$Grain\,Size = \frac{Sieve\,weight}{Total\,weight} \times 100 \tag{1}$$

3.2 Sample Digestion and Analysis of Metal Ions

1 g of sediment sample was digested with a solution of concentrated $HClO_4$ (2 ml) and HF (10 ml) to near dryness. Subsequently, a second addition of $HClO_4$ (1 ml) and HF (10 ml) were made and the mixture was evaporated to near dryness. Finally, $HClO_4$ (1ml) was added and the sample was evaporated until white fumes appeared. The residue was dissolved in concentrated HCl and diluted to 25 ml prior for heavy metals analysis using Atomic Absorption Spectroscopy (Model: SpetrAA B65).The total organic carbon content was evaluated using the Walkey and Black titration method (Walkey & Black, 1934). The sediment pH was determined using JENWAY 370 pH meter. A buffer solution with pH 7.0 was added to a beaker and the pH electrode was then inserted, the pH meter was calibrated to a pH of 7.0. 20 g of each sample was placed in a beaker; 50ml of distilled water was added to each sample and stirred for 30 minutes before inserting the probe into the system. Also, 50 ml of filtered water samples were placed in beakers after which the pH probe was inserted and the values were then read off from the electronic meter attached to the probe and data obtained recorded (Bascomb, 1994).

4. Results and Discussion

The total organic carbon (orgC) in sediments of Qua Iboe River estuary and associated creeks ranged from 0.05% to 1.36%, with an average of 0.38% during the dry season and 0.08% and 1.03% with an average of 0.28% during the wet season (Table 1). The higher values observed during dry season may be as a result of high anthropogenic activity during this season. Also, co-precipitation with carbonate minerals is another important source of organic carbon (Fortsner & Wittmann, 1983; Alloway, 1990). pH values ranged from 6.71 to 9.69 with an average of 7.82 during dry season and 5.99 to 7.67 with an average of 6.59 during wet season. The higher concentrations of Cd, Cr, Cu, Fe, Ni, Pb, Zn, and Hg during this season were attributed to lower pH levels in the sediments. The lower pH values may have been from run-off from bush land areas which particularly introduces tannic acid (tannins) which are found naturally in leaves which also account for a tea-like colour of the seas (Barnes, Meyer, & Freeman, 1998).

Table 1. Descriptive statistics

Variable	Dry Season				Wet Season			
	Mean	StDev	Min	Max	Mean	StDev	Min	Max
pH	7.82	0.85	6.71	9.65	6.59	0.61	5.99	7.65
orgC	0.38	0.41	0.05	1.36	0.28	0.31	0.08	1.03
Sand	43.6	17.39	18.62	75.32	43.6	17.39	18.62	75.32
Silt	33.58	8.37	20.42	49.68	33.58	8.37	20.46	49.68
Clay	24.07	14.72	4.22	45.24	24.07	14.72	4.22	45.24
Cd	0.08	0.04	0.02	0.16	0.21	0.36	0.04	1.1
Cr	0.1	0.04	0.01	0.13	0.06	0.04	0.03	0.14
Cu	0.09	0.03	0.01	0.13	0.07	0.03	0.03	0.13
Fe	42.2	1.9	39.54	44.63	28.53	1.14	26.05	29.4
Pb	0.08	0.03	0.03	0.13	0.07	0.03	0.03	0.12
Zn	7.51	0.29	7.06	7.85	6.44	0.34	5.86	6.9
Ni	12.51	0.49	11.43	13.03	10.86	1.07	9.2	11.95
Hg	0.0005	0.0005	0.00	0.001	0.000125	0.00035	0	0.001

The descriptive statistics of pH, orgC (%), sand (%), silt (%), clay (%) and heavy metals content (mg/g) in sediment of Qua-Iboe estuary and associated creeks during wet and dry seasons.

The percentage of sand fraction ranged from 18.62% to 75.32% with an average of 43.06%, while silt fraction ranged from 20.42% to 49.68% with an average of 33.08% and the percentage fraction of clay ranged from 4.22% to 45.28% with an average of 24.07%.

From the data presented on Table 1, it is observed that both physical and chemical parameters vary between seasons (wet and dry), and these may be as a result of differences in anthropogenic inputs due to intense seasonal variations, influence of tides and salt water intrusion (Asuquo, 1998; Ekwere et al., 1992). However, mean concentrations of Cd (0.08 ± 0.004 mg/g d·wt), Cr (0.1 ± 0.04 mg/g d·wt), Cu (0.09 ± 0.003 mg/g d·wt), Fe (42.2 ± 1.9 mg/g d·wt), Pb (0.08 ± 0.003 mg/g d·wt), Zn (7.51 ± 0.29 mg/g d·wt), Ni (12.51 ± 0.49 mg/g d·wt) obtained during dry season were higher than the mean concentrations of Cd (0.21 ± 0.03 mg/g d.wt), Cr (0.06 ± 0.004 mg/g d·wt), Cu (0.07 ± 0.003 mg/g d·wt), Fe (28.53 ± 1.14 mg/g d·wt), Pb (0.07 ± 0.003 mg/g d·wt), Zn (6.44 ± 0.34 mg/g d·wt) and Ni (10.86 ± 1.07 mg/g d·wt) obtained during the wet season. Apparently, the mean values of Cd and Zn were higher than marine sediment quality standards (MSQS) of 5.1 and 410 ppm respectively. The lower concentrations of these metals in sediment during wet season may be attributed to the lower pH levels dissolving these metals into the water column.

The analyses of the data using Pearson's correlation matrix among the levels of orgC, pH, Sand, Silt, Clay, Cd, Cr, Cu, Fe, Pb, Zn, Ni, and Hg in sediment samples during wet and dry seasons are presented in Tables 2 (a & b). During wet season, strong positive correlations were observed between Cd and Clay (r = 0.87), Pb (r = 0.65), Zn (r = 0.91), Hg (r = 0.61), Fe (r = 0.76), Cu with Clay (r = 0.80), Fe with Clay (r = 0.88), Cu (r = 0.82), Pb with Clay (r = 0.72), Fe (r = 0.64), Zn with Clay (r = 0.81), Fe (r = 0.83), Pb (r = 0.55), Ni with pH (r = 0.65), Hg with Zn (r = 0.56).

Table 2. Pearson's correlation matrix

(a)

	pH	orgC	Sand	Silt	Clay	Cd	Cu	Fe	Pb	Zn	Ni
orgC	-0.315										
Sand	0.547b	-0.028									
Silt	-0.105	0.037c	-0.650								
Clay	-0.683	-0.038	-0.822	0.135c							
Cd	-0.856	-0.053	-0.731	0.238c	0.865a						
Cu	-0.515	0.218c	-0.698	0.105c	0.797a	0.496b					
Fe	-0.631	-0.181	-0.863	0.332c	0.879a	0.756a	0.824a				
Pb	-0.335	-0.144	-0.867	0.559b	0.718a	0.646a	0.392c	0.643a			
Zn	-0.776	-0.324	-0.632	0.096c	0.806a	0.906a	0.481b	0.828a	0.551b		
Ni	0.654a	0.305c	0.382c	0.126c	-0.734	-0.839	-0.393	-0.597	-0.313	-0.827	
Hg	-0.908	0.482b	-0.239	-0.166	0.467b	0.611a	0.429b	0.390c	0.042b	0.558b	-0.511

(b)

	pH	orgC	Sand	Silt	Clay	Cd	Cr	Cu	Fe	Pb	Zn	Ni
orgC	-0.227											
Sand	-0.350	-0.079										
Silt	-0.081	0.663a	-0.650									
Clay	0.420b	-0.310	-0.822	0.135c								
Cd	-0.534	0.036c	0.138c	0.057c	-0.242							
Cr	-0.145	-0.257	0.022c	-0.416	0.446b	-0.221						
Cu	0.553b	-0.033	-0.848	0.349c	0.789a	-0.351	-0.049					
Fe	0.223c	-0.033	-0.810	0.543b	0.716a	-0.379	0.222c	0.532b				
Pb	0.237c	-0.210	-0.739	0.401b	0.682a	-0.447	0.148c	0.671a	0.766a			
Zn	0.206c	-0.179	-0.820	0.326c	0.913a	-0.223	0.497b	0.634a	0.830a	0.809a		
Ni	-0.241	0.165c	0.655a	-0.040	-0.902	0.236c	-0.627	-0.706	-0.500	-0.610	-0.856	
Hg	-0.164	-0.107	-0.041	-0.266	0.440b	-0.190	0.980a	-0.061	0.282c	0.157c	0.157c	-0.626

Pearson's correlation matrix of pH, Organic carbon, Silt, Clay and heavy metals in sediment of Qua-Iboe River estuary and associated creeks during wet (a) and dry (b) seasons.

a[*] significant at $p < 0.01$.

b[*] significant at $p < 0.05$.

c[*] not significant.

During dry season, low positive correlations were observed between Cd and orgC ($r = 0.04$), Sand ($r = 0.14$), Silt ($r = 0.06$), Ni ($r = 0.24$), Cr with ($r = 0.02$), Clay ($r = 0.45$), Fe ($r = 0.22$), Pb ($r = 0.15$), strong positive correlation with Zn ($r = 0.50$), Hg ($r = 0.98$), Strong significant correlations between Cu and pH ($r = 0.55$), Clay (0.79), Fe (0.53), Pb ($r = 0.67$) and Zn ($r = 0.63$) were observed.

There was positive correlation between Fe and pH ($r = 0.22$), Hg ($r = 0.28$) but strong positive correlations with Silt ($r = 0.54$), Clay ($r = 0.72$), Pb ($r = 0.77$) and Zn ($r = 0.83$), strong positive correlations between Pb and Clay ($r = 0.68$), Zn ($r = 0.81$), Zn with Clay (0.91) and Hg ($r = 0.54$), Ni with Sand ($r = 0.66$). Significant correlation among the variables indicates that there are linear relationships between the parameters. It is well established that organic matter content is an important controlling factor in the abundance of heavy metals (Rubio, Nombela, & Vilas, 2000). The poor correlation between Cd, Cr, Cu, Fe, Zn, Pb, Ni, Hg, and organic carbon, sand, silt fractions indicates that organic carbon and silicates are not the main geochemical carriers of the metal in sediments of the study area (Chatterjee et al., 2006).

Apparent difference of the anthropogenic inputs of heavy metals from the geogenic sources is important in evaluating the extents of heavy metal pollution. Enrichment factor (EF), Geo-accumulation index (Igeo), Modified degree of contamination (mC_d) was used to assess and interpret the pollution status of the estuary in different stations. Several kinds of refractory metals such as Al, Fe, Mg, Ti, Sc, Li and Cs have been used to normalize the grain size effect for heavy metal concentrations in sediments (Habes & Nigem, 2006; Baptista Neto, Smith, & McAllister, 2000; Schiff & Weisberge, 1999). In this study iron was used as a conservative tracer to differentiate natural from anthropogenic components. Although Fe and the heavy metals in the sediments showed discrepancies depending on the type of elements, significant correlations from Pearson correlation analysis were observed. Heavy metal concentrations were normalized to Fe to account for differences in grain size and mineralogy and then normalized by background values from the study carried out by Ekwere et al. (1992) who studied the geochemistry of sediments in Qua-Iboe estuary and associated creeks, to assess the anthropogenic input of metals in the study area. The advantage of using enrichment factor (EF) analysis is that it is possible to establish a contamination guideline. This technique has been well applied in several studies to assess metal contamination in marine sediments (Khaled, El-Neme, & El-Sikkaily, 2006; Acevedo-Figueroa, Jimenez, & Rodriguez-Sierra, 2006; Ghrefat & Yusuf, 2006; Barakat, Baghdadi, Rais, & Nadem, 2012; Mashiatullah, Chaudhary, Ahmed, Javel, & Ghaffar, 2013).

Enrichment factor is a convenient measure of geochemical trends and is used for comparison between areas. It is applied widely in sediment geochemical studies (Abraham, 1998; Soto-Jimenez & Pacz-Osuna, 2001; Kamau 2002; Qu, Chen, Yang, & Lu, 1993; Kehing, Pinto, Moreira, & Malm, 2003; Barakat et al., 2012).

According to Ergin, Saydam, Basturk, Erdem, and Yoruk (1991), the metal enrichment factor (EF) is defined by the equation below:

$$EF = \frac{\dfrac{M}{Fe}\,Sample}{\dfrac{M}{Fe}\,Background} \qquad (2)$$

$\dfrac{M}{Fe}\,Sample$ is the ratio of metal and Fe concentration of the sample, and $\dfrac{M}{Fe}\,Background$ is the ratio of the metal and Fe concentration of the background value.

The formula below was applied to the studied heavy metals in the study to assess the anthropogenic and lithogenic contributions:

$$\lfloor M \rfloor Lithogenic = \lfloor Fe \rfloor Sample \times \left(\frac{M}{Fe} \right) Lithogenic \qquad (3)$$

where $\left(\dfrac{\lfloor M \rfloor}{\lfloor Fe \rfloor} \right) Lthogenic$ corresponds to the average background ratio. The anthropogenic heavy metals can be estimated by the formula shown below:

$$\lfloor M \rfloor Anthropogenic = \lfloor M \rfloor Total - \lfloor M \rfloor Lithogenic \qquad (4)$$

Birch (2003) divided contamination into different categories based on EF values. EF<1 demonstrates "no enrichment", 1<EF<3 is "minor enrichment", EF=3-5 is "moderate enrichment", EF=5-10 is "moderately severe enrichment", EF=10-25 is "severe enrichment", EF=25-50 is "very severe enrichment" and EF>50 is "extremely severe enrichment". The enrichment of heavy metals in sediments of Qua Iboe River estuary and associated creeks is shown in Table 3.

The factor obtained for the studied area revealed that there were extreme enrichment of cadmium in all the stations during dry and wet seasons, severe enrichment of zinc, minor to moderate enrichment of lead, minor enrichment of copper, minor enrichment of nickel, no enrichment of chromium, mercury and iron. This is similar to work reported by Joseph (2002) in sediments of Port-Reitz creek, Mombasa; Rezaee, Saion, Yab, Abdi, and Riyahi (2010) in sediments cores from South China Sea; Habes and Nigem (2006), in bottom sediments of Wadi

Al-Arab Dam, Jordan.

Table 3. The result of enrichment factor (EF)

Stations	WET SEASON								DRY SEASON							
	Cd	Cr	Cu	Fe	Pb	Zn	Ni	Hg	Cd	Cr	Cu	Fe	Pb	Zn	Ni	Hg
Ikotlwang	56	0.6	4.2	1	1	11.5	1.1	5.1	32.1	0.5	6.2	1	7.3	15.3	1.2	0
Iwoachang	47.3	0.5	3.8	1	3.7	11.4	1.1	0	45.4	0.4	4.4	1	4.3	14.3	1.1	0
Ukpenekang	39.5	0.55	2.9	1	3.6	11.3	1.1	0	65	0.3	3.6	1	4.4	14.6	1.5	0
Atasi	43	0.4	1	1	3.8	12.1	1.3	0	71.9	0.4	1.5	1	4.4	14.6	1.6	0
Ukpenekang	9.5	0.7	2.5	1	1.5	11.8	1.3	0	107.4	0.5	2.3	1	2	14.8	1.7	0
Douglas creek	41.4	0.6	3.5	1	2	11.6	1.2	5	7.5	0.4	2	1	2.6	15	1.6	0
Egerton Port	48	0.8	2.7	1	2	11.9	1.1	4	45.8	0.3	2.5	1	3.1	14.1	1.6	0
Stubbs creek	69	2.3	3.3	1	3.5	11.9	1	0	25.7	1.3	2.9	1	4.9	15.5	1.2	0
Mean	44.26	0.81	2.98	1	2.64	11.7	1.2	1.8	50.1	0.5	3.2	1	4.13	14.8	1.4	0

The result of enrichment factor (EF) for both wet and dry seasons in the study area.

According to these researchers, possible enrichment of Cd in bottom sediments was attributed to anthropogenic inputs from fertilizers and pesticides used in agricultural activities. Manaf, Samah, and Zukki (2009) reported that domestic wastes is the primary source of the generation of solid wastes as a result the high concentration of Cd in Malaysia and its coast.

Sediment bacteria may also assist in the partitioning of cadmium from water to sediments. Studies indicate that concentrations of cadmium in sediments are at least one order of magnitude higher than in the overlying water. The mode of sorption of cadmium to sediments is important in determining its deposition and remobilization into water column (Harisson & De Mora, 1996). Cadmium, associated with carbonate minerals or co-precipitated with hydrous iron oxides, is less likely to be mobilized by re-suspension of sediments or biological activity. Cadmium that is adsorbed to mineral surfaces such as clay, or to organic materials, is more easily bioaccumulated or released in the dissolved state when the sediment is disturbed. Cadmium may re-dissolve from sediments under varying ambient conditions of pH, salinity, and redox potential (Di Toro et al., 1991). Cadmium is not known to form volatile compounds in the aquatic environment, so partitioning from water to the atmosphere does not occur. The geo-accumulation index was introduced by Muller in 1979. The model was used to assess metal pollution in sediments from Qua Iboe river estuary besides the enrichment factor. Geo-accumulation Index is expressed as:

$$Igeo = Log_2\left(\frac{c_n}{1.5B_n}\right) \tag{5}$$

where c_n is the measured concentration of the heavy metal (n) in the <63μm fraction of the sediment, B_n is the geochemical background value in average shale (Turekian & Wedepohl, 1961) of element n, and 1.5 is the background correction factor due to lithogenic effects. The index of geo-accumulation includes seven grades which show various degrees of enrichment above the background value ranging from unpolluted to much polluted water and sediment quality (Table 3). The highest grades (class six) reflect 100-fold enrichment above the background values (Sing, Hasnain, & Baneriee, 2003).

Table 4. Muller's classification for geo-accumulation index

Igeo value	Class	Sediment quality
≥0	0	Unpolluted
0-1	1	From unpolluted to moderately polluted
1-2	2	Moderately polluted
2-3	3	Moderately polluted to strongly polluted
3-4	4	Strongly polluted
4-5	5	strongly polluted to extremely polluted
>5	6	Extremely polluted

Muller's classification for geo-accumulation index.

Table 5. Results of geo-accumulation index (Igeo)

	WET SEASON								DRY SEASON							
Stations	Cd	Cr	Cu	Fe	Pb	Zn	Ni	Hg	Cd	Cr	Cu	Fe	Pb	Zn	Ni	Hg
IkotIwang	8.2	1.3	1.0	-0.5	1.2	1.9	6.5	1.2	2.6	0.7	1.0	-0.9	1.1	1.9	6.3	0.0
Iwoachang	3.4	1.1	0.9	-0.5	0.8	2.0	6.5	0.0	2.9	0.5	0.6	0.9	0.6	1.8	6.4	0.0
Ukpenekang	3.2	1.1	0.6	-0.5	0.8	1.9	6.6	0.0	3.3	0.2	0.5	-0.9	0.6	1.8	6.4	0.0
Atasi	3.2	0.7	-0.5	-0.6	0.8	1.9	6.0	0.0	3.4	0.5	-0.5	-0.9	0.6	1.8	6.5	0.0
Mkpanak	1.7	1.4	0.4	-0.6	-0.2	1.9	6.6	0.0	3.6	0.7	-0.2	-1.1	-0.3	1.7	6.5	0.0
Douglas creek	3.2	0.6	0.7	-0.5	0.2	1.9	6.6	1.2	5.7	0.5	-0.2	-0.9	0.03	1.8	5.4	0.0
Egerton Port	3.4	1.6	0.5	-0.5	0.2	2.0	6.5	1.2	2.9	0.2	0.01	-0.9	0.2	1.8	6.5	0.0
Stubbs creek	3.8	2.6	0.7	-0.5	0.8	2.0	6.5	1.2	2.4	1.7	0.2	-0.9	0.7	1.9	6.2	1.2
Mean	3.8	1.3	0.5	-0.5	0.6	1.9	6.5	0.6	3.4	0.6	0.2	-0.9	0.4	1.8	6.3	0.2

Results of geo-accumulation index (Igeo) for both wet and dry seasons in the study area.

Results from the mean geo-accumulation index (Equation 5. and Table 4) reveal the following trend Ni > Cd > Zn > Cr > Cu > Pb > Hg >Fe (Figure 7). Based on Muller (1979) classification (Table 3), marine sediment for geo-accumulation of metals, Ni belongs to class 6 (extremely polluted), Cd belongs to class 4 (strongly polluted), Zn belongs to class 2 (moderately polluted), Cr, Cu, Pb, and Hg belongs to class 1 (unpolluted to moderately polluted) and Fe belongs to class 0 (background concentration) making it a suitable normalizer for sediments of the study area. The geo-accumulation is important because on a weight per square meter basis, the uppermost superficial sediments serve as the largest heavy metal sinks in marine ecosystem. Once these heavy metals accumulate in sediments, they tend to pose threats to aquatic life as a result of re-suspension into the water column from geochemical cycling, bioaccumulating in benthic organisms that feed on substrate nutrient and also biomagnified through the aquatic food web.

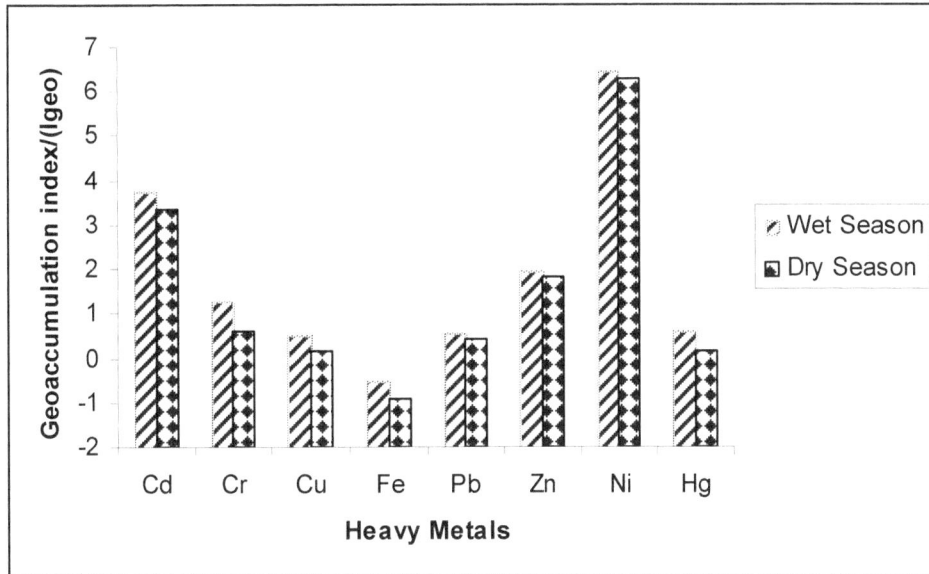

Figure 2. Seasonal variations in geo accumulation index of heavy metals in sediments of the study area

Modified degree of contamination (mC_d) is based on the calculation for each pollutant of a contamination factor (Cf). However, the Cf requires that at least five surficial sediment samples are averaged to produce a mean pollutant concentration which is then compared to baseline pristine reference level, according to the equation below:

$$CF = \frac{C_{Sample}}{C_{background}} \qquad (6)$$

where C_{sample} and $C_{background}$ respectively refer to the mean concentration of a pollutant in the contaminated sediments and the pre-industrial "baseline" sediment or average shale. The numeric sum of the K specifies contamination factors which express the overall degree (Hakanson, 1980) of sediment contamination (C_d) using the following formula:

$$C_d = \sum_{i=1}^{k} cf_i \qquad (7)$$

The C_d is aimed at providing a measure of the degree of overall contamination in surface layers of the sediment. Furthermore, all n species must be analyzed in order to calculate the correct C_d for the range of classes defined by (Hakanson, 1980). The modified formula is generalized by defining the degree of contamination m_{Cd} as the sum of all the contamination factors Cf for a given set of estuarine pollutants divided by the number of analyzed pollutants. The modified equation for a generalized approach to calculating the degree of contamination is given:

$$m_{Cd} = \frac{\sum_{i=1}^{k} Cf_i}{n} \qquad (8)$$

where n is the number of analyzed elements and (i) is i_{th} element (or pollutant) and Cf is formula to calculate M_{cd}. It allows the incorporation of as many metals as the study may analyze with no upper limit. Table 5 shows the model for classifying estuarine sediment.

Table 6. Hakanson (1980) classification of the modified degree of contamination (Abrahim et al., 2007)

m_{Cd} values	Sediment quality
$mCd < 1.5$	Nil to low degree of contamination
$1.5 \leq mCd < 2$	Low degree of contamination
$2 \leq mCd < 4$	Moderate degree of contamination
$4 \leq mCd < 8$	High degree of contamination
$8 \leq mCd < 16$	Very high degree of contamination
$16 \leq mCd < 32$	Extremely high degree of contamination
$mCd > 32$	Ultra high degree of contamination

Table 7. Results of modified degree of contamination

Stations	mCd	
	Dry season	Wet season
IkotIwang	40.00	36.00
Iwoachang	7.20	2.90
Ukpenekang	0.91	0.88
Atasi	2.73	2.10
Mkpanak	0.80	0.61
Douglass creek	3.08	2.69
Egerton Port	10.50	8.98
Stubbs creek	10.40	8.99
Mean	9.34	7.78

Modified degree of contamination using pristine values (Ekwere et al., 1992) for heavy metals in bottom sediments from Qua-Iboe River estuary and associated creeks, South Eastern Nigeria.

The earlier indices (EF and Igeo) gave diverse classification of sediment quality of the study area. The modified degree of contamination has an advantage over other indices and provides a measure of the overall degree of contamination of all the chemical parameters in each sampling station. Based on Hakanson classification (Equation 8 and Table 5), Ikot Iwang shows an ultra high degree of contamination during both seasons, Iwoachang is moderately contaminated from the wet season results but highly contaminated from the dry season resuts, Ukpenekang shows low degree of contamination, Atasi moderately polluted while Mkpanak shows low degree of contamination during both seasons, Douglass Creek is moderately polluted, Egerton Port and Stubbs Creek show very high degree of contamination. However, on the average, the sediments of Qua Iboe River estuary and some associated creeks were found to be highly contaminated during both seasons.

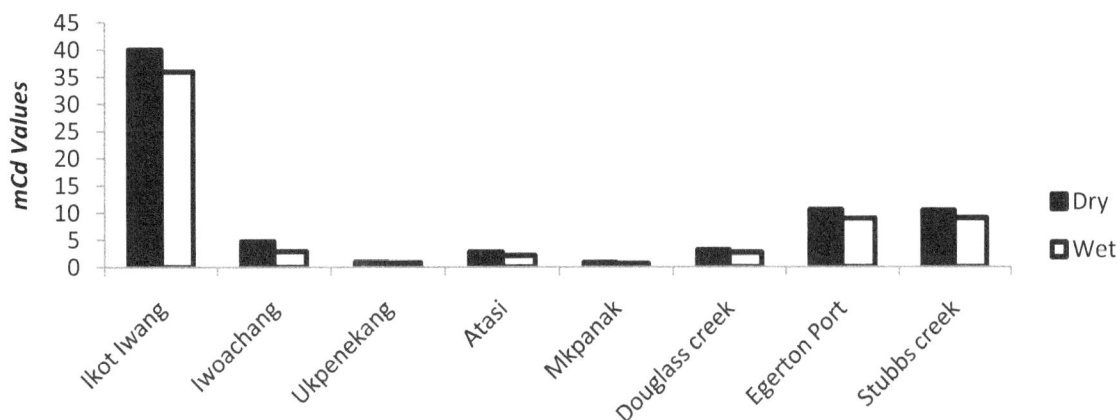

Figure 3. Modified degree of contamination

5. Conclusion

Identification and quantification of heavy metal sources, as well as their enrichment in marine sediments are important environmental scientific issues. Descriptive statistics showed that the mean concentrations of Cr, Cu, Fe, Pb, Ni, and Hg were lower than marine sediment quality standard (MSQS) but the mean concentration of Cd and Zn were significantly higher than MSQS. There were significant higher concentrations of heavy metals during the dry season than the wet season. Analysis of the data using Pearson correlation matrix showed significant correlation between the metals, Fe and Clay suggesting that Fe is the best chemical normalizer for other metals. Different metal assessment indices were applied in order to interpret the sediment's quality of Qua Iboe River estuary and associated creeks. All the EF's values of Cr, Fe and Hg were less than 1.5. However, the EF's values for Cd, Cu, Pb, Zn, and Ni were greater than 1.5 signifying greater percentages from anthropogenic inputs. The sources of pollution were mainly high surface runoffs, domestic effluents from the coastal dwellers and possibly from the oil exploration in the area. The high Igeo values for Ni may be attributed to oil spill and the slow water movement along the creeks. With regards to an overall degree of contamination as proposed by Hakanson (1980) sediment quality equation, the sediments of Qua Iboe River estuary and associated creeks fall between $8 \leq mCd < 16$ indicating high degree of contamination.

Acknowledgements

The authors would like to thank the following for assisting during sampling process, Eno Sampson, Evelyn Sampson and Patience Ekpenyong of the Department of Pure & Applied Chemistry. We are also grateful to Mr Effiom and Mr Charles, the laboratory staff who assisted immensely in the analytical work.

References

Abdullah, M. H., Sidi, J., & Aris, A. Z. (2007). Heavy metals (Cd, Cu, Cr, Pb, and Zn) in meretrix Roding, water and sediments from Estuaries in Sabah, North Borneo. *International Journal of Environmental & Science Education, 2*(3), 69-74.

Abraham, J. (1998). Spatial distribution of major and trace elements in Shaw Reservoir Sediments: An example from Lake Waco. Texas. *Environmental Geology, 36*, 3-4. http://dx.doi.org/10.1007/s002540050351

Acevedo-Figueroa, D., Jiménez, B. D., & Rodríguez-Sierra, C. J. (2006). Trace metals in sediments of two estuarine lagoons from Puerto Rico. *Environmental Pollution, 141*, 336-342. http://dx.doi.org/10.1016/j.envpol.2005.08.037

Alloway, B. J. (1990). *Heavy Metals in Soils*. Glasgow, London: John Wiley and Sons.

Ankley, G. T., Lodge, K., Call, D. J., Balcer, M. D., Brooke, L.T., Cook, P. M., . . . McAllister, J. J. (1992). Heavy metal concentrations in surface sediments in a nearshore environment, Jurujuba Sound, Southeast Brazil. *Environmental Pollution, 109*, 1-9.

Asuquo, F. E. (1998). Water pollution monitoring of Great Kwa River at Calabar (pp. 1-12). *In proceedings of the 1st conference of Nigeria water and sanitation*, Port Harcourt.

Babtista, J. A., Smith, B. J., & McAllister, J. J. (2000). Heavy metal concentrations in surface sediments in a nearshore environment, Jurujuba Sound, Southeast Brazil. *Environ. Pollut., 109*, 1-9. http://dx.doi.org/10.1016/S0269-7491(99)00233-X

Barakat, A., Baghdadi, M. E., Rais, J., & Nadem, S. (2012). Assessment of heavy metal in surface sediments of Day River at Beni–Mellal Region, Morocco. *Research Journal of Environmental and Earth Sciences, 4*(8), 797-806.

Barnes, K. H., Meyer, J. L., & Freeman, B. J. (1998). Sedimentation and Georgia's Fishes: An analysis of existing information and future research. *1997 Georgia Water Resources Conference*, March 20-22, 1997, the University of Georgia, Athens Georgia.

Bascomb, C. L. (1994). Physical and chemical analysis of < 2 mm samples. Soil survey laboratory methods. In B. W. Averg, & C. L. Bascomb (Eds.), *Soil Survey Technology Monographs, 6*, 14-41

Berg, S. B. (1999). Iron as a reference element for determining trace metal enrichment in Southern California coast shelf sediments. *Mar. Environ. Res., 48*, 161-176.

Calmano, W., Hong, J., & Forstner, U. (1993). Binding and mobilization of heavy metals in contaminated sediments affected by pH and redox potential. *Water Sci. Technol., 28*(8-9), 223-235.

Chtterjee, M., Silva-Filho, E. V., Sarkar, S. K., Sella, S. M., Bhattachrya, K., Satpaty, K. K., . . . Bhattachrya, B. D. (2006). Distribution and Possible Source of Trace Element in Sediment Core of Tropical Macrotidal Estuary and Their Ecotoxicological Significance. *Environment International, 3*, 346-356.

Dahilia, A., Apodaca, C., Emerson, J., Tui, N. G., & Allyn, S. V. (2003). Remediation of Pb^{2+}, Ca^{2+} and Zn^{2+} metal ions in waste water samples using iota-carrageenan. *Research Journal of Chemistry and Environment, 7*(2), 23-45.

Di Toro, D. M., Zarba, C. S., Hansen, D. J., Berry, W. J., Swartz, R. C., Coman, C. E., . . . Pasquin, R. P. (1991). Technical basis for establishing sediment quality criteria for non ionic organic chemicals by using equilibrium partitioning. *Environ. Toxicol. Chem., 10*, 1541-1583. http://dx.doi.org/10.1002/etc.5620101203

Ekwere, S. J., Akpan, E. B., & Ntekim, E. E. (1992). Geochemical studies of sediments in Qua Iboe estuary and associated creeks, South Eastern Nigeria. *Tropical Journal of Applied Science, 2*, 91-95.

Ergin, M., Saydam, C., Basturk, O., Erdem, E., & Yoruk, R. (1991). Heavy metal concentrations in surface sediments from the two coastal inlets(Golden Horn Estuary and Izmit Bay) of the northeastern Sea of Marmara. *Chem. Geo., 91*, 269-285. http://dx.doi.org/10.1016/0009-2541(91)90004-B

Forstner, U., & Wittmann, G. T. W. (1981). *Metal pollution in aquatic environment*. New York: Springer. http://dx.doi.org/10.1007/978-3-642-69385-4

Forstner, U., & Wittmann, G. T. W. (1983). *Metal Pollution in the Aquatic Environment*. Berlin, Heidelberg, New York, Tokyo: Springer-Verlag.

Ghrefat, H., & Yusuf, N. (2006). Assessing Mn, Fe, Cu, Zn and Cd pollution in bottom sediments of Wadi Al-Arab Dam, Jordan. *Chemosphere, 65*, 2114-2121. http://dx.doi.org/10.1016/j.chemosphere.2006.06.043

Gibbs, R. J. (1973). Water chemistry of the Amazon River. *Geochim. Cosmochim Acta, 36*, 1006-1066.

Habes, G., & Nigem, Y. (2006). Assessing Mn, Fe, Cu, Zn, and Cd pollution in bottom sediments of wadi Al-rab Dam, Jordan. *Chemosphere, 1-6*.

Hakanson, L. (1980). Ecological risk index for aquatic pollution control, a sedimetological approach. *Water Research, 14*, 975-1001. http://dx.doi.org/10.1016/0043-1354(80): 90143-8

Harrison, R. M., & De Mora, S. J. (1996). *Introduction chemistry for the Environmental Sciences* (2nd ed.). Cambridge: University Press.

Kamau, J. N. (2002). Heavy metals distribution and enrichment at port–Reitz Creeks, Mombasa Western Indian Ocean. *Journal Marine Science, 1*, 65-70.

Kehing, H. A., Pinto, F. N., Moreira, I., & Malm, O. (2003). Heavy metals and methyl mercury in a tropical coastal estuary and a mangrove in Brazil. *Organic Geochemistry, 34*, 661-669. http://dx.doi.org/10.1016/S0146-6380(03)00021-4

Khaled, A., El-Nemr, A., & El-Sikaily, A. (2006). An assessment of heavy-metal contamination in surface sediments of the Suez Gulf using geo-accumulation indices and statistical analysis. *Chem. Ecol., 22*(3), 239-252. http://dx.doi.org/10.1080/02757540600658765

Leivouri, M. (1998). Heavy metal contamination in surface sediment in theGulf of Finland and comparison with the Gulf of Bothnia. *Chemosphere, 36*, 43-59. http://dx.doi.org/10.1016/S0045-6535(97)00285-3

Maher, W. A., & Aislabe, J. (1992). Polycyclic aromatic hydrocarbons in nearshore marine sediments of Australia. *Sci. Total Environ., 11*(3), 143-164. http://dx.doi.org/10.1016/0048-9697(92)90184-T

Manaf, L. A., Samah, M. A. A., & Zukki, N. I. M. (2009). Municipal Solid Wastes Management in Malaysia: Practices and Challenges. *Wastes Management, 29*, 2902-2906. http://dx.doi.org/10.1016/j.wasman.2008.07.015

Manahan, S. E. (1991). *Environmental chemistry* (5th ed., pp. 565). Lewis Publishers.

Mashiatullah, A., Chaudhary, M. Z., Ahmed, N., Javel, T., & Ghaffar, A. (2013). Metal pollution and ecological risk assessment in marine sediments of Karachi Coast, Pakiatan. *Environmental Monitoring and Assessment, 185*(2), 1555-65. http://dx.doi.org/10.1007/s10661-012-2650-9

Muller, G. (1979). Schwermemalle in den sedimenten des Rheine-Vera Enderungenseit. *Umschau, 79*, 778-783.

Nriagu, J. O. (1989). A global assessment of the natural sources of atmosphere trace metals. *Nature, 338*, 47-49. http://dx.doi.org/10.1038/338047a0

Qu, C. H., Chen, C. Z., Yang, L. Z., & Lu, Y. L. (1993). Geochemistry of dissolved and Particulate Elements in the Major Rivers of China (The Huanghe, Changjiang, and Zhunjiang Rivers). *Estuaries, 16*, 475-467. http://dx.doi.org/10.2307/1352595

Razaee, K. H., Saion, E. B., Yap, C. K., Abdi, M. R., & Riyahi, B. A. (2010). Vertical Distribution of heavy Metals and Enrichment in the South China Sea Sediment Cores. *International Journal of Environment, 4*(4), 877-886.

Rubio, B., Nombela, M. A., & Vilas, F. (2000). Geochemistry of major and trace elements in sediments of the Ria de Vigo (NW Spain): An assessment of metal pollution. *Marine pollution Bulletin, 40*(11), 968-980. http://dx.doi.org/10.1016/S0025-326X(00)00039-4

Schiff, K. C., & Weisberg, S. B. (1999). Iron as a reference element for determining trace metal enrichment in Southern California coast shelf sediments. *Mar. Environ. Res., 48*, 161-176. http://dx.doi.org/10.1016/S0141-1136(99)00033-1

Sieve analysis. (2013). In Wikipedia, the free encyclopedia. Retrieved July 17, 2013 from en.wikipedia.org/wiki/sieve_analysis

Singh, A. K., Hasnain, S. I., & Baneriee, D. K. (2003). Grain Size and Geochemical Portioning of heavy metals in sediments of the Damodar River – A tributary of the lower Ganga, India. *Environmental Geology, 39*, 90-98. http://dx.doi.org/10.1007/s002540050439

Soto-Jimenez, M. F., & Pac-Osuna, F. (2001). Distribution and Normalization of heavy metal concentrations in mangrove and lagoonal sediments from Mazatlan Habour (S.E. Gulf of California). *Estuarine, Coastal and Shelf Science, 53*, 259-274. http://dx.doi.org/10.1006/ecss.2000.0814

Turekian, K. K., & Wedepohl, K. H. (1961). Distribution of the elements in some major units of the earth's crust. *Bulletin Environ.*

Wen, X. A., & Allen, H. E. (1999). Mobilization of heavy metals from Len River sediments. *Sci. Total Environ., 227*, 27-38. http://dx.doi.org/10.1016/S0048-9697(99)00002-9

6

Metal Contents in Sediments (Cd, Cu, Mg, Fe, Mn) as Indicators of Pollution of Palizada River, Mexico

Carlos Montalvo[1], Claudia A. Aguilar[1], Luis E. Amador[2], Julia G. Cerón[1], Rosa M. Cerón[1], Francisco Anguebes[1] & Atl V. Cordova[1]

[1] Universidad Autónoma del Carmen, Dependencia de Ciencias Químicas y Petrolera, México

[2] Universidad Autónoma del Carmen, Centro de Investigación de Ciencias Ambientales, México

Correspondence: Claudia A. Aguilar, Universidad Autónoma del Carmen, Dependencia de Ciencias Químicas y Petrolera, Calle 56 No.4, Ciudad del Carmen, Campeche, México. E-mail: caguilar@pampano.unacar.mx

Abstract

Some heavy metals and trace metals reach aquatic ecosystems from natural and anthropogenic sources, and are considered some of the most important environmental contaminants due to their toxicity, persistence and tendency to accumulate in aquatic organisms. Thus, their study is needed due to the environmental risk they pose. Concentrations of Cu, Cd, Mg, Fe and Mn in recent sediments of the deltaic lagoon-river system of the Palizada river, Campeche, Mexico were determined for three climatic seasons on the 2010 annual cycle. The results confirmed that the climatic season has great influence over the results variability. The highest levels of Cu, Fe and Mn were found during dry season, which may suggest significant evaporation phenomena in the area. Both Fe and Mn are abundant elements in the Earth crust; their concentrations could be related to the study area's characteristics, given the conjunction of two sedimentary provinces: terrigenous in the western portion and carbonated in the eastern. On the other hand, the results suggest a high relationship of Fe-Mn (r = 0.5131), Fe-clay (r = 0.5978), Cu-Mn (r = 0.8707), Cu-clay (0.8501) and Mn- clay (0.9311). The latter confirms the high dependence of these elements and the great affinity of some metallic elements for finer sediments. In conjunction, the climatic season and the sediment's characteristics are essential for metal mobilization and transport. Likewise, the Cd and Cu levels reported are lower than international parameter, indicating value ranges that could cause effects in exposed organisms.

Keywords: trace metals, Palizada river, Gulf of Mexico

1. Introduction

Coastal ecosystems are currently exposed to high amounts of contaminants coming from industrial and urban activities being poured discharged into the systems, which contributes to increasing the concentration of certain contaminants (Cuong et al., 2005; Vane et al., 2009).

Most heavy metals released into the environment reach aquatic systems through direct discharge, wet and dry deposition and erosion. Sediments can accumulate the heavy metals reaching the lake environment; on the other hand, changes in physicochemical conditions can remobilize and release metals into the water column These contributions can be transferred through the food chain to individuals.

The study of sediments in an aquatic ecosystem allows a comprehensive assessment of a site contamination, since they are the main recipients of most of the contaminants deposited from the water column by precipitation. The metals associated to the organic fraction can form solutions and become available for aquatic life, making it possible to establish a relationship between the sediments heavy metals and the living organisms from a given environment (Aguilar et al., 2012).

In this sense, Santos et al. (2003) & Lima et al. (2005) proved the importance of sediments as water pollution indicators. Characteristics such as size of grain sediments and the amount of organic matter are of great importance as they determine the presence and availability of certain contaminants such as heavy metals and trace metals (Sauvé et al., 2000).

1.1 Background of the Study Area

Terminos lagoon is located in the Gulf of Mexico; it is one of the most studied ecosystems in Mexico due to its importance as a nesting site for many bird species and holding many flora and fauna species. It covers an area of 705,016 hectares, making it in one of the largest protected natural areas of Mexico (INE/SEMARNAP, 1997). The northern continental shelf of the Laguna is highly productive in oil extraction, and it has been shown that the latter can be a source of contaminants (hydrocarbons, heavy metals and others). Besides the rivers (Palizada, Chumpán and Candelaria), some inland lakes such as Pom, Atasta, del Corte, San Carlos, del Este, Balchacah y Panlau –which flow into the lagoon in the south– are a source of agrochemicals and other contaminants.

The oil extraction industry located in this area emits different contaminants into the air, water and soil, including nickel, chromium, lead and cadmium (RETC, 2011).

Considering the area's characteristics, it is important to monitor metal concentrations in order to detect the contamination source and the sediment quality.

2. Method

2.1 Sediment Recollection

Three sampling campaigns were conducted during the rainy, dry and north winds climatic seasons. With that purpose, ten sampling stations were established, which were identified in a previous tour using a global positioning system (GPS). Surface sediment samples were collected at each point using a 0.1-m^3 Van Veen dredge. All sediment samples collected were stored in polypropylene containers previously washed with a solution of 10% HCL.

Figure 1 shows the study area, highlighting the sampling points.

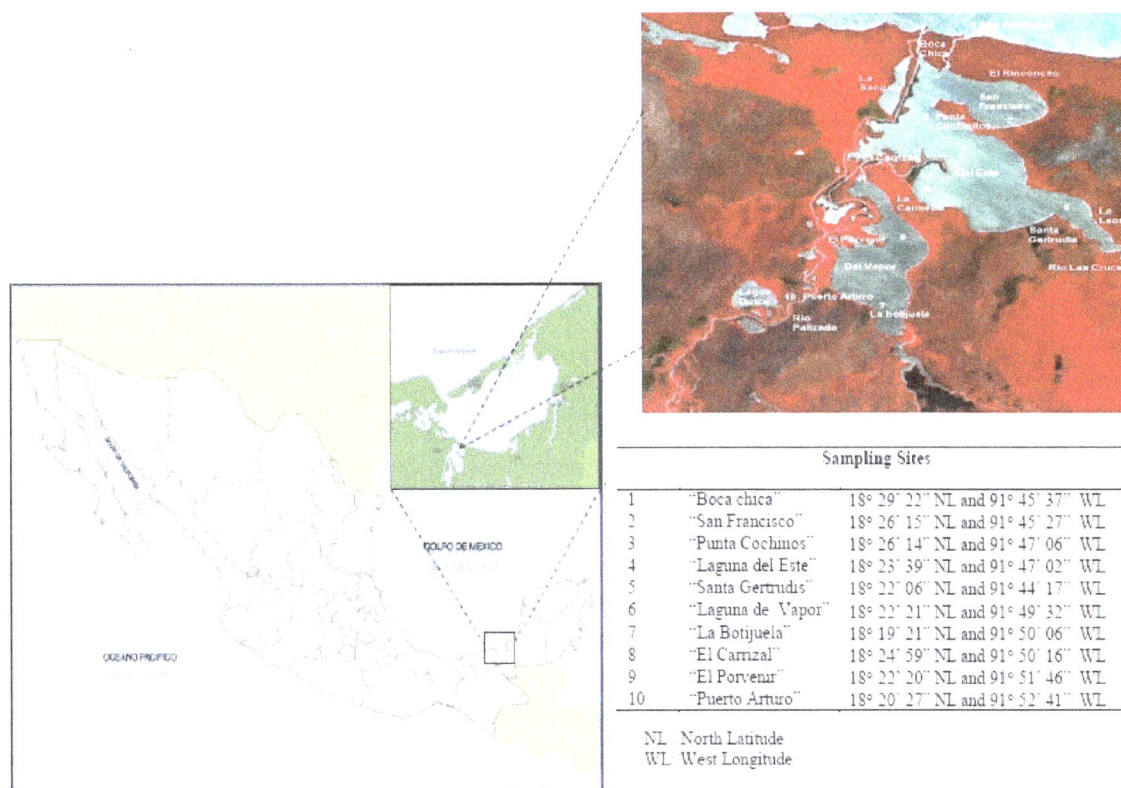

	Sampling Sites	
1	"Boca chica"	18° 29' 22" NL and 91° 45' 37" WL
2	"San Francisco"	18° 26' 15" NL and 91° 45' 27" WL
3	"Punta Cochinos"	18° 26' 14" NL and 91° 47' 06" WL
4	"Laguna del Este"	18° 23' 39" NL and 91° 47' 02" WL
5	"Santa Gertrudis"	18° 22' 06" NL and 91° 44' 17" WL
6	"Laguna de Vapor"	18° 22' 21" NL and 91° 49' 32" WL
7	"La Botijuela"	18° 19' 21" NL and 91° 50' 06" WL
8	"El Carrizal"	18° 24' 59" NL and 91° 50' 16" WL
9	"El Porvenir"	18° 22' 20" NL and 91° 51' 46" WL
10	"Puerto Arturo"	18° 20' 27" NL and 91° 52' 41" WL

NL : North Latitude
WL : West Longitude

Figure 1. Study area, showing the three sampling areas

2.2 Treatment of Samples

Collected sediments were dried by lyophilization for 24 hours, and homogenized using an agate mortar. A portion of the sediments was used to evaluate organic matter using the technique suggested in the AA-034-NMX-SCFI-2001, which is based on the weight loss on ignition.

Analyses of sediment texture were performed through the Bouyoucos-scale hydrometer technique proposed by Buchanan & Kain (1971). For metal analysis the analytical technique suggested by EPA in method 3050B, consisting of an acid digestion with HNO_3 in a sediment sample was used, the digestion is completed with the addition of H_2O_2, the volume is reduced, finally filtered sample and deionized water to volume for subsequent analysis.

2.3 Analysis of Metals in Sediments Samples

All samples were analyzed by Flame Atomic Absorption Spectrophotometer (AVANTA GBC-A4509), using standard solutions at different concentrations of recognized analytical grade (J.T Baker). The solutions were prepared according to the working ranges of equipment operation, (Table 1). All samples wer analyzed by duplicate.

Table 1. Operating limits in detecting elements

Element	Wavelength (nm)	Detection range ($\mu g\ L^{-1}$)*
Cadmiun	228.0	0.01 – 5.00
Iron	386.0	5.00 - 300
Cooper	222.0	0.10 -60
Magnesium	322.5	0.1- 150
Manganese	219.0	0.2-300

* Data manual operation of the equipment AVANTA GBC-A4509

2.4 Statistical Analyses

All data were evaluated to determine normality by applying the Shapiro-Wilks test, which is used for testing normality of a data set. It states as null hypothesis that a sample $x1$, ..., xn comes from a normally distributed population (Shapiro & Wilk, 1965). The Statgraphics software was used for this purpose, with a confidence level of 95%, which confirmed the normality of the experimental data.

To determine the heavy metal levels and soil texture variation per sampling site and climatic season, an analysis of variance (ANOVA) test was applied. Finally, a Pearson correlation was applied to determine the relationship between metals and the characteristics of sediment, as well as organic matter.

3. Results and Discussion

In marine systems, heavy metals attach to the sediment by processes such as adsorption and co-precipitation by iron and manganese hydroxides and oxides, adsorption in mineral clays, precipitation with organic matter, hydrolytic reactions of ions and dissolved complexes and other natural mechanisms related to the sediments physicochemical characteristics (Bruder et al., 2002; Sutherland & Tack, 2002).

Marine sediments act as metal integrators and concentrators (García et al., 2004). Depending on the metals physical and chemical properties, they can be mobilized and transported and even pose serious hazards for an ecosystem health. Therefore, their monitoring and control is a necessary action for the conservation of the environment, as well as to define mitigation policies and actions.

3.1 Manganese and Iron

Result obtained (Table 2) of the analysis of heavy metals in sediments of the Palizada river clearly showed that one of the most abundant elements is iron; other studies have found atypical iron concentrations, which are considered indicators of an increase in the rate of fine-material sediments contributed by the rivers (Zarazúa et al., 2011). Likewise, it was considered that there is an influence by the neighboring industries for this metal atypical increase (Mora et al., 2013).

Table 2. Ranges of metals concentrations (annual) in µg g^{-1}

Climatic season	Elements				
	Cu	Cd	Fe	Mg	Mn
I Dry	5.8-10.48	1.00 -1.71	216.6 -232.47	3.28– 4.61	127.51- 246.58
Mean	9.658	1.426	225.89	4.365	187.903
*SD	4.0065	0.260	5.2512	0.389	41.407
II Rainy	0.19 -0.256	1.01–1.60	215.4-222.54	14.69-17.91	6.89-13.44
Mean	0.1954	1.335	219.948	16.904	9.50477
*SD	0.056	0.2144	2.93225	1.1129	2.1569
III North winds	1.1-1.23	2.19-2.34	12.9-34.6	5.18-22.16	2.77-15.09
Mean	1.154	2.261	23.769	12.238	6.4262
*SD	0.04409	0.0427	6.846	6.93024	3.0634

*Standard deviation

On the other hand, manganese is another of the most abundant elements in this study and its concentrations increases during the dry season (range 127.1- 246.58 µg g^{-1}), which suggests significant evaporation phenomena in the area as well as low mobility of sediments. On the analysis of the results it was determined that there is a high relationship (Table 5) between Fe-Mn ($r = 0.5131$). This feature can be instrumental in their association and pre-disposition to transport other metals, along with their biogenic origin. Thus, it was considered that their levels are proper to the area characteristics, the natural movement of water bodies and sediment transport during different climatic seasons.

It has been shown that most of the iron in coastal systems is bound to the oxidation of organic matter. On the other hand, its presence in sediments increases productivity at shallow coastal regions (Hutchins & Bruland, 1998). Decomposition of organic material produces iron flows into water bodies and near the coast. In this study, the relationship between Fe and the organic matter (Fe-MO), were not significant ($r = 0.1770$) as shown in Table 5.

The statistical analyses show that the climatic season does influence the Fe and Mn concentrations ($p <0.05$), as shown in Table 3. The highest Fe, Mn and organic matter concentrations were determined during dry season. The same behavior has been identified in other studies, in which the most abundant metals were Fe and Mn, considered as natural input elements (Zarazúa et al., 2011).

Table 3. Statistical p values from the ANOVA and normality tests

Variation sources	Analyzed element				
	Cd	Cu	Fe	Mg	Mn
Site	0.2099	0.4753	0.8541	0.641	0.4467
Season	0.0001*	0.0002*	0.0000*	0.0000*	0.0000*
Annual mean	1.674	3.6673	156.55	11.69	69.81
Standard deviation	0.4685	4.925	95.68	6.68	91.70
Normality test (p)	0.00194**	0.0000**	0.00002**	0.0063**	0.0000**

* Significant values with a confidence value of 95%

** For the normality test, it means that data is normally distributed and the confidence value is less than 0.05

Fe and Mn are essential in marine sediments and constitute a source of minerals for different flora and fauna marine species. Although it is highly difficult to distinguish between anthropogenic and natural source, concentration increases between seasons are an important element for analysis.

These results of the influence of soil texture on transport, mobility and association with sediments are shown in Table 4.

Table 4. Ranges in percentages of sediment texture and organic matter content in µg g^{-1}

Climatic season	Sand	Silt	Clay	Organic matter
I Dry	27.29-38.2	10-26.09	34.8-57.94	1.46-13.33
Mean	31.755	17.372	50.883	4.589
*SD	3.535	5.995	8.272	4.3392
II Rainy	7.6-73.2	22.4-80.8	2.4-11.6	2.22-4.54
Mean	44.086	50.377	5.937	3.66722
*SD	21.412	20.2202	3.3814	0.84433
III North winds	15.2-85.6	10.4-76.4	2.4-11.6	2.02-4.18
Mean	50.633	42.717	6.65	2.99503
*SD	24.3834	22.546	3.0188	0.70282

*Standard deviation (SD)

The concentrations of sand and clay are predominant in the area and in the dry season show the highest level, showing certain relationship between metal levels detected and the climatic season. These results show high correlation values between iron and manganese with the sediment clay particles (Table 5). In particular, the iron relationships with clay particles showed high correlation (r = 0.5978).

Table 5. Correlation matrix of variables

	Analyzed elements							
	Cd	Cu	Fe	Mg	Mn	Sand	Silt	Clay
Cu	-0.2477							
Fe	-0.8953	0.3908						
Mg	0.0273	-0.6637	-0.1488					
Mn	-0.0404	0.8707*	0.5131*	-0.7129				
Sand	0.1552	-0.3060	-0.3234	0.4609	-0.3541			
Silt	0.1836	-0.5574	-0.1925	0.3137	-0.5929	-0.4717		
Clay	-0.3371	0.8501*	0.5978*	-0.7424	0.9311*	-0.4370	-0.5866	
Organic matter	-0.3024	0.1630	0.1770	-0.1421	0.2087	-0.0957	-0.1315	0.2214

* Significant values with a confidence value of 95%

Several studies have reported this behavior (Aguilar et al., 2012), which constitute significant factors in the mobility of heavy metals.

3.2 Cooper

Other studies have reported significant relationships between manganese and copper with organic matter (Karbassi et al., 2011); nevertheless, those relationships did not show in this study. Still, there is high affinity between Mn and Cu (r = 0.8707). These results are shown in Table 5 and may indicate association phenomena between these elements or common sources.

Regarding copper, levels can be considered low compared to other studies that report up to 18 µg g^{-1} in the sediments of the rivers flowing into Terminos lagoon (Aguilar et al., 2012). Previous studies have reported up to 23 µg g^{-1} of Cu in surface sediments at Baja California, Mexico with an increasing trend compared to areas considered free of contamination (Villaescusa et al., 2002). In this study copper showed a high relationship with the sediment finer particles, (Cu-clay r = 0.8501), which shows its high affinity to sediment providing greater absorption and ionic attraction due to negatively charged clay particles. This causes permanence in deposited sediments and bioavailability to the animal species whose habits make them susceptible to ingest these contaminants. Mn and Cu show high affinity with the sediment clay particles, which may suggest that there are association phenomena of Cu-Mn and the sediment finer material. This increases the possibility of retention and accumulation of copper in the sediments.

The high copper concentrations have been related to wastewater discharges and hydrocarbons (González et al., 2006). Concurrently, other studies attributed high levels of determined Cu and Cd to wastewater discharges in lakes in China (Zeng & Wu, 2013). On the other hand, none of the other metals analyzed showed a relationship with organic matter, which might suggest that the source is not biogenic. Recently, Vázquez et al. (2002) and Vázquez and Sharma (2004) related high copper and other elements concentrations in the sediments at the "Sonda de Campeche" to the combustion of gasoline and exploration, hydrocarbon production and shipping in the area. In this sense, the studies of Turner (2013), confirm that the geosolids derived from fine particles in paint used for boat maintenance increase the concentration of heavy metals such as copper in sediments and surrounding areas, even exceeding the limits established by the UK Environment Agency.

Recent studies (Mora et al., 2013), determined copper concentrations above those found in sediments of the Palizada river. Comparing these results with the standard for the protection and management of aquatic sediment by the Ontario Ministry of the Environment (Persaud et al., 1993), the values for copper and cadmium in this study are below those for the effects range low (ERL), defined as the concentration after which the first adverse effects are observed on benthic organisms.

Moreover, the results showed that the levels of heavy metals are below the ERL in the case of copper (ERL and ERM reference values for heavy metals can cause biological effects on exposed marine organisms). The effects incidence has been estimated in ranges lower than 25% at concentrations below ERL values (Long et al., 1995). The estimated limits for considering the risk values are 34 $\mu g\ g^{-1}$ for Cu, but the ERL for cadmium establish a risk value of 1.2 $\mu g\ g^{-1}$, which is exceeded in this study.

3.3 Magnesium

Another element that comes into attention is magnesium, whose values were higher during the norths season (maximum concentration of 22.16 $\mu g\ g^{-1}$), as shown in Table 2. This behavior may be due to the particular circulation mechanisms at the coastal area in this season, compared to other climatic seasons (Villanueva & Botello, 1992). The ANOVA confirms that there is significant statistical difference only by climatic season (p = 0.0000), but not by site (Table 3).

These determined concentrations may be derived from the amount of water mobilized during this climatic season, as the study area has the highest values in the country, with average annual precipitation of 1169 mm (Villalobos & Mendoza 2010). Concurrently, runoff from agricultural soils is high in fertilizers, which could be a source of increased Mg in sediments.

Additionally, this area is characterized for being highly productive with regards to agriculture with the use of ammonium phosphate fertilizers (Medina et al., 2009). Different studies have found that the use of fertilizers increases the concentration of heavy metals that can even be transferred to food and enhance adverse effects on aquatic organisms through runoff from agricultural soils to water bodies (Conceição et al., 2013). Like many metals, magnesium's highest concentrations are seen in surface sediments, which tend to decrease with depth (Higgins & Schrag, 2010). In this study, magnesium showed no significant relationship with any other element or with sediment characteristics.

3.4 Cadmium

Regarding cadmium, it is considered one of the most toxic element to the environment. Recent studies indicate that cadmium poses a serious ecological threat and contributes greatly to the toxicity response rates, as it is even more toxic than arsenic and lead (Min et al., 2013, Wei & Yang, 2010; Chabukdhara & Nema, 2013; Zeng & Wu, 2013). The results are shown in Table 2.

The concentration profile in relation to the other elements is low. This is consistent with Chakraborty & Owens (2013), who determined cadmium concentrations of up to 125 $\mu g\ g^{-1}$ along the Australian coast. On the other hand, previous studies in the Terminos lagoon have reported cadmium concentrations between 1.2 and 1.5 $\mu g\ g^{-1}$ in surface sediments (Ponce & Botello, 1991); likewise, concentrations of 0.40 $\mu g\ g^{-1}$ in river sediments were recently determined (Aguilar et al., 2012). These levels are below the maximum values reported in this study, which are 2.34 $\mu g\ g^{-1}$ in climatic season of norths.

The presence of cadmium in marine sediments, it is considered totally extraneous to life and its presence in sediments is mainly due to human action. In the sediments, the main solid cadmium species occurring under oxidizing conditions are CdO, CdCO, while under reducing conditions it is CdS (Villanueva & Botello, 1992). The oxyanionic species exist at high pH, whereas at low pH the predominant ion is Cd .

Cadmium solubility is controlled by the presence of organic matter and Fe-Mn hydrous oxides, which could be a determining factor in the retention of cadmium in the sediments as our results show high Fe-Mn relationships (r = 0.5131). Shows that the latter can predispose the presence of cadmium. In addition to low cadmium concentrations, the determining factor for its mobility is the absorption in several soil constituents, which could be Mn oxides > Fe hydrated oxides > Fe crystalline oxides. This also suggests association phenomena with Fe. However, the results show that there is no significant relationship between Cd-Fe.

Cadmium values exceed the limits established by the ERL (ERL and ERM reference values for heavy metals can cause biological effects on exposed marine organisms). The effects incidence has been estimated in ranges lower than 25% at concentrations below ERL values (Long et al., 1995). The estimated limits for considering the risk values are 1.2 $\mu g\ g^{-1}$ for cadmium.

The ANOVA results, shown in Table 3, confirm that only the season influences the concentration variations of the elements analyzed (p <0.05). However, for the sediment texture fractions, the site is the factor promoting concentration variability as shown in table 6 for the fractions of silt and Clay.

Table 6. Statistical p values from the ANOVA and normality tests

Variation sources	Sand	Silt	Clay	Organic matter
Site	0.0779	0.0006*	0.000*	0.4095
Climatic Season	0.0953	0.0395	0.3055	0.3370
Annual mean	41.627	35.7007	22.82	3.75065
Standard deviation	20.84	23.35	22.68	2.69968
Normality test (p)	0.005374**	0.002252**	0.00007**	0.000009**

* Significant values with a confidence value of 95%

** For the normality test, it means that data is normally distributed and the confidence value is less than 0.05

Also the organic matter content is usually high as compared to that of recent studies in the surrounding areas (Aguilar et al., 2012) since the area is influenced by the extensive mangroves, several macrophyte species and human settlements, thus contributing substantially to the variations of organic matter. This greatly predisposes both retention and mobility of heavy metals, in this study organic matter shows no seasonal climate variability or sampling site.

4. Conclusion

The absence of effective monitoring and control of contaminants at the Mexican coastal environment, the growing industrialization and urbanization, and mostly the lack of real implementation of environmental regulations have caused rivers and lagoons at the Gulf of Mexico to be at risk due to the presence of several contaminants.

It is very difficult to make the difference between human and natural sources, but both have influence on the levels of iron, magnesium and manganese, which were the most abundant in this study, likewise these elements are part of the earth's crust and may form partnerships with the characteristics of sediments primarily with the finest fractions. Magnesium, manganese and iron are natural components of the sediments the atypical increase in their concentrations reveals the human impact on the Palizada river, by the use of fertilizers and other products for agriculture, which is directly related to high concentrations in climatic seasons of dry. Also this fact clearly shows that the amount of metals deposited sediments from climatic seasons and that these levels are lower than those detected in the rainy season climate can infer what the dilution of contaminants.

The study area has a large human influence agricultural activities taking place in the Palizada river, this area also forms part of the Terminos lagoon and the "Sonda of Campeche", where it extracts more than 70 % of production hydrocarbon Mexico, according to current studies, this industry generates numerous emissions into the atmosphere including heavy metals, this may be one of the generation sources of heavy metals as due to the physical and chemical processes that occur these are deposited in water and sediments are precipitated as evidenced by numerous studies in the area.

Making a comparison with other studies , the levels of metals were detected in the Palizada river do not represent a risk factor for the ecosystem, however cadmium levels are slightly higher than international standards and if there could be considered a risk factor because cadmium does not come from any natural source, which may suggest that there are anthropogenic discharges into the Palizada river.

The seasonal climate has a major influence on the concentration and distribution of heavy metals, being in the case of iron, copper and manganese dry season the most significant climate for magnesium season was the most influential of showers. For cadmium the climatic season Norths was the most significant. Organic matter levels increase during the dry season in the study area are given numerous weather events with the rainy season with very high levels of precipitation and very extreme dry seasons with temperatures close to 40 ° C.

The sediments represent a reservoir of heavy metals, some of which may form associations with sediment texture as shown in this study with high correlation values between copper, iron and manganese clays, thus can be made available for other life forms.

Also the high values of correlations between the metals copper and manganese, iron and manganese may suggest that these elements may have a common source of generation, being the most abundant that were determined in this study.

References

Aguilar, C. A., Montalvo, C., Rodríguez, L., Cerón, J. G., & Cerón, R. M. (2012). American oyster (*Crassostrea virginica*) and sediments as a coastal zone pollution monitor by heavy metals. *International Journal of Environmental Science and Technology, 9*, 579-586. http://dx.doi.org/10.1007/s13762-012-0078-y

Bruder-Hubscher, V., Lagarde, F., Leroy, M. J. F., Coughanowr, C., & Enguehard, F. (2002). Application of a sequential extraction procedure to study the release of elements from municipal solid waste incineration bottom ash. *Analytica Chimica Acta, 451*, 285-295. http://dx.doi.org/10.1016/S0003-2670(01)01403-9

Buchanan, J. B., & Kain, J. M. (1971). Measurement of the physical and chemical environment. In Holmes N. A., & McIntyre A. D. (Eds.), Methods for the study of the marine benthos (pp. 30-58). Blackwell Scientific Publications, Oxford.

Chabukdhara, M., & Nema, A. K. (2013). Heavy metals assesment in urban soil around industrial clusters in Ghaziabad, India: Probabilistic health risk approach. *Ecotoxicology and Environmental Safety, 87*, 57-64. http://dx.doi.org/10.1016/j.ecoenv.2012.08.032

Chakraborty, S., & Owens, G. (2013). Metal distributions in seawater, sediment and marine benthic macroalgae from the South Australian coastline. *International Journal of Environmental Science and Technology.*

Conceição, F. T., Navarro, G. R. B., & Silva, A. M. (2013). Anthropogenic influences on Cd, Cr, Cu, Ni, Pb and Zn concentrations in soils and sediments in a watershed with sugar cane crops at São Paulo State, Brazil. *International Journal of Environmental Research, 7*, 551-560.

Cuong, D. T., Bayen, S., Wurl, O., Subramanian, K., Wong, K. K., Sivasothi, N., & Obbard, J. P. (2005). Heavy metal contamination in mangrove habitats of Singapore. *Marine Pollution Bulletin, 50*, 1732-1738. http://dx.doi.org/10.1016/j.marpolbul.2005.09.008

García, L., Soto, M. S., Jara, M. E., & Gómez, A. (2004). Fracciones geoquímicas de Cd, Cu y Pb en sedimentos costeros superficiales de zonas ostrícolas del estado de Sonora, México. *Revista Internacional de Contaminación Ambiental, 20*, 159-167.

González, M. C., Méndez, L., López, D., & Botello, A. (2006). Evaluación de la contaminación en sedimentos del área portuaria y zona costera de Salina Cruz, Oaxaca, México. *Interciencia, 31*, 647-656.

Higgins, J. A., & Schrag, D. P. (2010). Constraining magnesium cycling in marine sediments using magnesium isotopes. *Geochimica et Cosmochimica Acta, 74*, 5039-5053. http://dx.doi.org/10.1016/j.gca.2010.05.019

Hutchins, D. A., & Bruland, K. W. (1998). Iron-limited diatom growth and Si:N uptake ratios in a coastal upwelling regime. *Nature, 393*, 561-564. http://dx.doi.org/10.1038/31203

INE/SEMARNAP. (1997). Programa de manejo del área de protección de flora y fauna Laguna de Términos, México. Instituto Nacional de Ecología, Secretaría del Medio Ambiente, Recursos Naturales y Pesca, Mexico.

Karbassi, A. R., Torabi, F., Ghazban, F., & Ardestani, M. (2011). Association of trace metals with various sedimentary phases in dam reservoirs. *International Journal of Environmental Science and Technology, 8*, 841-852. http://dx.doi.org/10.1007/BF03326267

Lima, L., Olivares, S., Columbie, I., De la Rosa, D., & Gil, R. (2005).Niveles de plomo, zinc, cadmio y cobre en el Río Almendares, Ciudad Habana, Cuba. *Revista Internacional de Contaminación Ambiental, 21*, 115-124.

Long, E., Macdonald, D., Smith, S., & Calder, F. (1995). Incidence of adverse biological effects within ranges of chemical concentrations in marine and estuarine sediments. *Environmental Management, 19*, 81-97. http://dx.doi.org/10.1007/BF02472006

Medina, J., Volke, V. H., Galvis, A., González, J. M., Santiago, M. J., & Cortés, J. I. (2009). Propiedades químicas de un luvisol después de la conversión del bosque a la agricultura en Campeche, México. *Agronomia Mesoamericana, 20*, 217-235. http://dx.doi.org/10.15517/am.v20i2.4939

Min, X., Xie, X., Chai, L., Liang, Y., Li, M., & Ke, Y. (2013). Environmental availability and ecological risk assessment of heavy metals in zinc leaching residue. *Transactions of Nonferrous Metals Society of China, 23*, 208-218. http://dx.doi.org/10.1016/S1003-6326(13)62448-6

Mora, A., Alfonso, J. A., Baquero, J. C., Handt, H., & Vásquez, Y. (2013). Elementos mayoritarios, minoritarios y traza en muestras de sedimentos del medio y bajo Río Orinoco, Venezuela. *Revista Internacional de Contaminación Ambiental, 29*, 165-178.

Persaud, D., Jaagumagi, R., & Hayton, A. (1993). *Guidelines for the protection and management of aquatic sediment quality in Ontario.* Ministry of Environment and Energy, Ontario.

Ponce, G., & Botello, A. V. (1991). Aspectos geoquímicos y de contaminación por metales pesados en la Laguna de Términos, Campeche. *Hidrobiológica, 1*(2), 1-10.

RETC. (2011). *Registro de Emisiones y Trasferencia de Contaminantes.* SEMARNAT. Retrieved from http://app1.semarnat.gob.mx/retc/index.html

Santos, J. C., Beltrán, R., & Gómez, J. (2003). Spatial variations of heavy metals contamination in sediments from Odiel river (Southwest Spain). *Environment International, 29*, 69-77. http://dx.doi.org/10.1016/S0160-4120(02)00147-2

Sauvé, S., Hendershot, W., & Allen, H. (2000). Solid-solution partitioning of metals in contaminated soils: dependence of pH, total metal burden and organic matter. *Environmental Science & Technology, 34*, 1125-1131. http://dx.doi.org/10.1021/es9907764

Shapiro, S. S., & Wilk, M. B. (1965). An analysis of variance test for normality (complete samples). *Biometrika, 52*, 591-611. http://dx.doi.org/10.1093/biomet/52.3-4.591

Sutherland, R. A., & Tack, F. M. G. (2002). Determination of Al, Cu, Fe, Mn, Pb and Zn in certified reference materials using the optimized BCR sequential extraction procedure. *Analytica Chimica Acta, 454*, 249-257. http://dx.doi.org/10.1016/S0003-2670(01)01553-7

Turner, A. (2013). Metal contamination of soils, sediments and dusts in the vicinity of marine leisure boat maintenance facilities. *Journal of Soil Sediments, 13*, 1052-1056. http://dx.doi.org/10.1007/s11368-013-0686-2

Vane, C. H., Harrison, I., Kim, A. W., Moss-Hayes, V., Vickers, B. P., & Hong K. (2009). Organic and metal contamination in surface mangrove sediments of South China. *Marine Pollution Bulletin, 58*, 134-144. http://dx.doi.org/10.1016/j.marpolbul.2008.09.024

Vázquez, F. G., & Sharma, V. K. (2004) Major and trace elements in sediments of the Campeche Sound, Southeast Gulf of Mexico. *Marine Pollution Bulletin, 48*, 87-90. http://dx.doi.org/10.1016/S0025-326X(03)00328-X

Vázquez, F. G., Sharma, V. K., & Pérez, L. (2002). Concentrations of elements and metals in sediments of the southeastern Gulf of Mexico. *Environmental Geology, 42*, 41-46. http://dx.doi.org/10.1007/s00254-001-0522-7

Villaescusa, J. A, Gutiérrez, E. A., & Flores, G. (2000). Heavy metals in the fine fraction of coastal sediments from Baja California (Mexico) and California (USA). *Environmental Pollution, 108*, 453–462. http://dx.doi.org/10.1016/S0269-7491(99)00222-5

Villalobos, G. J., & Mendoza, J. (2010). La biodiversidad en Campeche: estudio de estado. Comisión Nacional para el Conocimiento y Uso de la Biodiversidad (CONABIO); Gobierno del Estado de Campeche; Universidad Autónoma de Campeche; El Colegio de la Frontera Sur; Campeche, México.

Villanueva, S., & Botello, A. (1992). Metales pesados en la zona costera del Golfo de México y caribe mexicano:

una revisión. *Revista Internacional de Contaminación Ambiental, 8,* 47-61.

Wei, B., & Yang, L. (2010). A review of heavy metals contaminations in urban soils, urban road dust and agricultural soils from china. *Microchemical Journal, 94,* 99-107. http://dx.doi.org/10.1016/j.microc.2009.09.014

Zarazúa, G., Tejeda, S., Ávila, P., Carapia, L., Carreño, C., & Balcázar, M. (2011). Metal content and elemental composition of particles in cohesive sediments of the Lerma River, México. *Revista Internacional de Contaminación Ambiental, 27,* 181-190.

Zeng, H., & Wu, J. (2013). Heavy metal pollution of lakes along the mid-lower reaches of the Yangtze River in China: intensity, sources and spatial patterns. *Int J Environ Res Public Health, 10,* 793-807. http://dx.doi.org/10.3390/ijerph10030793

7

Occupational Exposure to Atmospheric Emissions Produced During Live Gun Firing

Bernadette Quémerais[1]

[1] School of Medicine, University of Alberta, Edmonton, Canada

Correspondence: Bernadette Quémerais, Division of Preventive Medicine, School of Medicine, University of Alberta, Edmonton, AB., T6G 2T4, Canada. E-mail: quemerai@ualberta.ca

Abstract

A pilot study was performed in Quebec in 2006 and 2007 to estimate occupational exposure of soldiers during live gun firing. For this project three different weapons were tested; the C3 105 mm howitzer, the M777 155 mm howitzer, and the Carl Gustav 84 mm anti-tank. Only area samples were collected and, for safety reasons, samples were collected from 8 to 22 m away from the weapons approximately 90 cm above ground. Results showed that concentrations of total particulates were 1.25 mg/m^3, 4.02 mg/m^3 and 32.1 mg/m^3 for the 105 mm howitzer, the 155 mm howitzer, and the Carl Gustav anti-tank respectively. In addition, estimation on the size distribution determined that most particles were smaller than 4 μm. Hydrogen cyanide was detected from the 105 mm howitzer and the Carl Gustav anti-tank, and formaldehyde was detected from the 155 mm howitzer and the Carl Gustav anti-tank. Although observed concentrations were low for both compounds, it is believed that concentrations around the guns are higher than what was measured during this study. In addition, ambient temperature during trials seemed to have an influence of the dispersion of gases. Although concentrations observed were low, further investigations are needed to better determine soldiers' exposure during live gun firing, and the influence of environmental conditions on this exposure.

Keywords: occupational exposure, artillery, infantry, howitzer, atmospheric emissions, live gun firing

1. Introduction

During the course of their career, members of the Canadian Forces spend time on firing ranges for training purposes. Most soldiers train with small arms but infantry, armored and artillery soldiers train with heavier weapons such as howitzers, anti-tanks, and weapons mounted on armoured vehicles. Exposure to various chemicals has been described in the literature for people practicing in small arms firing ranges either recreationally or for occupational reasons (Bonnano, Robson, Buckley, & Modica, 2002; Gulson, Palmer, & Bryce, 2002; Mancuso, McCoy, Pelka, Kahn, & Gaydos, 2008; Demmeler, Nowak, & Schierl, 2009; Di Lorenzo et al., 2010; Diaz, Sariks, Viebig, & Saldiva, 2012).

Most of the studies looked at lead exposure in indoor ranges (Gulson et al., 2002; Mancuso et al., 2008; Demmeler et al., 2009; Di Lorenzo et al., 2010), although a few studies looked at outdoor ranges (Bonnano et al., 2002; Gulson et al., 2002; Mancuso et al., 2008). Mancuso et al. (2008) also looked at other chemicals such as silica, benzene, ethylbenzene, xylenes, toluene, carbon monoxide, nitrogen dioxide, and hydrogen cyanide. Diaz and Poulin (2012) looked at lead, antimony and barium in a ballistic laboratory.

All studies that measured personal samples showed elevated levels of lead (Bonnano et al., 2002; Mancuso et al., 2008; Diaz et al., 2012). In addition, Bonnano et al. (2002), Mancuso et al. (2008) and Demmeler et al. (2009) found elevated levels of lead in the blood of the shooters, some exceeding the ACGIH recommended BEI value of 30 μg/100 ml.

It has been shown that soils and pore water at military ranges are contaminated with various types of contaminants such as metals and explosive materials (Bennett, Kaufman, Koch, Sova, & Reimer, 2007; Berthelot, Valton, Auroy, Trottier, & Robidoux, 2008; Clausen & Korte, 2009; Martel et al., 2009; Laporte-Saumure, Martel, & Mercier, 2011; Lewis, Sjöström, Skyllberg, & Hägglund, 2010; Etim & Onianwa, 2012; Laporte-Saumure, Martel, & Mercier, 2012; M. R. Walsh, M. E. Walsh, & Ramsey, 2012). Recent work from Du et al. (2011) and Gillies et al. (2007) showed that artillery back blast resulted in fugitive emission of particulate matter ($PM_{2.5}$ and PM_{10}) from soils.

Since 2000, Defence Research and Development Canada Valcartier (DRDC Valcartier) in conjunction with the Army Corps of Engineers has done intensive work on the contamination of firing ranges in Canadian Forces and US Army Bases (Ampleman, Thiboutot, Desilets, Gagnon, & Marois, 2000; Ampleman et al., 2003; Dube, Thiboutot, Ampleman, Marois, & Bouchard, 2006; Jenkins et al., 2006; Marois, Gagnon, Thiboutot, Ampleman, & Bouchard, 2004; Thiboutot et al., 2004; Walsh et al., 2012) and have found that firing positions are contaminated with explosive materials.

Considering that energetic materials can result in health effects such as headaches, DRDC Valcartier researchers felt that gunners may be affected by gaseous emissions produced by live gun firing. In addition, the researchers were concerned about the size of the particles emitted during this activity. It was decided to characterize gaseous emissions, as well as particle size distribution and composition during live artillery and infantry gun firing. In 2006, DRDC Valcartier initiated a joint project with DRDC Toronto researchers to evaluate potential exposure and health risks for infantry and artillery soldiers.

This paper presents the main results obtained during this pilot study of measured area samples from live artillery gun firing. Subsequent studies have been performed by DRDC Valcartier on airborne emissions produced during live gun firing and collected at the muzzle of the gun (Diaz & Poulin, 2012) or from open burning of artillery propellant (Thiboutot, Ampleman, Pantea, Whitwell, & Sparks, 2012) but they were looking only at the environmental aspect. To our knowledge this is the only publication and study that looked at the occupational health aspect of live heavy gun firing.

2. Methods

2.1 Trials

In total three trials were conducted in Quebec, Canada, from September 2006 to February 2007. The first trial was conducted at the Munitions Experimental Testing Centre (METC) in Nicolet, Quebec, on the C3 howitzer 105 mm using the C60 Squash Head Practice projectile. The facility is equipped with a muffler which allows for accumulation of concentrated emissions making the determination of toxic substances easier. The second trial was conducted at Canadian Forces Base Valcartier (CFB Valcartier), Quebec, on 12th January 2007 during a live firing training exercise on the M777 howitzer 155 mm. The final trial was conducted at CFB Valcartier on 7th February 2007 during a live firing training exercise on the Carl-Gustav anti-tank 84 mm.

Detailed descriptions of the facilities and set-up, the weapons, the ammunitions, and physical and chemical composition of the propellants are given in the references from Table 1.

Table 1. Information on the trials

Date	Location	Weapon	Reference
19/20 September 2006	METC Nicolet	C3 Howitzer 105 mm	Quémerais et al. (2007)
12 January 2007	CFB Valcartier	M777 Howitzer 155 mm	Quémerais, Diaz, Poulin, and Marois (2007a)
7 February 2007	CFB Valcartier	Carl Gustav 84 mm	Quémerais, Diaz, Poulin, and Marois (2007b)

In the September trial, tests were conducted indoor as well as outdoor. The indoor set-up at the METC is shown in Figures 1 and 2 while the outdoor set-up is shown is Figure 3. Because of the pressure inside the muffler during gun firing it was impossible to leave the sampling equipment inside the muffler; a sampling hatch was designed specifically to allow sampling of gases and airborne particles inside the muffler (Figure 2). In each case the sampling equipment was installed on a table approximately 90 cm above ground. The C60 Squash Head Practice uses M67 propellant in up to six charge bags. Both indoor and outdoor tests were conducted at charge 4 and 6. Charge 4 contains 475 g of propellant while charge 6 contains 850 g of propellant.

Indoor sampling was carried out on 19th September 2006. The sampling valve was opened after firing two rounds at each charge and before ventilation was activated. Sample collection lasted only for few minutes to avoid clogging of the sampling media since concentration of emissions inside the muffler were extremely high. Outdoor sampling was carried out on 20th September 2006. To avoid damage from the blast effect the sampling station had to be located at a certain distance from the gun. Sampling location was selected according to wind direction in order to capture the plume of atmospheric emissions from the gun muzzle. Sample collection was performed continuously while firing ten rounds at each charge. Sampling times varied from 35 to 43 minutes. All indoor and outdoor sampling was carried out in duplicates.

Figure 1. Side view of the muffler

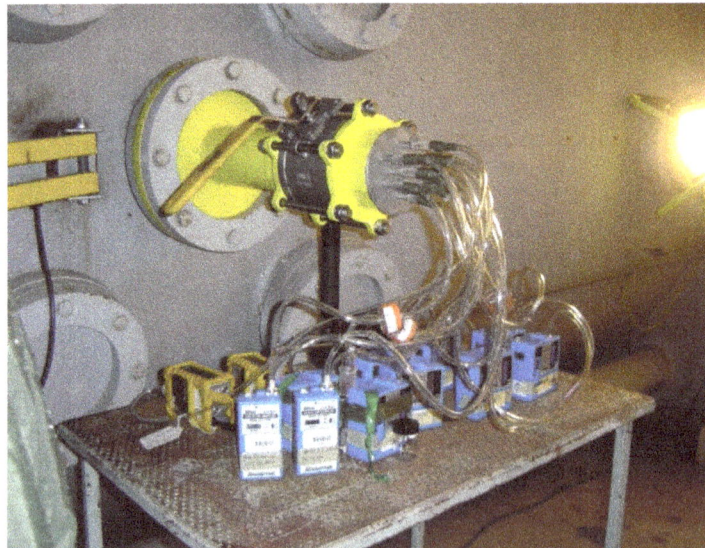

Figure 2. Sampling hatch with pressure valve and sampling equipment

Figure 3. Outdoor set-up with C3 howitzer 105 mm and sampling equipment

Sampling set-up for the second trial on January 2007 is shown in Figure 4. Two tables were setup with the sampling equipment: one on the left side of the gun at approximately 8 m, and 90 cm above ground (identified as Table #1), and the second one in the firing direction at approximately 22 m, and 90 cm above ground (identified as Table #2). The M777 155 mm projectile uses M1 propellant in up to five charge bags. Samples were collected continuously for a total of 170 minutes. In total, 72 rounds were fired, 69 at charge 4 plus 3 rounds at charge 5. Charge 4 represents 1.814 kg of propellant while charge 5 represents 2.523 kg of propellant.

Figure 4. Location of the sampling equipment around the M777 155 mm howitzer

Sampling set-up for the third trial in February 2007 is shown in Figure 5. Two coolers were located in line with firing bay #2 at 8 m (noted Station #1) and at 13 m (noted Station #2) from the firing bay. Sampling was carried out approximately 30 cm above ground. Coolers were protected from the back blast by a pile of salt bags. When firing full calibre ammunition, the danger zone created by the back blast is approximately 60 m. The propellant used is the AKB 204 and has a weight of 380 g. Sampling was carried out continuously for 105 minutes. In total 71 rounds were fired: 39 at bay #1 and 32 at bay #2.

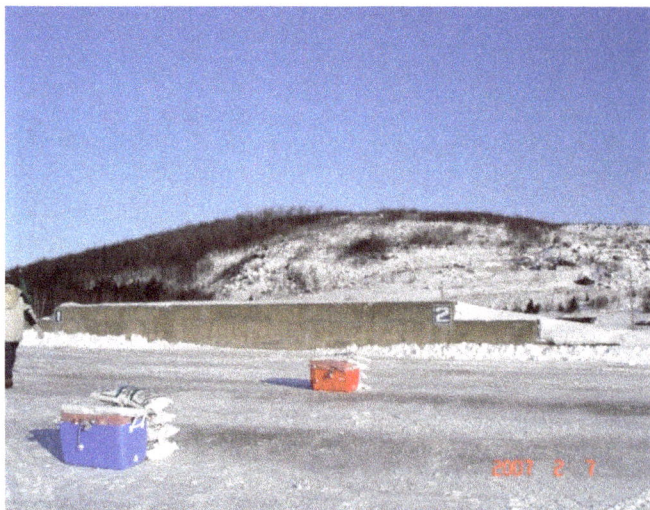

Figure 5. Positions of the sampling stations around the firing bays

2.2 Analytes

Atmospheric emissions were analyzed for total particulates, size distribution of particles, hydrogen cyanide, dinitrotoluene compounds, benzene, toluene, ethylbenzene, xylene, nitrogen oxide and ioxide, sulphur dioxide for all trials. Hydrogen sulphide, nitric acid, polycyclic aromatic hydrocarbons, metals, and aldehydes were added for trials #2 and #3. Nitroaromatic compounds were analyzed during the first trial but since they were never detected they were not analyzed in the following trials.

2.3 Sampling and Analytical Methods

Parameters were sampled and analyzed according to standardized NIOSH and OSHA methods using appropriate filters and sorbent tubes. Detailed methods are given in Quémerais et al. (2007) and Quémerais, Diaz, Poulin, and Marois (2007a). Blank filters and sorbent tubes were analyzed for each trial as controls.

Size distribution of particles was analyzed by scanning electron microscopy for trial #1 and then using a cascade impactor for trials #2 and #3. The impactor used was the Maple Personal Cascade Impactor (Series 290) from Thermo Electron Corporation and it was connected to a GilAir5 sampling pump from Sensidyne operating at a flow rate of 2 L/min. Since only one impactor was available it was located on Table #1 and Station #1 for the second and third trials respectively.

For all trials sampling pumps were calibrated prior to and just after sampling to ensure sampling flow remained constant during sample collection.

Analyses for dinitrotoluene compounds were performed at DRDC Valcartier. Size distribution for the first trial was determined directly on the filter collected for total particulates using a TSI DS-130 scanning electron microscope (SEM) equipped with a Gresham light element detector and an integrated X-Ray fluorescence (IXRF) digital imaging system calibrated with magnification standards on the SEM. These analyses were performed in a private laboratory accredited by the American Industrial Hygiene Association (AIHA) in the United States.

For the last two trials morphology and chemical composition of the particles were analyzed using a JEOL LSM-840A SEM equipped with a NORAN energy dispersive X-ray (EDX) spectrometer. All other analyses were performed in a private AIHA accredited laboratory in the United States.

3. Results

Many of the analytes were detected only during the September trial inside the muffler where emissions were concentrated. The analytes detected discussed in this paper comprise hydrogen cyanide, formaldehyde, and particulates which were identified most commonly.

Blanks were never detected for all analytes.

3.1 Weather Data

Weather data for each sampling day are shown in Table 2. There was approximately 30 °C difference between the warmest and the coldest day. Relative humidity was the lowest when the temperature was also the lowest. Wind speed was lowest during the 155 mm howitzer trial and highest during the 105 mm howitzer trial.

Table 2. Weather data (average during the sampling period)

	Temperature (°C)	Relative humidity (%)	Pressure (kPa)	Wind speed (km/hr)	Wind direction
20/09/2006	16	70	100.1	22	SW
12/01/2007	-0.5	100	100.8	10	WSW
07/02/2007	-15	65	99.8	18	WSW

3.2 Pump Calibration

Differences between pre and post calibration flow for each trial were below 15%. These differences were considered as acceptable and average pump flow was used to calculate sample concentration.

3.3 Concentration of Particles

Results for particle concentrations are shown in Table 3.

Table 3. Particle concentrations measured for each trial

Date	Test	Concentration (mg/m^3)
19/09/2006	Muffler-6 bags	320.00
		310.00
	Muffler-4 bags	180.00
		150.00
20/09/2006	Outdoor-6 bags	1.50
		0.99
	Outdoor-4 bags	1.10
		1.40
12/12/2007	Table #1	3.42
	Table #2	4.62
07/02/2007	Station #1	31.6
	Station #2	32.6

Duplicate samples collected during the first trial in Nicolet showed experiments were reasonably reproducible (2 to 20% difference between duplicates). Total particle concentrations inside the muffler potentially included fugitive emissions of residues present in the muffler before our tests. However this contribution should be minimal since the first section of the muffler was cleaned prior to each test (Ampleman et al., 2008). Outdoor tests gave average concentrations of 1.25 mg/m^3 for both charges 4 and 6. The fact that there was no difference between charges 4 and 6 is probably due to rapid dispersion and dilution of the plume.

During the trial with the M777 howitzer, particle concentration was higher at Table #2, located in line with the muzzle, than at Table #1, located on the left side of the gun. Average particle concentration was 4.04 mg/m^3 for this trial. Concentrations observed are higher than for the C3 105 mm howitzer likely because 72 rounds were fired in the second trial as compared to only 10 rounds in the first one, and consequently considerably more propellant was burned during the second trial. However, particle concentrations remained low considering that approximately 117 kg of propellant was burned during the exercise. This is probably due to the fact that the gun muzzle was higher than in the first trial (Figures 3 and 4) and that the plume was dispersed. However, smoke was observed at the gun muzzle directly after firing but also each time the breech was opened to insert a new round.

Results were quite different for the Carl Gustav anti-tank weapon. Particle concentrations were much higher than in the previous two trials giving an average of 32.1 mg/m^3. It is possibly due to the fact that the emissions produced by combustion of the propellant are expelled at the back of the weapon. The samplers were located 8 and 13 m directly behind one of the firing bay therefore, they were in-line with the emissions. Particle concentrations at each station are similar suggesting that atmospheric emissions were not diluted between stations and that sampling stations were located in the plume.

3.4 Size Distribution of Particles

Results for size distribution obtained using scanning electron microscopy are shown in Table 4. Results for the outdoor 4 bags in trial #1 and all results for trials #2 and #3 have been discarded since there were discrepancies between the weigh and the electron microscopy analysis of the filters. It was concluded that the filters may have been affected by humidity and that weighing was not accurate, therefore results obtained were discarded.

Table 4. Size distribution of particles (% of total number of particles)

Date	Test	< 4 µm (%)	< 10 µm (%)	< 50 µm (%)
19/09/2006	Muffler-6 bags	79.8	98.8	100.0
		78.5	98.3	99.8
	Muffler-4 bags	82.4	98.2	100.0
		76.6	97.9	99.7
20/09/2006	Outdoor-6 bags	98.7	100.0	100.0
		90.6	99.3	100.0

For the first trial, the mass concentration of each fraction (i.e. < 4 µm, < 10 µm, and < 50 µm) was estimated from the total number distribution. For this calculation, it was assumed that all particles in the smaller fraction had a diameter of 4 µm and that all particles in the second fraction had a diameter of 10 µm. Particles were assumed to all be spherical with a mass density of 1 g/cm^3. Calculations for the mass distribution were done according to the following equations (Walter, 2011).

$$v(d_p) = n(d_p) \times (\pi d_p^3)/6$$

$$m(d_p) = v(d_p) \times \rho$$

Where $n(d_p)$ is the number distribution, $v(d_p)$ is the volume distribution, d_p is the particle diameter, $m(d_p)$ is the mass distribution, and ρ is the particle density.

Using these calculations the mass distribution of particles with a diameter of 4 µm gives averages of 26.6% and 31.7% for the indoor and outdoor tests respectively.

Particles inside the muffler showed a distribution that tends slightly to larger particles than outside. Using an average of 31.7%, the mass concentrations for the fraction below 4 µm were estimated for the first trial outdoor test at charge 4, and the second and third trials based on the total concentration of particles (Table 5). Since they are not relevant in terms of occupational exposure, data from inside the muffler were not included.

Table 5. Estimated mass concentrations for the fraction below 4 μm

Date	Test	< 4 μm (mg/m^3)
20/09/2006	Outdoor-6 bags	0.5
		0.3
	Outdoor-4 bags	0.4
		0.4
12/01/2007	Table #1	1.1
	Table #2	1.5
07/02/2007	Station #1	10.0
	Station #2	10.3

Results obtained using scanning electron microscopy for the various stages of the impactor showed that there were no particles on the first 6 stages of the impactor for the trial performed in January (Poulin, Diaz, & Quémerais, 2008a). Since the cut-off point of the stage #6 of the impactor is 1.55 μm, this observation suggests that most particles were smaller than 2 μm, and therefore mostly respirable. Electron microscopy analysis of all stages of the impactor showed that on stage #7 particles originated mainly from soot and potassium sulphate (Poulin et al., 2008a). Stage #8 was composed primarily of soot (Poulin et al., 2008a). The last stage did not show any particle.

For the trial with the Carl Gustav, the first stages of the impactor (i.e. stages #1 to #5, particles over 3.5 μm) showed very few particles when analyzed using scanning electron microscopy. However, the last stages were covered with small particles. Stage #6 showed that many particles were composed of NaCl, likely originating from the salt bags. However, the last three stages of the impactor were covered with submicron particles composed mainly of propellant combustion residues and aluminum. Aluminum is believed to originate from the warhead (Poulin, Diaz, & Quémerais, 2008b). The last stage of the impactor showed a fairly large area (40 μm) covered with ultrafine particles (<100 nm) (Poulin et al., 2008b).

3.5 Hydrogen Cyanide and Formaldehyde

Results for hydrogen cyanide and formaldehyde are shown in Table 6. Average concentrations for hydrogen cyanide inside the muffler were 17.4 and 20.9 mg/m^3 for charges 4 and 6 respectively. The ratio of charge 4 to charge 6 of 83.3% is much higher than the ratio of propellant suggesting that the sampler may have overloaded at least while sampling at charge 6. There was no significant difference between charge 4 and charge 6 during the outdoor test suggesting dilution of the emissions. Hydrogen cyanide was not detected during the trial with the M777 howitzer 155 mm, and showed much lower concentrations during the trial with the Carl Gustav anti-tank. In addition, the level observed at Station #1 was slightly higher than level at Station #2 suggesting a negative gradient concentration from the firing bay, contrarily to results observed with particles.

Unfortunately, formaldehyde was not measured during the first trial. Formaldehyde concentration was higher at Table #1 (left side of the gun) than at Table #2 (in front of the gun muzzle) during the test with the M777 howitzer 155 mm. For the test with the Carl Gustav anti-tank weapon, Station #1 showed a higher concentration than Station #2 suggesting again a negative gradient concentration from the firing bay. It is interesting to note that this behaviour is opposite to the behaviour of particles (Table 3).

Table 6. Concentrations of hydrogen cyanide and formaldehyde

Date	Test	Hydrogen cyanide (mg/m^3)	Formaldehyde (µg/m^3)
19/09/2006	Muffler-6 bags	21.2	NA
		20.5	NA
	Muffler-4 bags	17.87	NA
		16.83	NA
20/09/2006	Outdoor-6 bags	0.21	NA
		0.17	NA
	Outdoor-4 bags	ND	NA
		0.16	NA
12/01/2007	Table #1	ND	7.1
	Table #2	ND	3.6
07/02/2006	Station #1	0.027	8.2
	Station #2	0.022	5.8

NA: non available.

ND: non detected.

4. Discussion

This study has many limitations. Since it was a pilot study, personal exposure samples were not collected and TWA were not calculated, it is quite difficult to estimate health risks for soldiers. In addition, it was not possible to install the sampling stations close to the guns for safety reasons and emissions were diluted and dispersed before reaching the samplers. Finally, only exposure to atmospheric emissions was evaluated. Combustion residues are deposited on the gun itself, on unused projectiles and on soldiers' uniforms. Therefore there is also a risk for dermal exposure that was not evaluated here.

4.1 Concentrations and Size Distribution of Particles

Total average particle concentrations were 1.25, 4.02, and 32.1 mg/m^3 for the 105 mm howitzer, the 155 mm howitzer, and the 84 mm Carl Gustav respectively. These concentrations are approximately three orders of magnitude higher than the 22 µg/m^3 annual average PM$_{10}$ concentration in Quebec City in 2008 (WHO, 2013) and than the average ambient level of PM$_{2.5}$ which varied between 6 and 10 µg/m^3 in Southern Quebec between 2000 and 2010 (Environment Canada, 2012). It confirms that the particles collected were originated from the combustion of the propellant and not from ambient particles. However, there is a possibility that some of these particles were produced by fugitive emissions produced by the back blast as noted by Du et al. (2011) and Gillies et al. (2007).

Detailed composition of particles was not analyzed during the first trial therefore it is impossible to know what proportion of collected particles may have originated from fugitive emissions in this case. During the last two trials, the ground was frozen and covered with snow therefore fugitive emissions of particles from the ground were unlikely. This was confirmed by the electron microscopy analysis of the filters showing that particles originated mainly from soot and potassium sulfate (Poulin et al., 2008a, 2008b). During the last trial, scanning electron microscopy analysis showed that some of the particles were composed of NaCl and likely originated from the salt bags used to protect the coolers. They may have contributed significantly to the overall mass (Poulin et al., 2008b).

The analysis performed using scanning electron microscopy of the different stages of the impactor (trials #2 and #3) showed that most particles collected were below 2 µm, and a significant amount of ultrafine particles (<100 nm) were also observed in the last trial. Estimated average mass concentrations of particles below 4 µm were 0.4 mg/m^3 and 1.3 mg/m^3 for the 105 mm and 155 mm howitzers respectively.

It is safe to assume that soldiers may be exposed to high concentration of fine and ultrafine particles. During the second trial with the M777 155 mm howitzer, smoke was observed each time the gunner was opening the breech to insert a new round. In a recent literature review on the health effects on air pollution, Anderson and Thundiyil (2012) noted that increases as low as 10 $\mu g/m^3$ in PM_{10} and $PM_{2.5}$ concentrations were found to have significant cardiovascular and respiratory health effects, even during short-term exposure.

Considering that, during this study, soldiers were exposed to higher concentrations of particulates, personal exposure should be collected to clarify potential health risks to the soldiers. Size distribution should also be determined more precisely to know the importance of fine (<2.5 μm) and ultrafine particles (<100 nm) since these are the most important in terms of health effects (Martins et al., 2010).

4.2 Hydrogen Cyanide and Formaldehyde

Results for these gases show a different behaviour than the behaviour of particles. The fact that hydrogen cyanide was detected at much higher concentrations in September than in February contrarily to what was observed for particles suggest that temperature may have an influence on the behaviour of gases. Combustion of propellants produces extremely high temperatures (Lengellé, Duterque, & Trubert, 2002). Gases expelled from the weapons are at a much higher temperature than the ambient air.

It is then safe to assume that the density of gases is lower than the density of air and that these gases likely rise. This will be even more pronounced in winter when the temperatures are below zero. This could explain why the concentration of hydrogen cyanide was one order of magnitude lower in February than in September since the temperature in February was 31 °C lower than in September. It also explains the negative gradient in concentration between Station #1 and #2 in the February trial. In contrast to particles that were projected at the back of the weapon, gases appear to have risen away from the gun due to their low density. This implies that concentrations close to the weapons are probably much higher, at least for a short period of time. It also means that concentrations of gaseous compounds around the guns may be quite different in winter and in summer.

Formaldehyde is an irritant of the upper respiratory tract and is considered a human carcinogen (IARC, 2012). Background levels in rural areas are around 1 $\mu g/m^3$, levels slightly lower than what was measured during our study (European Commission, 2004). The European Union recommends a level of 30 $\mu g/m^3$ for a 30-min exposure (European Commission, 2004) and ACGIH recommends a level of 370 $\mu g/m^3$ for a 15-min exposure (ACGIH, 2012). Levels measured here are lower than the recommended limits but they were not personal exposure samples and it is likely that soldiers were exposed to values higher than those measured during the study.

According to the US Environmental Protection Agency, most people would experience discomfort at concentrations of hydrogen cyanide of 1.44 mg/m^3 for a 4-hours exposure (USEPA, 2013). Although these levels are much higher than the concentrations measured during the study, it would be important to know the exact exposure level of the soldiers. In addition, since both formaldehyde and hydrogen cyanide are irritants of the upper respiratory tract, exposure to both of these compounds may have a combined effect (ACGIH, 2012).

5. Conclusion

Although this study had many limitations, it underlines potential health risks for artillery and infantry soldiers. The combustion of propellant produces many toxic compounds that are released through the muzzle of the gun or when the breech is opened to insert a new projectile. Soldiers working around these guns are exposed to these compounds.

This pilot study showed that there is a potential for significant exposures to particles, hydrogen cyanide, and formaldehyde. In fact, results showed that most particles produced during live gun firing are below 4 μm and that soldiers may be exposed to high concentrations of these particles. Fine (<2.5 μm) and ultrafine particles (<100 nm) are associated with premature death, aggravated asthma, and chronic bronchitis (Martins et al., 2010). For gaseous emissions, both hydrogen cyanide and formaldehyde are irritants of the upper respiratory tract (ACGIH, 2012). Hydrogen cyanide may also give headaches, nausea and has effect on the thyroid while formaldehyde is a human carcinogen (IARC, 2012).

Therefore it is essential to characterize soldiers' exposure to fine and ultrafine particles as well as to gaseous emissions by collecting personal exposure samples. Further investigations are needed to better estimate soldiers' exposure to these chemicals. Since it would be difficult to collect samples for all compounds, the use of biomarkers may be an alternative for this type of study when they exist.

6. Funding

This study was funded by DRDC Land Force R&D Program Sustain Thrust (12SG04) and the Strategic Environmental Research and Development Program (SERPD, project ER-1481).

Acknowledgements

The author would like to thank the researchers from DRDC Valcartier and Dr David Halton for their useful comments on the paper and without whom this study would not have been possible. Special thanks are given to Emmanuela Diaz and Isabelle Poulin for analyzing all the samples for energetic compounds and for performing all the scanning electron microscopy.

References

American Conference of Government Industrial Hygienists. (2012). *TLVs® and BEIs®*. Cincinnati: ACGIH.

Ampleman, G., Thiboutot, S., Désilets, S., Gagnon, A., & Marois, A. (2000). *Evaluation of the soils contamination by explosives at CFB Chilliwack and CFAD Rocky Point*. Quebec City: Defence Research Establishment Valcartier, TR 2000-103.

Ampleman, G., Thiboutot, S., Lewis, J., Marois, A., Gagnon, A., Bouchard, M., . . . Pennington, J. (2003). *Evaluation of the impacts of live fire training at CFB Shilo (Final Report)*. Quebec City: Defence Research and Development Canada, TR 2003-066.

Ampleman, G., Thiboutot, S., Marois, A., Gamache, T., Poulin, I,. Quémerais, B., & Melanson, L. (2008). *Analysis of propellant residues emitted during 105-mm Howitzer live firing at the muffler installation in Nicolet, Lac St. Pierre, Canada*. Quebec City: Defence Research and Development Canada, TR 2007-514.

Anderson, J. O., & Thundiyil, J. G. (2012). Clearing the air: a review of the effects of particulate matter air pollution on human health. *Journal of Medical Toxicology, 8*, 166-175. http://dx.doi.org/10.1007/s13181-011-0203-1

Bennett, J. R., Kaufman, C. A., Koch, I., Sova, J., & Reimer, K. J. (2007). Ecological risk assessment of lead contamination at rifle and pistol ranges using techniques to account for site characteristics. *Science of the Total Environment, 374*, 91-101. http://dx.doi.org/10.1016/j.scitotenv.2006.12.040

Berthelot, Y., Valton, E., Auroy, A., Trottier, B., & Robidoux, P. Y. (2008). Integration of toxicological and chemical tools to assess the bioavailability of metals and energetic compounds in contaminated soils. *Chemosphere, 74*, 166-77. http://dx.doi.org/10.1016/j.chemosphere.2008.07.056

Bonanno, J., Robson, M. G., Buckley, B., & Modica, M. (2002). Lead exposure at a covered outdoor firing range. *Bulletin of Environmental Contamination and Toxicology, 68*, 315-23. http://dx.doi.org/10.1007/s001280256

Clausen, J. L., Bostick, B., & Korte, N. (2011). Migration of lead in surface water, pore water, and groundwater with a focus on firing ranges. *Critical Review in Environmental Science and Technology, 41*, 1397-448. http://dx.doi.org/10.1080/10643381003608292

Clausen, J., & Korte, N. (2009). The distribution of metals in soils and pore water at three U.S. military training facilities. *Soil and Sediment Contamination, 18*, 546-63. http://dx.doi.org/10.1080/15320380903085683

Demmeler, M., Nowak, D., & Schierl, R. (2009). High blood lead levels in recreational indoor-shooters. *International Archives of Occupational and Environmental Health, 82*, 539-42. http://dx.doi.org/10.1007/s00420-008-0348-7

Di Lorenzo, L., Borraccia, V., Marisa, C., Mantineo, G. A., Tiziana, C., Marina, M., & Soleo, L. (2010). Lead exposure in firearms instructors of the Italian state police. *Medicina del Lavoro, 101*, 30-7.

Diaz, E., Sarkis, J. E. S., Viebig, S., & Saldiva, P. (2012). Measurement of airborne gunshot particles in a ballistics laboratory by sector field inductively coupled plasma mass spectrometry. *Forensic Science International, 214*, 44-7. http://dx.doi.org/10.1016/j.forsciint.2011.07.016

Diaz, E., Savard, S., & Poulin, I. (2012). Air residues from live firings during military training. *WIT Transactions on Ecology and the Environment, 157*, 399-410. http://dx.doi.org/10.2495/AIR120351

Du, K., Yuen, W., Wang, W., Rood, M. J., Varma, R. M., Hashmonay, R. A., . . . Kemme, M. R. (2011). Optical remote sensing to quantify fugitive particulate mass emissions from stationary short-term and mobile continuous sources: Part II. field applications. *Environmental Science and Technology, 45*, 666-72. http://dx.doi.org/10.1021/es101906v

Dubé, P., Thiboutot, S., Ampleman, G., Marois, A., & Bouchard, M. (2006). *Preliminary assessment of the dispersion of propellant residues from the static live firing of 105 MM howitzer*. Quebec City: Defence Research and Development Canada, TM 2005-284.

Environment Canada. (2012). *Ambient levels of particulate matter*. Retrieved from http://www.ec.gc.ca/indicateurs-indicators/default.asp?lang=en&n=029BB000-1#r2

Etim, E. U., & Onianwa, P. C. (2012). Lead contamination of soil in the vicinity of a military shooting range in Ibadan, Nigeria. *Toxicology and Environmental Chemistry, 94*, 895-905. http://dx.doi.org/10.1080/02772248.2012.678997

European Commission. (2004). *The Index Project–Summary on recommendations and management options*. Joint Research Centre, Institute for Health and Consumer Protection, Physical and Chemical Exposure Unit.

Gillies, J. A., Kuhns, H., Engelbrecht, J. P., Uppapalli, S., Etyemezian, V., & Nikolich, G. (2007). Particulate emissions from U.S. department of defense artillery backblast testing. *Journal of the Air and Waste Management Association, 57*, 551-60. http://dx.doi.org/10.3155/1047-3289.57.5.551

Gulson, B. L., Palmer, J. M., & Bryce, A. (2002). Changes in blood lead of a recreational shooter. *Science of the Total Environment, 293*, 143-50. http://dx.doi.org/10.1016/S0048-9697(02)00003-7

International Agency for Research on Cancer. (2012). *IARC Monographs on the Evaluation of Carcinogenic Risks to Humans*. IARC Monographs, Volume 100 (F), ISBN-13 9789283213239.

Jenkins, T. F., Hewitt, A. D., Grant, C. L., Thiboutot, S., Ampleman, G., Walsh, M. E., . . . Pennington, J. C. (2006). Identity and distribution of residues of energetic compounds at army live-fire training ranges. *Chemosphere, 63*, 1280-1290. http://dx.doi.org/10.1016/j.chemosphere.2005.09.066

Laporte-Saumure, M., Martel, R., & Mercier, G. (2011). Characterization and metal availability of copper, lead, antimony and zinc contamination at four Canadian small arms firing ranges. *Environmental Technology, 32*, 767-81. http://dx.doi.org/10.1080/09593330.2010.512298

Laporte-Saumure, M., Martel, R., & Mercier, G. (2012). Pore water quality in the upper part of the vadose zone under an operating Canadian small arms firing range backstop berm. *Soil and Sediment Contamination, 21*, 739-55. http://dx.doi.org/10.1080/15320383.2012.691576

Lengellé, G., Duterque, J., & Trubert, J. F. (2002). *Combustion of Solid Propellants*. Bruxelles: North Atlantic Treaty organization, RTO-EN-023

Lewis, J., Sjöström, J., Skyllberg, U., & Hägglund, L. (2010). Distribution, chemical speciation, and mobility of lead and antimony originating from small arms ammunition in a coarse-grained unsaturated surface sand. *Journal of Environmental Quality, 39*, 863-70. http://dx.doi.org/10.2134/jeq2009.0211

Mancuso, J. D., McCoy, J., Pelka, B., Kahn, P. J., & Gaydos, J. C. (2008). The challenge of controlling lead and silica exposures from firing ranges in a special operations force. *Military Medicine, 173*, 182-6.

Marois, A., Gagnon, A., Thiboutot, S., Ampleman, G., & Bouchard, M. (2004). *Caractérisation des sols de surface et de la biomasse dans les secteurs d'entraînement, Base des Forces Canadiennes, Valcartier*. Quebec City: Defence Research and Development Canada.

Martel, R., Mailloux, M., Gabriel, U., Lefebvre, R., Thiboutot, S., & Ampleman, G. (2009). Behavior of energetic materials in ground water at an anti-tank range. *Journal of Environmental Quality, 38*, 75-92. http://dx.doi.org/10.2134/jeq2007.0606

Martins, L. D., Martins, J. A., Freitas, E. D., Mazzoli, C. R., Gonçalves, F. L. T., Ynoue, R. Y., . . . Andrade, M. F. (2010). Potential health impact of ultrafine particles under clean and polluted urban atmospheric conditions: a model-based study. *Air Quality, Atmosphere and Health, 3*, 29-39. http://dx.doi.org/10.1007/s11869-009-0048-9

Poulin, I., Diaz, E., & Quemerais, B. (2008a). *Particulate matter emitted from the M777 howitzer during live firing*. Quebec City: Defence Research and Development Canada, TR 2008-215.

Poulin, I., Diaz, E., & Quemerais, B. (2008b). *Airborne contaminants in two anti-tank weapons back blast plume - Carl Gustav 84-mm and M72 66-mm*. Quebec City: Defence Research and Development Canada, TR 2008-242.

Quemerais, B., Diaz, E., Poulin, I., & Marois, A. (2007). *Characterization of atmospheric emissions during live gun firing–Test on the M777 howitzer 155 mm.* Toronto: Defence Research and Development Canada, TR 2007-102. Retrieved from http://www.dtic.mil/cgi-bin/GetTRDoc?AD=ADA477180

Quemerais, B., Diaz, E., Poulin, I., & Marois, A. (2008). *Characterization of atmospheric emissions during live gun firing–Test on the Carl Gustav anti-tank, 84 mm weapon.* Toronto: Defence Research and Development Canada, TR 2007-103. Retrieved from http://www.dtic.mil/cgi-bin/GetTRDoc?AD=ADA480051

Quemerais, B., Melanson, L., Ampleman, G., Thiboutot, S., Diaz, E., & Poulin, I. (2007). *Characterization of atmospheric emissions during live gun firing at the muffler installation in Nicolet, Lac St. Pierre, Canada–Test on howitzer 105 mm.* Toronto: Defence Research and Development Canada, TR 2007-060.

Thiboutot, S., Ampleman, G., Marois, A., Gagnon, A., Bouchard, M., Hewitt, A., . . . Ranney, T. A. (2004). *Environmental conditions of surface soils, CFB Gagetown training area: delineation of the presence of munitions related residues (phase III, final report).* Quebec City: Defence Research and Development Canada, TR 2004-205.

Thiboutot, S., Ampleman, G., Pantea, D., Whitwell, S., & Sparks, T. (2012). Lead emissions from open burning of artillery propellants. *WIT Transactions on Ecology and the Environment, 157,* 273-84. http://dx.doi.org/10.2495/AIR120241

US Environmental Protection Agency. (2013). *Acute Exposure Guideline Levels.* Retrieved from http://www.epa.gov/oppt/aegl/pubs/results6.htm

Walsh, M. R., Walsh, M. E., & Ramsey, C. A. (2012). Measuring energetic contaminant deposition rates on snow. *Water Air and Soil Pollution, 223,* 3689-99. http://dx.doi.org/10.1007/s11270-012-1141-5

Walsh, M. R., Walsh, M. E., Ampleman, G., Thiboutot, S., Brochu, S., & Jenkins, T. F. (2012). Munitions propellants residue deposition rates on military training ranges, *Propellants. Explosives, Pyrotechnics, 37,* 393-406. http://dx.doi.org/10.1002/prep.201100105

Walter, J. (2011). Size distribution characteristics of aerosols. In P. Kulkarni, P. A. Baron, & K. Willeke (Eds.), *Aerosol Measurement: Principles, Techniques, and Applications* (3rd ed., pp. 41-54). USA: John Wiley.

World Health Organization. (2013). *Urban air pollution database.* Retrieved from http://www.who.int/phe/health_topics/outdoorair/databases/en/index.html

8

Influence of Habitat Pollution on Organophosphate Esters and Polycyclic Aromatic Hydrocarbons in Cicadas

Haruki Shimazu[1]

[1] School of Science and Engineering, Kinki University, Osaka, Japan

Correspondence: Haruki Shimazu, School of Science and Engineering, Kinki University, Osaka, 577 8502, Japan. E-mail: hshimazu@civileng.kindai.ac.jp

Abstract

The present study examines the concentration levels of seven organophosphate esters (OPEs) and nine polycyclic aromatic hydrocarbons (PAHs) in cicadas and the influence factors of their habitats on the contamination of cicadas. Adult cicadas, nymphal exoskeletons, soils, and saps were sampled in Japan. The total concentrations of seven OPEs and nine PAHs for the adult cicadas ranged from 107 to 8940 ng/g-dw and from 58.9 to 1580 ng/g-dw, respectively. Some OPEs and PAHs were detected in heads, thoraxes, and abdomens of the adult cicadas. The concentrations were higher in the heads than in the other parts. The relationship between OPEs and PAHs in soils and those in cicadas was considerably positive. For some OPEs and PAHs, the concentrations in the saps tend to be higher as those in the cicadas increase. These tendencies indicate that cicadas intake OPEs and PAHs from soils and saps contaminated with these pollutants. The concentrations of OPEs and PAHs tend to increase with those in the atmospheric depositions. This probably shows that OPEs and PAHs in atmospheric depositions pollute soils, and cicadas intake the pollutants from the contaminated soils and saps.

Keywords: Japanese cicada, organophosphate esters, polycyclic aromatic hydrocarbons

1. Introduction

Organophosphate esters (OPEs) are a group of widely used commercial chemicals such as organic plasticizers, flame retardants, and photographic films. Polycyclic aromatic hydrocarbons (PAHs) are produced primarily as a result of incomplete combustion from anthropogenic sources such as cars, incinerators, and factories. Therefore, the pollutants have been observed in various environments, including surface water (Andresen et al., 2004), sediments (Kawagoshi et al., 1999), air (Nam et al., 2008), and indoor air (Hartmann et al., 2004). The United States Environmental Protection Agency (US EPA) includes some PAHs in the EPA's Priority Chemical List. Anthracene and tri-2-chloroethyl phosphate are included in the Candidate List by REACH in the European Commission. From many perspectives, some countries have regulated OPEs and PAHs because some OPEs and PAHs are considered to be neurotoxic, mutagenic, and carcinogenic (Jamal, 2002; Luch, 2005).

Rachel Carson (1962) pointed out the dangers of widespread use and bioconcentration of chemicals in Silent Spring. Many authors have hitherto reported bioconcentration, bioaccumulation, and biomagnification for various organisms in the laboratory and in the environments (Bruggeman et al., 1984; Muir et al., 1986; Gobas et al., 1989; Debruyn et al., 2004; Kelly et al., 2004; Kondo et al., 2005). For recent studies, Johnson-Restrepo et al. (2008) identified tetrabromobisphenol A and hexabromocyclododecanes in tissues of humans, dolphins, and sharks. Alava J. J. et al. (2009) reported pollutants, such as polychlorinated biphenyls and polybrominated diphenyl ethers in Galapagos sea lions. In relation to the current state, the bioaccumulation regulations including the bioconcentration factor (BCF) and bioaccumulation factor (BAF) have been introduced in the United States, Canada, and Europe. Biomagnification factor (BMF) and trophic or food web magnification factor (TMF) have been discussed as bioaccumulation assessment criteria (Gobas et al., 2009). These factors are associated with bioaccumulation that occurs when organisms are exposed to chemicals in air, water, and their diet. However, they are not associated with bioaccumulation of chemicals in soil. It is necessary to focus on soil contamination of chemicals because the concentrations in soil tend to be high. This study targets cicadas because they spend most of their lives underground as nymphs and may encounter many predators such as wasps and birds. Due to this, cicadas probably bring pollutants in the soil into the food web of the land.

The objectives of this paper are to determine the concentration levels of the OPEs and the PAHs in cicadas and to investigate the influence factors on the contamination of cicadas from their habitats. In this paper, we analyze OPEs and PAHs in cicadas, soils, and tree saps. Adult cicadas feed saps and nymph cicadas drink root juice. From the comparison between the concentrations, the degree of influence of OPEs and PAHs in soils and tree saps on cicada was evaluated.

2. Materials and Methods

2.1 Cicadas

Black cicadas (*Cryptotympana facialis*) were sampled in this study. The black cicada is an insect of the order Hemiptera, suborder Homoptera, in the superfamily Cicadoidea. The cicadas live from the west Japan to the south Japan. The range of the cicada distribution gradually extends to the north. An adult cicada is usually 60 to 70 mm long including its wings. It is one of the largest species of cicada in Japan.

After mating, the female deposits her eggs into the bark of a twig. When the eggs hatch, the cicada nymphs drop to the ground, where they burrow. Cicada nymphs go through a life cycle that lasts about five years as nymphs for most of their lives. They suck xylem from the roots of tree. Moles, mole crickets, and ground beetles are predators for cicada nymphs. In contrast, the life of an adult cicada ranges from two weeks to one month. Adult cicadas also drink plant sap. They are commonly eaten by spiders, mantises, wasps, and birds.

2.2 Sample Collection

Adult cicadas, exoskeletons, soils, and tree saps were collected in Osaka, Japan. Osaka is in a mixed traffic area as well as an industrial, and residential area. Osaka has about 20 000 manufacturing business establishments, and a population of 8 800 000. There are a lot of main roads such as Kinki expressway with approximately 90 000 cars for every twelve hours in the daytime. Seven sampling locations in Osaka were in the campus of Kinki University (34° 39' 05 N and 135° 35' 13 E), with an area of 469 000 m^2.

From August to September in 2011, adult black cicadas and exoskeletons were sampled (n = 30 and n = 39). Four soil samples were collected in November, 2011. They were collected from 0–20 cm depth near the trees from which the exoskeletons were sampled.

From August to September in 2012, adult black cicadas and exoskeletons were sampled (n = 10 and n = 6). Ten adult black cicadas were divided into heads, thoraxes, and abdomens to measure OPEs and PAHs in the individual parts. Three soil samples were collected from 0–20 cm depths in September, 2012. Saps from cherry trees were also sampled in 2012 (n = 7).

2.3 OPEs and PAHs

Seven OPEs and nine PAHs shown in Table 1 are the target compounds in this study because many studies have reported that they are frequently detected in aquatic and airborne environments. Some OPEs and PAHs, for example, TCEP, BaA, BaP, BbF, BkF, DahA, FL and PY, are also designated as hazardous substances that have the potential to harm humans through long-term exposure under Air Pollution Control Law in Japan.

Seven extrapure grade OPEs were purchased from Tokyo Chemical Industry Co., Ltd., Japan, and diluted with acetone and hexane (for pesticide residue and polychlorinated biphenyl analysis from Wako Pure Chemical Industries, Ltd., Japan) to make calibration standards. Nine PAHs of standard material grade were purchased from Wako Pure Chemical Industries, Ltd. and diluted with acetone and hexane to make calibration standards.

2.4 Analytical Methods and Instruments

The samples of adult cicada and exoskeleton were dried in a dark place for one month and broken into shatters in aluminum foil by hand. Then, each whole sample was put into a cellulose extraction thimble (Whatman, UK) and extracted with 40 mL of dichloromethane (for pesticide residue and polychlorinated biphenyl analysis from Wako Pure Chemical Industries, Ltd.) for 15 min by ultrasonic extraction. The extract was concentrated to 2 mL with a rotary evaporator. The extract was filtered with a disposable filter device (Whatman, PURADISCTM 25TF) and then the extract was concentrated to 0.1 mL under N$_2$ flow. Hexane was added to the extract to make 2 mL.

The samples of soil and sap were dried in a dark place for one month. Then, 1 g of the soil or 0.5 g of the sap was put into a cellulose extraction thimble and extracted with 40 mL of dichloromethane for 15 min by ultrasonic extraction. The extract was concentrated to 2 mL with a rotary evaporator. The extract was filtered with a disposable filter device and was concentrated to 0.1 mL under N$_2$ flow. Hexane was added to the extract to make 2 mL.

The contents of OPEs and PAHs in these extracts were analyzed using a GC/MS (5975B inert XL E/CI MSD; Agilent technologies, USA) equipped with a HP-5MS capillary column (30m × 0.25 mm i.d., 0.25μm film thicknesses; Agilent technologies, USA). The GC conditions were as follows: splitless injection of 2 μL; injection port temperature, 250°C; GC temperature program, 70 °C (hold 1.5 min) to 180 °C at 20 °C/min, and to 280 °C at 5 °C/min (hold 1 min); and carrier gas, helium.

The mass spectrometer was operated in the electron impact mode with an electron energy of 70 eV. After each pollutant was qualified using three representative fragment ions, it was quantified using the largest one. The quantification was performed by an external calibration method. The recoveries and the variation coefficients for OPEs and PAHs in the pretreatment process for the analysis ranged from 70% to 120% and from 7% to 20%, respectively.

Table 1. OPEs and PAHs measured in this study and their detection limits with the GC/MS

Chemicals	Abbreviations	logKow	Detection limits (pg)
OPEs			
Tributyl phosphate	TBP	4.00	0.72
Tri-2-butoxyethyl phosphate	TBXP	3.75	9.5
Tri-2-chloroethyl phosphate	TCEP	1.44	5.3
Tris (1,3-dichloroisopropyl) phosphate	TDCPP	3.65	3.7
Triethyl phosphate	TEP	0.80	1.4
Tris (2-ethylhexyl) phosphate	TEHP	9.49	0.76
Triphenyl phosphate	TPP	4.59	2.5
PAHs			
Anthracene	AN	4.45	3.5
Benzo(a)anthracene	BaA	5.76	1.6
Benzo(a)pyrene	BaP	6.13	1.5
Benzo(b)fluoranthene	BbF	5.78	1.6
Benzo(k)fluoranthene	BkF	6.11	1.8
Benzo(g,h,i)perylene	BghiP	6.63	4.0
Dibenzo(a,h)anthracene	DahA	6.75	4.2
Fluoranthene	FL	5.16	0.80
Pyrene	PY	4.88	1.0

Note. The abbreviations are used in this study. logKow values were obtained from the website of SRS Inc. The detection limits were calculated from values three times the signal-noise ratio at the baseline of the chromatogram with the GC/MS.

3. Results and Discussion

3.1 Sample Properties

For the adults of black cicada (n = 30), the arithmetic mean ± standard error (range) was 60.9 ± 3.7 mm (52–67 mm) for the length and 0.832 ± 0.222 g (0.49–1.24 g) for the dry weight. The arithmetic means of the length were 59.6 mm for males (n = 15) and 62.2 mm for females (n = 15). For the exoskeletons of black cicada (n = 45), the arithmetic mean ± standard error (range) was 34.0 ± 2.2 mm (29–38 mm) for the length and 0.275 ± 0.056 g (0.18–0.54 g) for the dry weight. The arithmetic means of the length were 34.5 mm for males (n = 23) and 33.6 mm for females (n = 22).

Ten adults of black cicada were divided into heads, thoraxes, and abdomens, for which the target compounds were measured. The arithmetic mean ± standard error (range) was 62.4 ± 2.3 mm (60–67 mm) for the length and 1.04 ± 0.200 g (0.77–1.38 g) for the dry weight. The average dry weight was 0.068 g for heads, 0.543 g for thoraxes, and 0.278 g for abdomens.

3.2 OPEs and PAHs in Cicadas

The analytical results for OPEs and PAHs in adult cicadas and exoskeletons are shown in Table 2. Seven OPEs and eight PAHs were detected in the adult cicadas. The median concentration and the range for the total of seven OPEs were 3340 ng/g-dw and from 107 to 8940 ng/g-dw, respectively. Those for the total of nine PAHs were 203 ng/g-dw and from 58.9 to 1580 ng/g-dw. All of the OPEs and PAHs were detected in the exoskeletons. The median concentration and the range for seven OPEs were 470 ng/g-dw and from 83.4 to 25 700 ng/g-dw, respectively. Those for nine PAHs were 103 ng/g-dw and from 13.9 to 3010 ng/g-dw. The total concentrations of OPEs were higher than those of PAHs in both the adult cicadas and the exoskeletons.

TBP, TEP, FL, and PY were detected in all of the adult cicadas. TEHP and BbF were detected frequently. TBP and FL were detected in all of the exoskeletons. TEP and PY were detected frequently. The highest median concentration value was 1160 ng/g-dw for TBXP in the adult cicadas and 220 ng/g-dw for TBP in the exoskeletons. The concentrations for the detected OPEs and PAHs in both the adult cicadas and exoskeletons were compared by sex. The difference of the concentrations between male and female was not observed.

Some OPEs and PAHs were detected in the heads, thoraxes, and abdomens of the adult cicadas. Figure 1 shows the median and the range between the 25th percentile and the 75th percentile for TBP, TEP, and TEHP. The OPE concentrations in the heads were higher than those in the other parts. However, the OPE contents were slightly higher in the thoraxes than in the other parts because the weight ratio of thorax in a whole body is about 60%.

Table 2. OPEs and PAHs detected in adult cicada and exoskeleton

	Adult cicada			Exoskeleton		
	Median*	Range	DR**	Median*	Range	DR**
TBP	132	49.1-2210	30/30	220	76.5-1400	45/45
TBXP	1160	N.D.-8030	21/30	N.D.	N.D.-25 300	8/45
TCEP	30.6	N.D.-449	19/30	N.D.	N.D.-192	21/45
TDCPP	N.D.	N.D.-2110	4/30	N.D.	N.D.-1280	7/45
TEP	47.1	23.6-256	30/30	48.4	N.D.-158	38/45
TEHP	285	N.D.-982	24/30	N.D.	N.D.-843	15/45
TPP	N.D.	N.D.-243	7/30	N.D.	N.D.-140	4/45
Σ7OPEs	3340	107-8940	30/30	470	83.4-25,700	45/45
AN	5.10	N.D.-26.9	18/30	11.9	N.D.-84.4	29/45
BaA	N.D.	N.D.-676	1/30	N.D.	N.D.-32.5	10/45
BaP	N.D.	N.D.-1310	8/30	N.D.	N.D.-481	9/45
BbF	89.4	N.D.-254	28/30	17.7	N.D.-718	28/45
BkF	N.D.	N.D.-19.8	1/30	N.D.	N.D.-305	5/45
BghiP	N.D.	N.D.	0/30	N.D.	N.D.-428	2/45
DahA	N.D.	N.D.-159	2/30	N.D.	N.D.	0/45
FL	23.7	9.55-69.5	30/30	40.9	13.9-379	45/45
PY	13.3	6.25-48.6	30/30	30.5	N.D.-317	43/45
Σ9PAHs	203	58.9-1580	30/30	103	13.9-3010	45/45

Note. *; N.D. (Not detected) data is included in the calculation of median concentrations. **; DR means the detection rate of OPEs and PAHs. ***; All units are ng/g-dw.

3.3 OPEs and PAHs in Soils and Saps

The analytical results for OPEs and PAHs for the soils and saps are shown in Table 3. Five OPEs and nine PAHs were detected in the soils. The median concentration and the range for the total of seven OPEs were 190 ng/g-dw and from 54.3 to 828 ng/g-dw, respectively. Those for the total of nine PAHs were 248 ng/g-dw and from 75.7 to

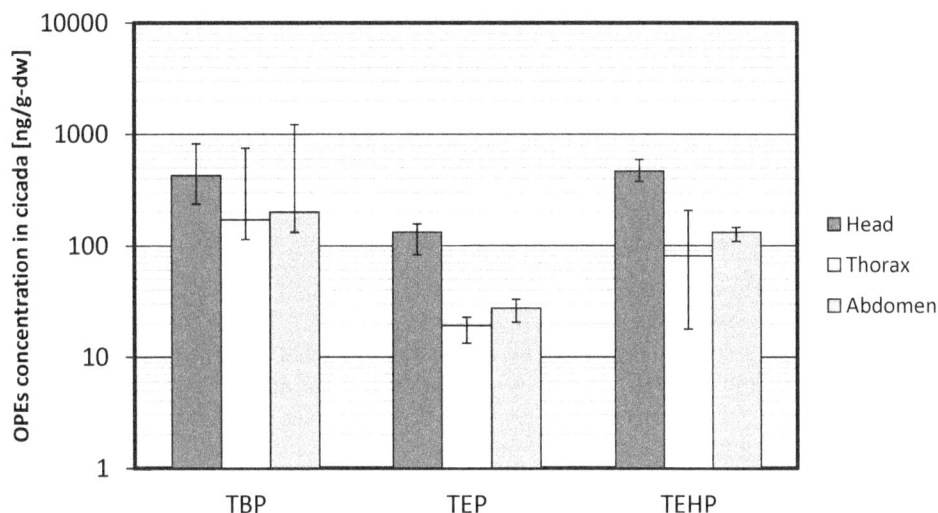

Figure 1. Comparison of OPEs in heads, thoraxes, and abdomens for adult black cicadas

Note. The number of samples is 10 for each part. The detection rates are seven or more. The bars show the median values of the measured concentrations. The ranges show between the 25th percentile and the 75th percentile values of the measured concentrations.

Table 3. OPEs and PAHs detected in soil and sap

	Soil			Sap		
	Median*	Range	DR**	Median*	Range	DR**
TBP	25.5	10.2-79.3	7/7	78.5	63.2-448	7/7
TBXP	75.4	N.D.-775	6/7	N.D.	N.D.-352	3/7
TCEP	5.55	N.D.-31.6	4/7	N.D.	N.D.	0/7
TDCPP	N.D.	N.D.	0/7	N.D.	N.D.	0/7
TEP	6.53	N.D.-7.18	6/7	13.7	9.93-30.2	7/7
TEHP	27.8	N.D.-58.6	6/7	47.6	N.D.-131	6/7
TPP	N.D.	N.D.	0/7	N.D.	N.D.	0/7
Σ7OPEs	190	54.3-828	7/7	338	92.2-608	7/7
AN	5.41	0.50-14.3	7/7	N.D.	N.D.	0/7
BaA	19.9	7.93-92.6	7/7	N.D.	N.D.	0/7
BaP	29.1	9.52-117	7/7	N.D.	N.D.	0/7
BbF	26.4	16.2-75.7	7/7	N.D.	N.D.	0/7
BkF	5.27	N.D.-37.6	4/7	N.D.	N.D.	0/7
BghiP	N.D.	N.D.-42.1	3/7	N.D.	N.D.	0/7
DahA	3.25	N.D.-66.4	4/7	N.D.	N.D.	0/7
FL	42.1	12.7-109	7/7	4.20	2.80-15.0	7/7
PY	39.4	12.1-104	7/7	N.D.	N.D.-6.40	2/7
Σ9PAHs	248	75.7-501	7/7	4.20	2.80-17.3	7/7

Note. *; N.D. (Not detected) data is included in the calculation of median concentrations. **; DR means the detection rate of OPEs and PAHs. ***; All units are ng/g-dw.

501 ng/g-dw, respectively. Four OPEs and two PAHs were detected in the saps. The median concentration and the range for the total of seven OPEs were 338 ng/g-dw and from 92.2 to 608 ng/g-dw, respectively. Those for the total of nine PAHs were 4.20 ng/g-dw and from 2.80 to 17.3 ng/g-dw, respectively. The total concentrations for nine PAHs were lower in saps than in the cicadas and soils.

TBP, AN, BaA, BaP, BbF, FL, and PY were detected in all of the soils. TBXP, TEP, and TEHP were detected frequently. TBP, TEP, and FL were detected in all of the saps. TEHP was detected frequently. The highest median concentration value was 75.4 ng/g-dw for TBXP in the soils and 78.5 ng/g-dw for TBP in the saps.

3.4 Influence Factors on Cicada Contamination

To determine the influence of OPEs and PAHs in the soils on cicada contamination, the relationships between the OPEs and PAHs in the soils and those in the cicadas are shown in Fig. 2 (A). The correlation coefficient for the adult cicada was 0.574. The OPEs and PAHs in the adult cicadas and nymphal exoskeletons tend to be higher as those chemicals in the soils increase. Though there are three chemicals in the adult cicada and three chemicals in the exoskeleton, the median concentrations in the cicadas tend to increase with those in the saps. These tendencies indicate that cicadas intake OPEs and PAHs from soils and saps contaminated with these pollutants.

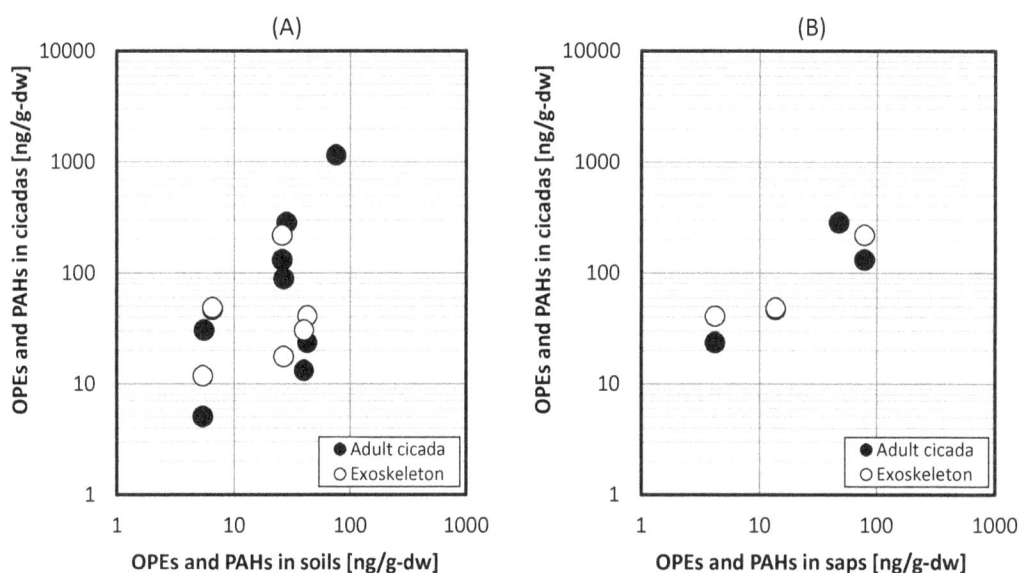

Figure 2. Relationships between the median concentrations for OPEs and PAHs in soils and those in cicadas (A). Relationship between the median concentrations for OPEs and PAHs in saps and those in cicadas (B)

Note. The median concentrations are calculated including N.D. (Not detected) data. Solid circles indicate the concentrations in adult cicadas. Open circles indicate the concentrations in exoskeletons.

Shimazu (2011) measured OPEs and PAHs in the air and the atmospheric deposition in Osaka, and calculated their surface loadings. The relationship between OPE and PAH surface loadings for non-rainfall time and those concentrations in the soils is shown in Fig.3. The correlation coefficients was 0.619. Significant relationship was observed between the OPE and PAH surface loadings and those concentrations in the soils. This indicates that the atmospheric depositions have an influence on soil contamination. The results of the present study suggest that OPEs and PAHs in the atmospheric depositions pollute soils, and cicadas intake OPEs and PAHs from soils and saps contaminated with these pollutants.

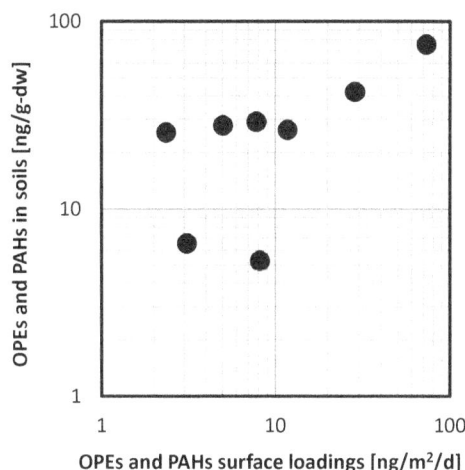

Figure 3. Relationship between the median values of OPE and PAH surface loadings, and the median concentrations for OPEs and PAHs in the soils

Note. The median concentrations are calculated including N.D. (Not detected) data. The OPE and PAH surface loadings are calculated with those concentrations in coarse particulate matter; particle matter <1.0 μm of atmospheric depositions for non-rainfall time (Shimazu, 2011).

4. Conclusion

The total concentrations of seven OPEs and nine PAHs for the adult cicadas ranged from 107 to 8940 ng/g-dw and from 58.9 to 1580 ng/g-dw, respectively. Those concentration levels for the exoskeletons were almost same. The patterns of the pollutants were similar between the adult cicadas and the exoskeletons. Some OPEs and PAHs were detected in heads, thoraxes, and abdomens of the adult cicadas. Although the concentrations were higher in the heads than in the other parts, the contents were slightly higher in the thoraxes than in the other parts. The relationship between OPEs and PAHs in soils and those in cicadas was considerably positive. For some OPEs and PAHs, the concentrations in the saps tend to be higher as those in the cicadas increase. These tendencies indicate that cicadas intake OPEs and PAHs from soils and saps contaminated with these pollutants. The concentrations of OPEs and PAHs in the soils tend to increase with those in the atmospheric depositions. This probably shows that OPEs and PAHs in atmospheric depositions pollute soils, and cicadas intake the pollutants from the contaminated soils and saps.

Acknowledgments

The author would like to thank Kinki University for providing the grant to complete this research. The author acknowledge the assistance given by Kazuya Nagahama and Masayoshi Nakashima.

References

Alava, J. J., Ikonomou, M. G., Ross, P. S., Costa, D., Salazar, S., Aurioles-Gamboa, D., & Gobas, F. A. P. C. (2009). Polychlorinated biphenyls and polybrominated diphenyl ethers in Galapagos sea lions (Zalophus wollebaeki). *Environmental Toxicology and Chemistry. 28,* 2271-2282. Retrieved from http://onlinelibrary.wiley.com/enhanced/doi/10.1897/08-331.1/

Andresen, J. A., Grundmann, A., & Bester, K. (2004). Organophosphorus flame retardants and plasticisers in surface waters. *Science of the Total Environment, 332,* 155-166. Retrieved from http://www.sciencedirect.com/science/article/pii/S0048969704003316#

Bruggeman, W. A., Opperhuizen, A., Wijbenga, A., & Hutzinger, O. (1984). Bioaccumulation of superlipophilic chemicals in fish. *Toxicological and Environmental Chemistry, 7,* 173-189.

Carson, R. L. (1962). *Silent spring.* New York (NY): Houghton Mifflin.

Debruyn, A. M. H., Ikonomou, M. G., & Gobas, F. A. P. C. (2004). Magnification and toxicity of PCBs, PCDDs, and PCDFs in upriver-Migrating Pacific Salmon. *Environmental Science and Technology, 38,* 6217-6224. Retrieved from http://pubs.acs.org/doi/pdf/10.1021/es049607w

Gobas, F. A. P. C., Bedard, D. C., Ciborowski, J. J. H., & Haffner, G. D. (1989). Bioaccumulation of chlorinated hydrocarbons by the mayfly hexagenia limbata in Lake St. Clair. *Journal of Great Lakes Research, 15*, 581-588. Retrieved from http://www.sciencedirect.com/science/article/pii/S0380133089715124

Gobas, F. A. P. C., Wolf, W. D., Burkhard, L. P., Verbruggen, E., & Plotzke, K. (2009). Revisiting bioaccumulation criteria for POPs and PBT assessments. *Integrated Environmental Assessment and Management,   5,   624-637.   Retrieved   from* http://onlinelibrary.wiley.com/doi/10.1897/IEAM_2008-089.1/full

Hartmann, P. C., Bürgi, D., & Giger, W. (2004). Organophosphate flame retardants and plasticizers in indoor air. *Chemosphere,   57,   781-787.   Retrieved   from* http://www.sciencedirect.com/science/article/pii/S0045653504007155

Jamal, G. A., Hansen, S., & Julu, P. O. O. (2002). Low level exposures to organophosphorus esters may cause neurotoxicity.   *Toxicology,   181-182,   23-33.   Retrieved   from* http://www.sciencedirect.com/science/article/pii/S0300483X0200447X#

Johnson-Restrepo, B., Adams D. H., & Kannan, K. (2008). Tetrabromobisphenol A (TBBTA) and hexabromocyclododecanes (HBCDs) in tissues of humans, dolphins, and sharks from the United States. *Chemosphere,   70,   1935-1944.   Retrieved   from* http://www.sciencedirect.com/science/article/pii/S0045653507012374

Kawagoshi, Y., Fukunaga, I., & Itoh, H. (1999). Distribution of organophosphoric acid triesters between water and sediment at a sea-based solid water disposal site. *Journal of Material Cycles and Waste Management, 1*, 53-61. Retrieved from http://link.springer.com/article/10.1007/s10163-999-0005-6

Kelly, B. C., Gobas, F. A. P. C., & Mclachlan, M. S. (2004). Intestinal absorption and biomagnification of organic contaminants in fish, wildlife, and humans. *Environmental Toxicology and Chemistry, 23*, 2324-2236. Retrieved from http://onlinelibrary.wiley.com/doi/10.1897/03-545/full

Kondo, T., Yamamoto, H., Tatarazako, N., Kawabe, K., Koshio, M., Hirai, N., & Morita, M. (2005). Bioconcentration factor of relatively low concentrations of chlorophenols in Japanese Medaka. *Chemosphere,   61,   1299-1304.   Retrieved   from* http://www.sciencedirect.com/science/article/pii/S004565350500473X

Luch, A. (2005). *The Carcinogenic Effects of Polycyclic Aromatic Hydrocarbons*. London: Imperial College Press.

Muir, D. C. G., Yarechewski, A. L., Knoll, A., & Webster, G. R. B. (1986). Bioconcentration and disposition of 1,3,6,8-tetrachlorodibenzo-p-dioxin and octachlorodibenzo-p-dioxin by rainbow trout and fathead minnows. *Environmental   Toxicology   and   Chemistry.   5,   261-272.   Retrieved   from* http://onlinelibrary.wiley.com/doi/10.1002/etc.5620050305/abstract

Nam, J. I., Thomas, G. O., Jaward, F. M., Steinnes, E., Gustafsson, O., & Jones, K. C. (2008). PAHs in background soils from Western Europe: Influence of atmospheric deposition and soil organic matter. *Chemosphere,   70,   1596-1602.   Retrieved   from* http://www.sciencedirect.com/science/article/pii/S0045653507009873

Shimazu, H. (2011) Polycyclic aromatic hydrocarbons and organophosphoric triesters in air and atmospheric deposition. *Journal of Japan Society of Civil Engineering, Ser.G. 67*(7), III_741-III_747 (in Japanese).

SRC Inc. Home page. (2014). http://esc.syrres.com/fatepointer/search.asp

9

Evaluation of Sunflower (*Helianthus annuus L.*), Sorghum (*Sorghum bicolor L.*) and Chinese Cabbage (*Brassica chinensis*) for Phytoremediation of Lead Contaminated Soils

Rodrick Hamvumba[1], Mebelo Mataa[1] & Alice Mutiti Mweetwa[2]

[1] Department of Plant Sciences, School of Agricultural Sciences, University of Zambia, Zambia

[2] Department of Soil Sciences, School of Agricultural Sciences, University of Zambia, Zambia

Correspondence: Alice Mutiti Mweetwa, Department of Soil Sciences, School of Agricultural Sciences, University of Zambia, Box 32379, Lusaka, Zambia. E-mail: alicemweetwa@yahoo.com

Abstract

The problems associated with heavy metal contamination are widespread and are especially common in developing countries. A pot study was carried out to evaluate the effectiveness of Sunflower (*Helianthus annuus*), Sorghum (*Sorghum bicolor*) and Chinese cabbage (*Brassica chinensis*) at removing lead from the soil. Lead contaminated soils were collected from Kabwe near the old Lead mine and characterized for total and extractable lead, pH, organic matter, texture and cation exchange capacity. The average total and extractable lead concentrations were 23 313 and 5876 mg/kg, respectively, in contaminated soil compared to 57.75 and 10.02 mg/kg in uncontaminated soil. The contaminated soil was then diluted with uncontaminated soil to achieve five contamination levels of 5876, 2500, 1000, 500 and 10.02 mg/kg. Test plants were grown for 10 weeks after which below and above ground dry biomass yields were determined and tested for lead concentration and uptake. Results from this study show that Chinese cabbage is more effective at lead uptake than Sunflower and Sorghum. Results also show that high soil lead concentration results in poor plant growth, low biomass yield and increased lead accumulation in plant tissue.

Keywords: Zambia, Kabwe, lead, phytoremediation, soil, contamination, mining, pollution

1. Introduction

Heavy metal contamination results mainly from anthropogenic activities such as mining and metal smelting. This contamination is a serious environmental problem that may threaten whole ecosystems and humans. Humans and wildlife are exposed to heavy metals through several pathways that may include contaminated drinking water and food, inhaling particulates and contaminated soil health (Qu, Ma, Yang, Bi, & Huang, 2012; Ikenaka et al., 2010). Exposure to heavy metals such as lead have been shown to have several negative effects in humans. Examples of such effects include problems with coordination of muscles, nerve damage, brain damage, impaired hearing, stunted growth and mental retardation in children (United States Environmental Protection Agency [USEPA], 2013).

Currently, most studies are aimed at finding means of removing these heavy metals from the soil, and thus various methods have been suggested that include both *ex-situ* and *in-situ* methods. The *ex-situ* methods require the removal of polluted soil for treatment on or off site, and subsequent return of the treated soil to the original site. This follows that the conventional *ex-situ* methods rely on excavating, detoxifying and/or destroying the polluting agent using physical or chemical means in order to stabilize, solidify, immobilize or completely destroy the contaminant (Ghosh & Singh, 2005). The *in-situ* methods which involve the destruction or transformation of the contaminant on site (Vandgrift, Reed, & Tasker, 1992) are preferred to the *ex-situ* techniques because they are cheaper and have potentially reduced impact on the ecosystem.

Phytoremediation, through the technology of phytoextraction, is a concept of using plants to clean the environment (Prasad & Freitas, 2003). This technology is suitable for *in situ* clean-up of toxic metal polluted sites (Vashegyi et al., 2005). Unlike organic compounds, metals cannot be degraded, and clean-up usually requires their removal from polluted sites. Phytoremediation has proved to be an effective way of removing the toxic metals from soil (Prasad & Freitas, 2003). The application of this concept takes into account the importance of biodiversity that encompasses several agricultural, horticultural and wild metal accumulating plants. More than 400 plants that hyperaccumulate heavy metals, from different families, have so far been tested for their potential use in the decontamination of polluted sites. Of these families, Brassicaceae has been reported to have the largest number of hyperaccumulaters (Prasad & Freitas, 2003). Phytoremediation strategies appear promising as a way to reduce lead contamination in soils and thereby minimize its potential hazardous effects.

In Zambia, lead was mined in Kabwe (Central Province). Mining activities started at the beginning of the 20th century and continued for 90 years until 1994 when the mine was closed, and smelting and sulphuric acid production operations conducted discontinued (The World Bank Group, 2013). Kabwe ranks among the 10 most polluted places in the world (Blacksmith Institute, 2007) with over 300,000 of the total population being affected by contaminated soil by 2003 (The World Bank Group, 2013). A lot of effort has been made to determine the extent of heavy metal contamination of the soils, aquatic systems and plants in Kabwe and other mining areas in Zambia (Ikenaka et al., 2010; Tembo, Sichilongo, & Cernak, 2006). On average, children's blood lead levels in Kabwe have been reported to be as high as 5 to 10 times the permissible WHO/EPA maximum of 10 µg/dL (Blacksmith Institute, 2007).

The objective of the study was to determine the effectiveness of *Sorghum bicolor* L., *Brassica chinensis* and *Helianthus annuus* L. in reducing concentrations of lead in contaminated soils from Kabwe. Specifically, the study evaluated the growth of and uptake of lead by *Sorghum bicolor* L., *Brassica chinensis* and *Helianthus annuus* L. in lead contaminated soils.

2. Method

2.1 Soil Sample Collection

The lead contaminated soil was collected from a location near the former Lead Mine in Kabwe District. The lead contaminated soil was then characterized to determine selected physical and chemical properties using the methods described below.

2.2 Characterization of the Soil

The following methods were used to characterize the soil used in the study.

2.2.1 Particle Size Distribution

Particle size analysis of the fine earth fraction was determined using the hydrometer method (Day, 1965); air dried soils (50 grams) were weighed in triplicate and placed in the dispersing cups. Fifty milliliters of calgon (sodium hexametaphosphate) was then added and the cups filled with tap water to the half way mark. The cups were stirred continuously for five minutes. The suspension was transferred into a sedimentation cylinder using a stream of tap water, the cylinder filled to the 1000 ml mark. A plunger was used to mix the contents thoroughly, after 20 seconds the hydrometer was lowered carefully into the cylinder and the density of the suspension read. The temperature was read using a thermometer. After two hours the density and temperature were read again and the particle size distribution analyzed.

2.2.2 Determination of Soil Organic Matter

The Walkley and Black Method was used to determine organic matter (Olsen & Sommers, 1982); One gram of air dried soil was weighed and placed in a conical flask, then 10 ml of 1N $K_2Cr_2O_7$ was added. Twenty milliliters of concentrated H_2SO_4 was rapidly added to the conical flask by directing the stream into the suspension using an automatic pipette. The conical flask was then swirled vigorously for a minute. The conical was stored in the fume hood for 30 minutes. Afterwards 150 ml of distilled water was added followed by 10 ml of concentrated H_3PO_4. Ten drops of diphenylamine indicator solution was added to the conical flask and it was titrated with Iron (II) Sulphate.

2.2.3 Determination of Soil pH

To determine pH, the electronic pH meter was used (Mclean, 1982); 10 g of air dry Lead contaminated soils was placed in a 50 cm^3 beaker and 25 cm^3 0.01 M $CaCl_2$ added. Each mixture was placed on shaker for 30 minutes and pH determined using a pH meter.

2.2.4 Determination of the Effective Cation Exchange Capacity of the Soil

2.2.4.1 Determination of the Exchangeable Acidity

To determine the exchangeable acidity, the NaOH titration method was used (Mclean, 1982); 10 g air dry soil was weighed and put into a 250 cm^3 Erlenmeyer flask, 100 cm^3 of 1 M KCl was then added and covered with a rubber stopper. The conical flask was shaken for 1 hour and then the suspension filtered and the filtrate collected in a beaker. A volume of 25 cm^3 of KCl extract was pipetted into a 250 cm^3 conical flask and approximately 100 cm^3 of distilled water was added. To the same flask, 5 drops of phenolphthalein indicator was added to the solution with a standard 0.01 M NaOH solution to a permanent pink end point. The amount of base used was equivalent to the exchangeable acidity in the aliquot taken.

2.2.4.2 Extraction of Exchangeable Bases

To determine the exchangeable bases, the ammonium acetate method was used (Thomas, 1982); 10 g of air dry soil was weighed and passed through a 2 mm sieve and put on a No 42 Whitman filter paper. Some 25 ml of 1N NH$_4$OAc was added four times, the filtrate was removed and the soil was then rinsed with Ethanol four times as well. The rinsing was repeated and 100 ml of 2 M KCl was added. Concentrations of K$^+$, Ca^{2+}, Mg^{2+} and Na$^+$ were determined using the Atomic Absorption Spectroscopy (AAS).

2.2.5 Determination of Total Lead in the Soil

To determine the total concentration of lead in the soil, the Aqua Regia method was used; 1.0 g of lead contaminated soil was weighed using an electronic balance and to which 18 ml of HCl and 6 ml HNO$_3$ were added. The sample was allowed to digest on a hot plate until enough of the solution had evaporated. The sample was then cooled, 18 ml of HCl and 6 ml of HNO$_3$ added again. The mixture was brought to 50 ml volume using distilled water and amount of total Lead determined using AAS.

2.2.6 Determination of Extractable Lead Using 0.5 M HNO$_3$

To determine extractable lead, the Nitric Acid method was used; 20 g of each soil sample was weighed using an electronic balance and to which 40 cm^3 of 0.5 M HNO$_3$ was added in order to extract only the Lead that may be soluble for plant uptake. The mixture was allowed to shake for 2 hours, afterwards it was filtered through No 42 Whitman filter paper and amount of extractable Lead determined using the AAS.

2.3 Green House Experiment

A greenhouse crop trial with 5 levels of lead contamination (Table 1) and 3 crop species (Chinese cabbage, Sunflower and Sorghum) was set up in a Completely Randomized Design (CRD) and replicated four times. The lead contaminated soil was diluted to 43, 20 and 9% of the original lead concentration using ordinary soils from the Field Station at the University of Zambia. The dilution was done to monitor the performance of the different plant species in varying levels of lead in the soil. The dilutions were made based on the concentration of extractable lead in both contaminated and ordinary soil samples. The soils (ordinary soil and lead contaminated soil) were thoroughly mixed to ensure uniform distribution of Lead in each pot.

Table 1. Dilutions of soil for greenhouse crop trials for Chinese cabbage, sunflower and sorghum

Soil Sample	Extractable Lead (mg/kg)	Total Lead (mg/kg)
Undiluted	5876	23 313
Dilution 1 (43%)	2500	11 500
Dilution 2 (20%)	1000	4600
Dilution 3 (9%)	500	2300
Control	10	58

The Milika, Chihili and Sima crop varieties of sunflower, Chinese cabbage and sorghum, respectively, were used for this trial. Seeds were sown directly to the appropriate depth. Basal and Top dressing fertilizer were applied at recommended rates for each plant species and in relation to plant density. The plants were grown for 10 weeks after which the shoots and roots were harvested and dried.

2.4 Determination of Lead in the Leaves of Sorghum, Sunflower and Chinese Cabbage

To determine lead concentrations in the leaves of all plants, approximately 1 g of each plant sample biomass was weighed using a sensitive electronic balance and ashed at 500 °C to which 10 ml 1 N HNO$_3$ was added, filtered and brought to volume of 100 ml by adding distilled water. Concentrations of lead were then determined by AAS. Lead uptake was calculated from the Lead concentrations by multiplying the concentration by the total dry mass of the crop.

2.5 Statistical Analysis

To assess whether there were significant differences among various treatment effects, Analysis of Variance was carried out.

3. Results

3.1 Soil Characterization

The characteristics of uncontaminated and lead contaminated soils are presented in Table 2. The lead contaminated soil was slightly acidic (pH 6.49) with a loamy sand texture and high concentration of Lead exceeding the US EPA's permissible soil Lead content in bare soil of 400 mg/ kg (USEAP, 2001). The ordinary soil had a neutral soil pH (pH 7.15) with a sandy loam texture, low organic matter content (1.4%) and low soil lead concentration. The organic matter content was higher (3.36%) in the contaminated soil than that of the ordinary soil, perhaps due to high presence of grass root during the time of soil collection. In both soils, the ECEC was moderately low, being 5.72 and 6.53 meq/100 g for the lead contaminated and the normal soils, respectively.

Table 2. Selected chemical and physical characteristics of soil

Soil sample	Soil pH	Extractable Lead (mg/kg)	Total Lead (mg/kg)	OM (%)	Texture	ECEC (meq/100g)
Contaminated	6.49	5876	23 313	3.36	Loamy sand	5.72
Uncontaminated	7.15	10.02	57.75	1.4	Sandy loam	6.53

3.2 The Effect of Lead on Plant Growth

In general, an inverse relationship was observed between plant growth and soil lead concentration (Figure 1). Poor plant growth in all species was observed at high soil lead concentrations as there was poor root and shoot growth (Figure 1). Complete failure of sunflower to grow at high lead concentrations was also observed.

3.3 Below Ground Biomass Yield at Harvest

Sunflower produced the highest below ground biomass followed by sorghum and lastly Chinese cabbage. In all the three plant species, an increase in biomass was observed as the concentrations of lead among treatments declined (Figure 2). The results show that sunflower, which has a bigger growth form, had a larger below ground biomass yield than the other two crops. Significant differences were observed for each crop species among the first three soil treatments containing 5867, 2500, and 1000 mg/kg of lead as shown in Figure 2.

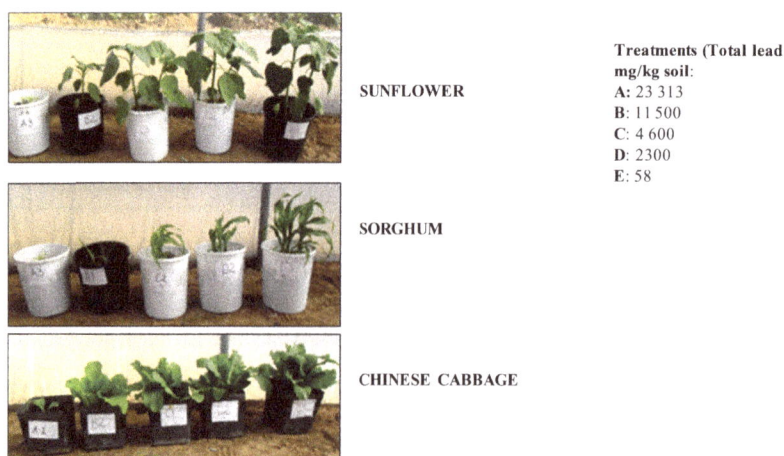

SUNFLOWER

SORGHUM

CHINESE CABBAGE

Treatments (Total lead mg/kg soil:
A: 23 313
B: 11 500
C: 4 600
D: 2300
E: 58

Figure 1. Plant growth of sunflower (*Helianthus annuus*), sorghum (*Sorghum bicolor*) and Chinese cabbage (*Brassica chinensis*) at 6 weeks after sowing

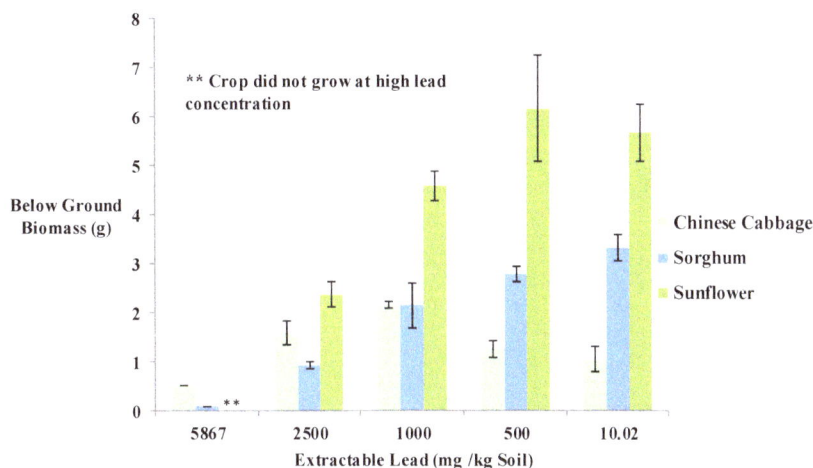

Figure 2. Below ground biomass yield at harvest of sunflower (*Helianthus annuus*), sorghum (*Sorghum bicolor*) and Chinese cabbage (*Brassica chinensis*)

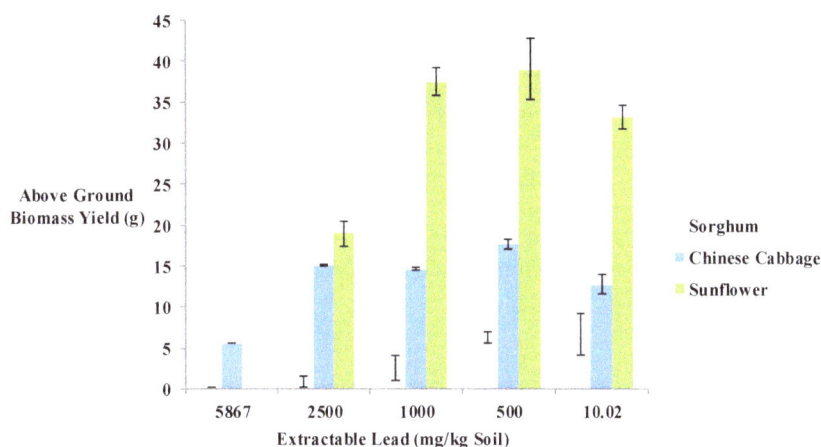

Figure 3. Above ground biomass yield at harvest of sunflower (*Helianthus annuus*), sorghum (*Sorghum bicolor*) and Chinese cabbage (*Brassica chinensis* var)

3.4 Above Ground Biomass Yield at Harvest

At soil lead concentrations where it was able to grow, sunflower produced the highest above ground biomass followed by Chinese cabbage and sorghum (Figure 3). Sorghum had the lowest biomass yield and showed severe signs of stunted growth due lead toxicity. Generally, an inverse relationship was observed as shoot biomass yield reduced with increase in soil lead concentrations among treatments.

3.5 Lead Concentrations in Below Ground Tissue at Harvest

Figure 4 shows the lead concentrations in below ground tissue for all the 3 plant species grown in soil with different lead levels. In all of the plants analyzed, the root portion of the plant showed the highest levels of lead, followed by the shoots, in which the levels were undetectable (for sunflower and sorghum). Chinese cabbage accumulated more lead in the roots than sunflower and sorghum at all the levels of lead contamination, except for the uncontaminated soil. With Chinese cabbage, there was a linear increase in lead concentration in the roots as the soil lead concentration increased. Sunflower showed a slight constant or uniform Lead accumulation from treatment 2500 to 500 mg/ kg.

The results from this study suggest that Chinese cabbage is more tolerant to high soil lead concentration than sunflower and sorghum. Sunflower proved to be very sensitive to high lead levels in soils as it failed to grow when extractable lead concentrations were high (5867 mg /kg) and its growth was very poor.

3.6 Lead Concentrations in Above Ground Tissue at Harvest- Chinese Cabbage

Lead concentrations in the above ground tissue in Chinese cabbage increased with increase in soil lead concentration (Figure 5). A drastic decrease of Lead content in the shoots was observed between 5876 to 2500 mg/kg showing a significant difference in the accumulation of lead between the two treatments. An exponential decrease of lead content in shoots of Chinese cabbage was observed in treatments 5876 to 10.02 mg/ kg as well. However, lead in the shoots of sunflower and sorghum was at levels below the experimental limit of detection which was 0.5 mg/kg.

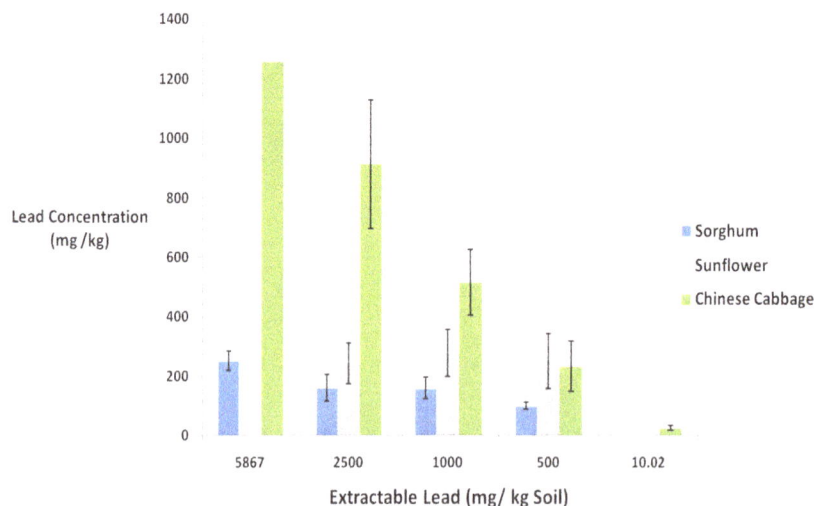

Figure 4. Lead concentrations in below ground tissue at harvest of sunflower (*Helianthus annuus*), sorghum (*Sorghum bicolor*) and Chinese cabbage (*Brassica chinensis* var)

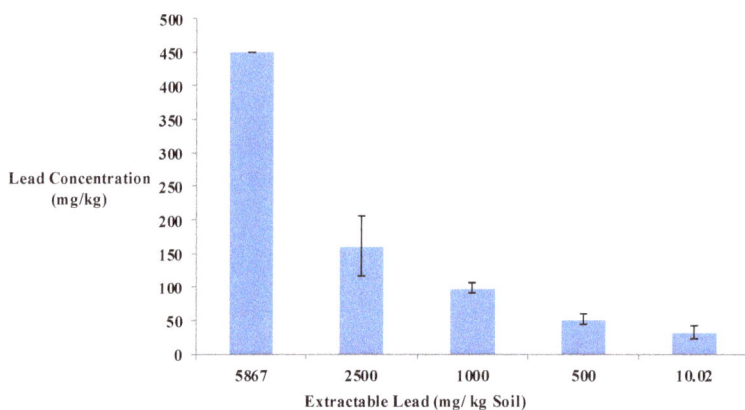

Figure 5. Lead concentrations in above ground tissue of cabbage (*Brassica chinensis)* at harvest

3.7 Lead Uptake in Above Ground Tissue at Harvest- Chinese Cabbage

Lead uptake was derived from the lead concentrations and biomass yields of the tissues (Figure 6). The results show that uptake of Lead by Chinese cabbage increased with the increase in soil extractable lead concentration. However, Chinese cabbage showed signs of uptake threshold as in treatment 5876 and 2500 mg/ kg the total lead uptake in both treatments was almost the same. Among treatments, there were significant differences with regards to total uptake of Lead. The mass of total Lead uptake in pots by Chinese cabbage among the different treatments is shown in Figure 6.

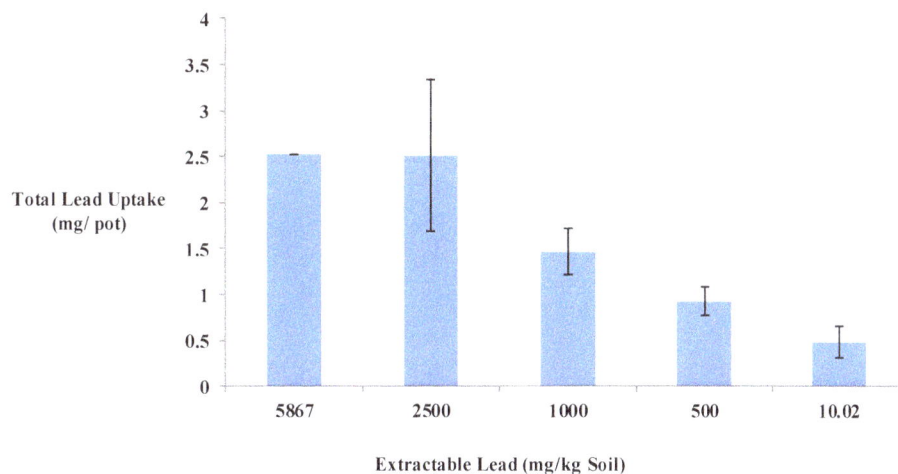

Figure 6. Lead uptake in above ground tissue at harvest (10 weeks) for Chinese cabbage (*Brassica chinensis*)

4. Discussion

Chinese cabbage, sunflower and sorghum all showed a poor growth pattern as the concentration of Lead in the soil increased among treatments. The observed inverse relationship between plant growth and soil lead concentration could be attributed to the direct negative effects of lead on plants (Figure 1). Symptoms of lead toxicity at higher lead levels included stunted growth and chlorosis. Previously, Sharma and Dubey (2005) noted that excess lead causes a number of toxicity symptoms in plants such as stunted growth, chlorosis and blackening of the root system, this agrees with the observations of this study. Alternatively, lead in the soil has been shown to be able to complex other plant nutrients such as phosphorus, thereby rendering them both unavailable for uptake (Xie, Wang, Sun, & Li, 2006). It is possible that plants suffered phosphorus deficiencies contributing to poor plant growth i.e. stunting, even if the classic purple discoloration of leaves was not observed. Sunflower was more sensitive to high concentrations of lead in the soil to the extent that plants failed to grow up to harvest. However, this observation was rather unexpected as sunflower together with other species of the family Asteraceae have been proposed to be hyperaccumulators of heavy metals (Prasad & Freitas, 2003). This observation could suggest that a threshold of lead contamination exists above which plants cannot survive. High soil concentrations of lead have been shown to reduce seed germination, root and shoot length, tolerance index and dry mass of roots and shoots of rice (Mishra & Choudhuri, 1998). This could explain the observed growth patterns and the complete failure of sunflower seeds to germinate in soils with high levels of lead.

According to Seregin and Ivanov (2001), lead phytotoxicity leads to inhibition of enzyme activities, disturbed mineral nutrition, water imbalance, change in hormonal status and alteration in membrane permeability. These disorders upset the normal physiological activities of the plant and would result in loss of biomass yield as observed in this study (Figure 3). According to Xiong (1998), *Brassica pekinensis* belonging to the same family as Chinese cabbage (Brassicaceae) accumulated unusually high contents of lead in the root and shoot tissues. The results from this study suggest that Chinese cabbage is more tolerant to high soil lead concentrations than sunflower and sorghum. Sunflower proved to be very sensitive to high lead levels in soils as it failed to grow when extractable lead concentrations were high (5867 mg /kg) and its growth was very poor.

According to Sharma and Dubey (2005), roots can take up significant quantities of lead whilst greatly restricting its translocation to above ground parts as observed in Sunflower and Sorghum. Therefore, it can be suggested that Chinese cabbage was more effective at accumulating lead in the shoots than sunflower and sorghum. Though this may seem to be a good characteristic of Chinese cabbage, most researchers recommend that plants appropriate for phytoremediation should accumulate metals only in the roots (Prasad & Freitas, 2003).

Chinese cabbage appeared to be more efficient at the uptake of lead from the soil than sorghum and sunflower. Prasad and Freitas (2003) noted that it is the combination of high metal accumulation and high biomass production that results in the most efficient metal removal. From the results obtained, soil lead concentration affects the amount of plant biomass produced; reducing with an increase in soil lead concentration. Lead accumulation in plant tissue increases with increase in soil lead concentration as observed in sunflower, sorghum and Chinese cabbage among the different treatments. It can be concluded that Chinese cabbage is capable of taking up lead from the soil and that it is more efficient than sunflower and sorghum. However, indications are that this would only be applicable to sites that contain low to moderate levels of lead contamination because plant growth is not sustained in heavily contaminated soils.

Based on the observations from the current study, and in order to get more insightful observations, plants should be allowed to grow to full maturity. This will allow for partitioning of plant tissue so as to determine how much lead would be accumulated in each plant part. Secondly, field experiments should be done to evaluate the response of plants grown directly in the sites of contamination. Greenhouse conditions are not truly representative of field conditions and the results from such studies are only indicative. In the field, plants are subjected to less than optimal conditions and thus may have different morphological and physiological responses.

Acknowledgements

We wish to acknowledge the technicians in the Department of Soil Science and Mr. V. Shitumbanuma for their technical input into this study.

References

Blacksmith Institute. (2007). *The world's worst polluted places* (pp. 1-69). The top ten of the dirty thirty, Blacksmith Institute, New York, USA. Retrieved from http://www.worstpolluted.org

Day, P. R. (1965). *Particle size fractionation and particle size analysis.* In C. A. Black (Ed.), Methods of soil analysis. *Agronomy, 9,* 545-567.

Ghosh, M., & Singh S. P. (2005). A Review on Phytoremediation of Heavy Metals and Utilization of its Byproducts. *Applied Ecology and Environmental Research, 3*(1), 1-18, Retrieved from http://ecology.kee.hu/pdf/0301_001018.pdf

Ikenaka, Y., Nakayama, S. M. M., Muzandu. K., Choongo, K., Teraoka, H., Mizuno, N., & Ishizuka, M. (2010). Heavy metal contamination of soil and sediment in Zambia. *African Journal of Environmental Science and Technology, 4*(11), 729-739. http://dx.doi.org/10.4314%2Fajest.v4i11.71339

Mclean, E. O. (1982). Soil pH and lime requirement. In A. L. Page, R. H. Miller, & D. R. Keeney (Eds.), Methods of soil analysis. Part 2. Chemical and microbiological properties. (2nd ed.). *Agronomy, 9,* 99-223.

Mishra, A., & Choudhuri, M. A. (1998). Amelioration of lead and mercury effects on germination and rice seedling growth by antioxidants. *Biologia. Plantarum, 41,* 469-473. http://dx.doi.org/10.1023/A:1001871015773

Olsen, S. R., & Sommers, L. E. (1982). Phosphorus. In A. L. Page, R. H. Miller, & D. R. Keeney (Eds.), *Methods of soil analysis: Part 2- Chemical and microbiological properties.* (2nd ed., pp. 403-430). Madison, WI: American Society of Agronomy.

Prasad, M., & Freitas, H. O. (2003). Metal hyperaccumulation in plants- Biodiversity prospecting for phytoremediation technology. *Electronic Journal of Biotechnology, 6*(3). http://dx.doi.org/10.2225/vol6-issue3-fulltext-6

Qu, C. S., Ma, Z. W., Yang, J., Bi, J., & Huang, L. (2012). Human exposure pathways of heavy metals in a lead-zinc mining area, Jiangsu Province, China. *PLoS ONE, 7*(11), e46793. http://dx.doi.org/10,1371/journal.pone.00467.

Seregin, I. V., & Ivaniov, V. B. (2001). Physiological aspects of cadmium and lead toxic effects on higher plants. *Russian Journal of Plant Physiology, 48,* 523-544. http://dx.doi.org/ 10.1023/A: 1016719901147

Sharma, P., & Dubey, R. S. (2005). Lead toxicity in plants. *Brazilian. Journal of Plant Physiology, 17*(1) 35-52. http://dx.doi.org/10.1590/S1677-04202005000100004

Tembo, B. D., Sichilongo, K., & Cernak, J. (2006). Distribution of copper, lead, cadmium and zinc concentrations in soils around Kabwe town in Zambia. *Chemosphere, 63*(3), 497-501. http://dx.doi.org/10.1016/j.chemosphere.2005.08.002

The World Bank Group. (2013). Copperbelt Environment Project tackles the lead and uranium danger in Zambia. Retrieved February 28, 2014, from http://web.worldbank.org/WBSITE/EXTERNAL/NEWS/0,,contentMDK:23169108~menuPK:141310~pagePK:34370~piPK:34424~theSitePK:4607,00.html

Thomas, G. W. (1982). Exchangeable cations. In A. L., Page (Ed.), Methods of soil anaylsis. Part 2. (2nd ed). Agronomy Monograph 9, American Society of Agronomy, Madison, WI.

USEPA. (2001). Lead; Identification of dangerous levels of lead. Final Rule. Federal register, 40 CRF Part 745. Retrieved February 28, 2014, from http://epa.gov/superfund/lead/products/rule.pdf

USEPA. (2013). *Human health and lead.* Retrieved February 28, 2014, from http://www.epa.gov/superfund/lead/health.htm

Vandgrift, G. F., Reed, D. T., & Tasker I. R. (Eds.). (1992). *Environmental remediation: Removing organic and metal ion pollutants.* ACS Symposium Series 509. Washington DC: ACS

Vashegyi, A., Mezôsi, G., Barta, K., Farsang, A., Dormány, G., Bartha, B., & Erdei, L. (2005). Phytoremediation of heavy metal pollution: A case study. *Acta Biologica Szegediensis, 49*(1-2), 77-79. Retrieved from http://www2.sci.u-szeged.hu/ABS/2005/Acta%20HP/4977.pdf

Xie, Z. M., Wang, B. L., Sun, Y. F., & Li, J. (2006). Field demonstration of reduction of lead availability in soil and cabbage (*Brassica chinensis* L.) contaminated by mining tailings using phosphorus fertilizers. *Journal of Zhejiang University Science B, 7*(1), 43-50. http://dx.doi.org/ 10.1631/jzus.2006.B0043

Xiong, Z. T. (1998). Lead uptake and effects on seed germination and plant growth in a lead hyperaccumulator-*Brassica pekinensis* Rupr. *Bulletin of Environmental Contamination and Toxicology, 60,* 285-291. http://dx.doi.org/ 10.1007/s001289900623

10

Hospitalizations for Respiratory Problems and Exposure to Industrial Emissions in Children

Rémi Labelle[1], Allan Brand[2], Stéphane Buteau[2,3] & Audrey Smargiassi[2,4]

[1] Université de Montréal. Département de Santé Environnementale et de Santé au Travail, Montréal, Québec, Canada

[2] Institut National de Santé Publique du Québec (INSPQ), Canada

[3] Department of Medecine, McGill University, Montréal, Québec, Canada

[4] Chaire sur la pollution de l'air, les changements climatiques et la santé, en partenariat avec la Direction de Santé publique de l'Agence de la santé et des services sociaux de Montréal et l'INSPQ, Université de Montréal, Montréal, Québec, Canada

Correspondence: Audrey Smargiassi, Chaire sur la pollution de l'air, les changements climatiques et la santé, en partenariat avec la Direction de Santé publique de l'Agence de la santé et des services sociaux de Montréal et l'INSPQ, Université de Montréal, Montréal, Québec, Canada. E-mail: audrey.smargiassi@umontreal.ca

Abstract

Industrial activities such as metal smelting, petroleum refining, and open mining emit air pollutants that can affect the health of surrounding communities. Few studies have assessed respiratory effects of acute exposure to industrial air emissions in children. In this study, we examined the association between daily exposure to air emissions from an industrial complex and hospitalizations for respiratory problems of children living nearby using a case crossover design. We used hospitalizations for respiratory problems of children under 5 years old living within 7.5 km of the industrial complex from January 1, 2001 to December 31, 2010. Pollutant exposure was estimated using daily mean and maximum concentrations of SO_2 and $PM_{2.5}$ at fixed monitoring stations located near the complex. We also calculated the daily percentage of hours that a child's residence was downwind of the industrial complex as an indicator of exposure to emissions. Odds-ratios were adjusted for temperature, relative humidity and wind speed, and calculated using conditional logistic regressions, reported by increases of interquartile range. A significant positive association was found between hospitalization for asthma or bronchiolitis and the percentage of hours downwind (OR: 1.11, 95% CI=1.01–1.22) but large statistical variability was noted for associations with all three exposure metrics (OR maximum SO_2 levels: 1.06, 95% CI=0.98–1.15; OR daily maximum $PM_{2.5}$ levels: 0.97, 95% CI=0.86–1.09). The results suggest that exposure to the mixture of air pollutant emissions from an industrial complex may induce respiratory health problems in children residing nearby.

Keywords: emissions, downwind exposure, case-crossover, odds-ratios, $PM_{2.5}$, SO_2

1. Introduction

Few studies have assessed the respiratory effects of exposure to air emissions in children from various types of industries including metal smelters, cement plants, power plants, petrochemical industries, open mining or wood processing plants (Aekplakorn et al., 2003; Bertoldi et al., 2012; Deger et al., 2012; Liu, Lessner, and Carpenter, 2012; Pless-Mulloli, Howell, and Prince, 2001; Rusconi et al., 2011; Smargiassi et al., 2009). Most studies that assessed the effects of industrial air emissions used a cross-sectional design and usually evaluated the respiratory outcome using a questionnaire. These studies have found some associations between respiratory health outcomes, e.g. increased respiratory symptoms, emergency department visits and hospitalizations or decreased pulmonary functions in children, and exposure to industrial air emissions. Nonetheless, only a few studies assessed the respiratory effects of short term exposure to industrial emissions in children (e.g. Aekplakorn et al., 2003; Lewin, Buteau, Brand, Kosatsky, and Smargiassi, 2013; Smargiassi et al., 2009).

The objective of the present study was to estimate the association between daily levels of sulphur dioxide (SO_2), fine particles ($PM_{2.5}$) and the percentage of daily hours downwind from an industrial complex with an aluminum smelter in the Saguenay region of Quebec, Canada and the hospitalizations of children aged 0 to 4 years old for asthma, bronchiolitis and other respiratory causes living in proximity to the industrial complex.

2. Methods

2.1 Study Period, Area and Population

The study area was located in the Saguenay region of the province of Quebec (Canada).It was centered on three point source emissions (referred to as the industrial complex) near residential neighborhoods. The first emission source (A), i.e. a cast house, emitted, in 2012, 7 569 tons of SO_2 and 1 413 tons of $PM_{2.5}$ (Environment Canada, 2013). The Söderberg process was used in the past at this plant, and air pollutant emissions were larger (Alain et al., 2011). Emission point B (alumina production plant) emitted much less than point A (86 tons of SO_2 and 151 tons of $PM_{2.5}$) in 2012 (Environment Canada [EC], 2013). The third plant, emission point C (an iron smelter) emitted in 2012, 639 tons of SO_2 and 17 tons of $PM_{2.5}$ (EC, 2013). Another industry is also present in the area of study, a paper mill that emitted less pollutant than the industrial complex: in 2012, it emitted 83 tons of $PM_{2.5}$ (EC, 2013). The study area was defined as the area within a buffer of 7.5 km centered on the industrial complex that included the three emission point sources (see Figure 1 for the location of the industrial complex). The location of the industrial complex was defined as the point between the three emission sources weighted by total emissions of SO_2 and $PM_{2.5}$ from each source during the study period. Approximately 6 985 children aged under 5 year old lived in the area in 2011 (Statistics Canada, 2012).

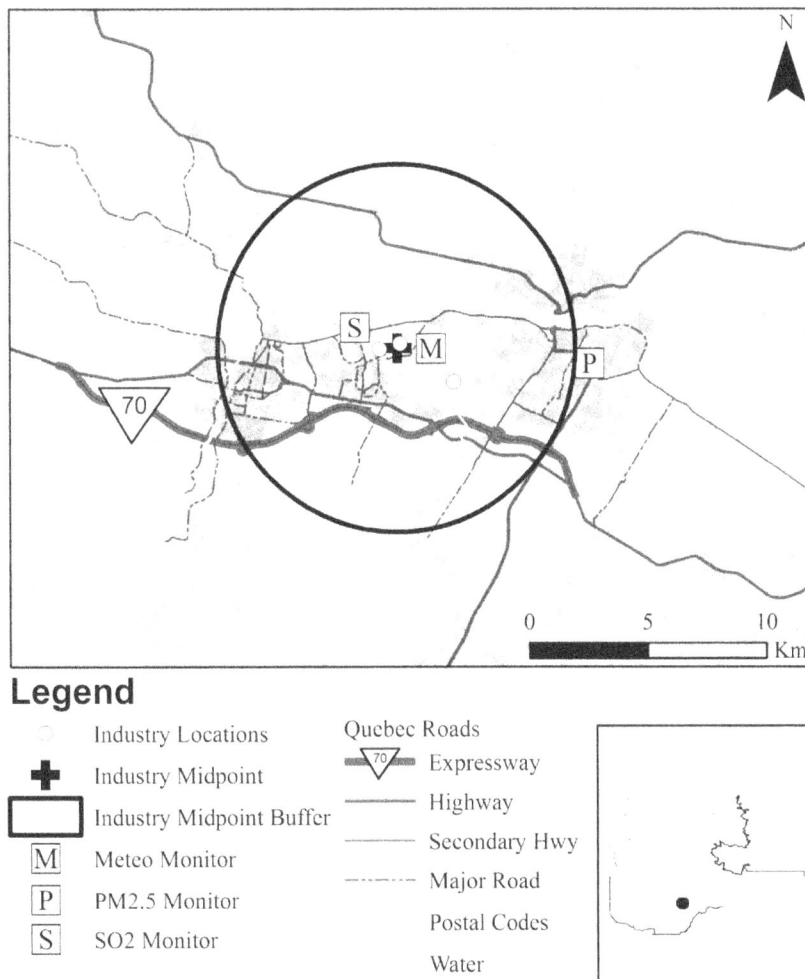

Figure 1. Map of an area of 7.5 km of radius located around an industrial complex comprising an aluminum smelter, where its location has been determinate by a central point located between the three factories, weighted by the quantities of emissions of SO_2 and $PM_{2.5}$ produced by each one of them, in the region of Saguenay, Quebec, Canada

2.2 Health Data

We used hospitalisations that occurred between January 1st 2001 and December 31th 2010 from the MED-ECHO database, for respiratory conditions in children aged 0 to 4 years, living within 7.5 km of the industrial complex (i.e. the weighed point between three emissions sources). This particular age group was chosen as it was reported more vulnerable to air emissions in previous studies (Bateson & Schwartz, 2008). The distance to industry was selected based on previous work in Quebec looking at respiratory hospitalizations near an aluminium smelter which found increased risks of hospitalizations for asthma and bronchiolitis in children aged 2 – 4 (Lewin et al., 2013). The database contains the sex of the children, the six digit postal code of the place of residence and the date and the cause of hospitalisation according to the international code of disease version 9 (ICD-9) or version 10 (ICD-10). The respiratory conditions considered were as follows: asthma (ICD-9: 493/ICD-10: J45-J46), bronchiolitis (ICD-9: 466\ICD-10: J21), other respiratory problems (ICD-9: 460-519/ICD-10: J00-J99). Hospitalisations occurring within 30 days of a prior event were discarded.

2.3 Exposure to Industrial Emissions

Three exposures variables were used: 1) the daily amount of time that the child's place of residence was downwind of the industrial complex, and the daily mean and maximum concentrations of 2) SO_2 and, 3) $PM_{2.5}$. The daily amount of time that the residence was exposed was computed as follows. We considered that a residence was exposed for an hour if its six digit postal code centroid was located within the area delimited by two angles of 22.5 degrees on both sides of a vector representing the direction of the wind originating from the industrial complex. Figure 1 presents the area of study and the location of the meteorological and pollution monitoring stations. Wind data and other meteorological variables (daily mean temperature, relative humidity and wind speed average) were from a meteorological monitoring station located at approximately 1.5 km from the industrial complex. The data was obtained from the Data Access Integration (DAI) team, a portal containing environmental data managed by Environment Canada. The number of downwind hours per day was divided by the number of hours with wind direction data available to produce the daily percentage of hours downwind from the industrial complex. Days with less than 18 hours of measurements recorded were discarded from the analyses. For hours with no wind (i.e. 0 km/h), all six digit postal codes centroids located within 2.5 km of the industrial complex location (as described above) were considered exposed. Daily mean and maximum concentrations of SO_2 and $PM_{2.5}$ were from the National Air Pollution Surveillance (NAPS) program (EC, 2013). The station measuring SO_2 with UV fluorescence was located at 1.77 km from the industrial complex. The $PM_{2.5}$ levels were measured for a subset of the study period, i.e. 2004 to 2008, with a TEOM sensor with dryer at a station located 8.32 km away from the industrial complex.

2.4 Statistical Analyses

We assessed the association between our three exposure variables (i.e. percent of time downwind, PM2.5 and SO2 pollution monitors) and hospitalisations for respiratory conditions in children, using a time-stratified case-crossover design, similar to what was done in Lewin et al. (2013). For each case date, control dates were selected by using the same days of the week within the same month as the case date (Lumley & Levy, 2000). For example if a subject was hospitalized Saturday December 11th 2004, the other Saturdays of the month (December 4, 18, 21) were its control dates. This design allows us to control for secular trends in our hospitalization data (Janes, Sheppard, & Lumley, 2005; Maclure, 1991).

We performed the analyses for children aged 0 to 4 years and for the 2-4 years subgroup. We also analysed hospitalisations for all respiratory conditions and for asthma and bronchiolitis separately. Analyses were also performed with lag 1 exposure (the day previous to the hospitalisation), and for only those individuals living within 2.5 km of the industrial complex. Analyses were also performed separately for the two emission point sources of the industrial complex. For these analyses, the hospitalisations considered were those within 7.5 km of each point source (not within 7.5 km of the industrial complex).

In order to assess associations between time spent downwind or exposure based on pollution monitors, and hospitalisations for respiratory problems in children, we estimated Odds ratios (OR) with conditional logistic regressions. In the conditional logistic regressions, each case-control date pairs was a strata and the ORs were estimated based on the difference in exposure on case and control days (Rothman, Greenland, and Lash, 2008). We report ORs with their 95% confidence intervals (95% CI) based on the inter-quartile range (IQR) of each exposure variable. When case dates were associated with less than three control dates, the strata and related case was removed from our analysis. Associations for our exposure variables were adjusted with the inclusion of daily mean temperature, relative humidity, and average daily wind speed (km/h) in our logistic regressions. All analyses were performed with the R software (http://www.r-project.org/).

3. Results

During the study period, there were 2 909 hospitalizations for respiratory problems in children aged 0 to 4 years old (1 006 for asthma and bronchiolitis). Table 1 details the exposure variables during the study periods. Residences in the study area were downwind from the industrial complex on an average 28.58 % of the time (IQR of 27.24). The daily mean SO_2 levels measured during the study period at the monitoring station was 8.91 ppb with a standard deviation of 14.88 (IQR of 9.22 ppb). For $PM_{2.5}$ data only available for the period 2004 to 2008, the average daily concentration was 5.52 $\mu g/m^3$ (standard deviation of 4.46 $\mu g/m^3$; IQR of 4.03 $\mu g/m^3$).

Table 1. SO_2 and $PM_{2.5}$ concentrations and percentage of hours per day that residential postal codes in a buffer of 7.5 km were downwind of the industrial complex

Pollutant	Years of data	Mean	Standard Deviation	IQR	Number of days of exposition
Percentage of hours downwind	2001-2010	28.58	15.41	27.24	15 025 385[a, b]
SO_2 (ppb)	2001-2010	8.91	14.88	9.22	3 652
$PM_{2.5}$ ($\mu g/m^3$)	2004-2008	5.52	4.46	4.29	578

[a] Number of receptor points 2001-2005 : 5077 six-character residential postal codes in the study area x 1722 days of wind data. 25 days were missing wind data.

[b] Number of receptor points 2006-2010 : 5264 six-character residential postal codes in the study area x 1802 days of wind data. 13 days were missing wind data.

Table 2 shows the results of the adjusted associations between hospitalizations for asthma and bronchiolitis in children aged between 0 to 4 years old living in the area of study and the three exposure variables at lag 0. A statistically significant positive association was found only with the percentage of hours downwind (OR: 1.11; 95% CI=1.01-1.22, for an IQR of 27% of daily hours downwind) at lag 0 but large statistical variability was noted for the associations with all three exposure metrics (i.e. percentage of hours downwind and levels of SO_2 and $PM_{2.5}$). Results with the percentage of hours downwind were similar at lag 1 (Table A1). Crude and adjusted associations were also similar (Table A2, supplemental material). When analyses were restricted to the small number of children living within 2.5 km of the industrial complex (Table A3), the association was no longer statistically significant for the percentage of hours downwind and very large statistical variability was noted with all exposure metrics.

Table 2. Adjusted[a] associations between hospitalizations for asthma and bronchiolitis in children aged less than 5 years of age, living in a 7.5 km buffer around the industrial complex and daily exposure variables lag 0

Pollutants	n day (n case days)	Daily mean		Hourly max	
		IQR (h/day)	OR (95%CI)	IQR	OR (95%CI)
Percentage of hours downwind[b]	4155 (957)	27.24% (6.48)	1.11 (1.01-1.22)	N/A	N/A
SO_2 (ppb)[c]	3967 (921)	9.22	1.03 (0.98-1.08)	47.00	1.06 (0.98-1.15)
$PM_{2.5}$ ($\mu g/m^3$)[d]	1775 (415)	4.29	0.94 (0.84-1.06)	10.00	0.97 (0.86-1.09)

[b] 16 cases (i.e., 16 case days + 72 control days) were excluded due to missing wind data in case-crossover strata

[c] 48 cases (i.e., 48 case days + 204 control days) were excluded due to missing SO_2 concentrations data in case-crossover strata

[d] 580 cases (i.e., 580 case days + 2519 control days) were excluded due to missing $PM_{2.5}$ concentrations data in case-crossover strata

Associations with SO_2 daily levels and with the percentage of hours downwind were slightly more pronounced for children aged 2-4 years of age than for children aged 0 to 4 years (see Table A4 in supplemental material).

The adjusted OR for the association with the daily mean SO_2 levels was 1.10 per an increase of 9.22 ppb (95% CI: 0.98-1.08), and with the daily hourly maximum SO_2 levels, it was 1.06 per increase of 47.00 ppb in IQR (95%CI: 0.98-1.15). Fewer hospitalizations were found in this age group and the statistical variability was high for the associations with $PM_{2.5}$ levels. No association was noted when hospitalizations for all respiratory symptoms were considered (supplemental material, Table A5).

When associations with hospitalizations for children living within 7.5 km of the industrial were performed separately for the three emission points instead of within 7.5 km of the industrial complex (defined as the point between the three emission sources, weighted by total emissions), the results were rather similar to the analyses that were performed using the midpoint of the three emission sources (Table A6 – A8 in supplemental material).

4. Discussion

In this study we noted small associations between hospitalisations for bronchiolitis and asthma in children aged less than 5 years old living in proximity to an industrial complex with an aluminum smelter and the daily percentage of hours downwind and SO_2 levels at lag 0 and lag 1, although large statistical variability was noted. No association was noted when all respiratory symptoms were included, suggesting that associations were specific to asthma and bronchiolitis and do not include respiratory infections or other respiratory issues.

It is interesting to note that statistically significant results were found between the percentage of hours downwind but not with SO_2 and $PM_{2.5}$ levels measured at monitoring stations. The percentage of hours downwind may capture effects of mixture of pollutants which otherwise may be difficult to assess. Furthermore in remote areas, air pollutant levels are not always monitored and the percentage of hours downwind could be an easy way to assess risks of acute effects in proximity to industrial sources.

Our results are concordant with those of previous studies that assessed the respiratory effects of acute exposure to industrial emissions in children. For example, Aekplakorn et al. (2003) observed in a panel study of Thai children aged 6 to 14 years exposed to air emissions of a coal power station that pulmonary functions inversely decreased with daily particulate levels. Our results can also be compared to those of Lewin et al. (2013), who assessed associations between hospitalisation for asthma in children younger than 5 years old exposed to the emissions of another Quebec aluminum smelter, and who used the same exposure variables as the ones in the present study. Lewin et al. (2013) reported greater risks with the percentage of hours downwind and daily $PM_{2.5}$, but only for children aged between 2 to 4 years old. While our shorter time period for $PM_{2.5}$ makes any conclusions on the association with $PM_{2.5}$ more limited, it is still possible that the difference noted between our study and the one of Lewin et al. (2013) is related to a different composition of the particles emitted by the two industrial complexes.

The estimation of exposure in an epidemiological study performed with secondary data is often difficult. In the present work, we assumed that children were consistently in the area under study and that they were always exposed when the exposure was estimated. These interpretations of the daily reality of the children are likely incorrect in some cases. It is indeed impossible to know if the children were outside or inside their place of residence on a giving day, or if they were always in the study area where the measures of exposure took place. Other limitations include the measures themselves. Even if we used $PM_{2.5}$ and SO_2 measurement located the closest to the residences of the hospitalised children as possible, they were sometimes located kilometers away from the residences. This is a particular concern for the $PM_{2.5}$ measurement station given that it is located a few hundred meters outside of the study area. Moreover only a few years of $PM_{2.5}$ data were available for analyses, which limit the reliability of the results with this pollutant in the present study.

In conclusion, our study was designed to estimate the respiratory effects of acute exposure to emissions from an industrial complex with an aluminum smelter complex in young children. Our results support those of previous studies that have shown that the risk of hospitalisation for respiratory problems in young children living in proximity to industrial air emitters increases with air pollutant levels. The use of the percentage of hours per day downwind from an industry appears as a valuable metric to represent exposure in remote areas. Nonetheless improved exposure measures are needed to a better estimate of the risks of respiratory effects associated with acute exposure to industrial emissions like aluminum smelters, as relatively few studies have pertained to this issue.

Competing Interests

The authors declare that they have no competing interests.

Authors' contributions

RL was involved in the modelling, the interpretation of results, and in the writing of the manuscript. AB was involved in data collection, in the interpretation of the results, and in the writing of the manuscript. SB was involved in the interpretation of the results and writing of the manuscript. AS conceived the study, directed the modelling and writing of the manuscript.

All authors reviewed and approved the final manuscript.

Acknowledgement

"The author(s) would like to acknowledge the Data Access Integration (DA) Team for providing the data and technical support. The DAI Portal (http://climat-quebec.qc.ca/CC-DEV/trunk/index.php/pages/dai) is made possible through collaboration among the Global Environmental and Climate Change Centre (GEC3), the Adaptation and Impacts Research Division (AIRD) of Environment Canada, and the Drought Research Initiative (DRI)."

References

Aekplakorn, W., Loomis, D., Vichit-Vadakan, N., Shy, C., Wongtim, S., & Vitayanon, P. (2003). Acute effect of sulphur dioxide from a power plant on pulmonary function of children, Thailand. *International journal of epidemiology, 32*, 854-861. http://dx.doi.org/10.1093/ije/dyg237

Bateson, T. F., & Schwartz, J. (2008). Children's response to air pollutants. *J. Toxicol. Environ. Health. A, 71*, 238-243.

Bertoldi, M. et al. (2012). Health effects for the population living near a cement plant: an epidemiological assessment. *Environment international, 41*, 1-7. http://dx.doi.org/10.1016/j.envint.2011.12.005

Deger, L. et al. (2012). Active and uncontrolled asthma among children exposed to air stack emissions of sulphur dioxide from petroleum refineries in Montreal, Quebec: A cross-sectional study. *Can. Respir. J., 19*, 97-102.

Environment Canada. (2013). *National Air Pollution Surveillance Network - Data.* Retrieved from http://www.ec.gc.ca/air-sc-r/default.asp?lang=En&n=9547191B-1

Janes, H., Sheppard, L., & Lumley, T. (2005). Case-crossover analyses of air pollution exposure data: referent selection strategies and their implications for bias. *Epidemiology (Cambridge, Mass.), 16*, 717-726. http://dx.doi.org/10.1097/01.ede.0000181315.18836.9d

Lewin, A., Buteau, S., Brand, A., Kosatsky, T., & Smargiassi, A. (2013). Short-term risk of hospitalization for asthma or bronchiolitis in children living near an aluminum smelter. *Journal of Exposure Science and Environmental Epidemiology, 23*, 474–480. http://dx.doi.org/10.1038/jes.2013.27

Liu, X., Lessner, L., & Carpenter, D. O. (2012). Association between residential proximity to fuel-fired power plants and hospitalization rate for respiratory diseases. *Environmental health perspectives, 120*, 807-810. http://dx.doi.org/10.1289/ehp.1104146

Lumley, T., & Levy, D. (2000). Bias in the case-crossover design: Implications for studies of air pollution. *Environmetrics, 11*, 689-704. http://dx.doi.org/10.1002/1099-095X(200011/12)11:6<689::AID-ENV439>3.0.CO;2-N

Maclure, M. (1991). The case-crossover design: a method for studying transient effects on the risk of acute events. *Am. J. Epidemiol., 133*, 144-153.

Pless-Mulloli, T., Howel, D., & Prince, H. (2001). Prevalence of asthma and other respiratory symptoms in children living near and away from opencast coal mining sites. *International journal of epidemiology, 30*, 556-563. http://dx.doi.org/10.1093/ije/30.3.556

Rothman, K. J., Greenland, S., & Lash, T. L. (2008). *Modern Epidemiology* (3rd ed). Philadelphia, PA: Lippincott Williams & Wilkins.

Rusconi, F. et al. (2011). Asthma symptoms, lung function, and markers of oxidative stress and inflammation in children exposed to oil refinery pollution. *The Journal of asthma : Official journal of the Association for the Care of Asthma, 48*, 84-90. http://dx.doi.org/10.3109/02770903.2010.538106

Smargiassi. A. et al. (2009). Risk of asthmatic episodes in children exposed to sulfur dioxide stack emissions from a refinery point source in Montreal, Canada. *Environmental health perspectives, 117*, 653-659. http://dx.doi.org/10.1289/ehp.0800010

Statistics Canada. (2012). Census subdivision of Saguenay, V - Quebec. Focus Geogr. Ser. 2011 Census. Retrieved from http://www12.statcan.gc.ca/census-recensement/2011/as-sa/fogs-spg/Facts-csd-eng.cfm?LANG=Eng&GK=CSD&GC=2494068

Appendix A

Table A1. Adjusted[a] associations between hospitalizations for asthma and bronchiolitis in children aged less than 5 years of age, living in a 7.5 km buffer around the industrial complex and daily exposure variables at lag 1

Pollutants	n day (n case days)	Daily mean		Hourly max	
		IQR (h/day)	OR (95%CI)	IQR	OR (95%CI)
Percentage of hours downwind[b]	4166 (958)	27.24% (6.48)	1.18 (1.07-1.29)	N/A	N/A
SO_2 (ppb)[c]	3965 (918)	9.22	0.99 (0.94-1.04)	47.00	0.97 (0.90-1.05)
$PM_{2.5}$ ($\mu g/m^3$)[d]	1802 (421)	4.29	1.02 (0.93-1.12)	10.00	1.02 (0.92-1.14)

[b] 14 cases (i.e., 14 case days + 61 control days) were excluded due to missing wind data in case-crossover strata

[c] 50 cases (i.e., 50 case days + 214 control days) were excluded due to missing SO_2 concentrations data in case-crossover strata

[d] 573 cases (i.e., 573 case days + 2491 control days) were excluded due to missing $PM_{2.5}$ concentrations data in case-crossover strata

Table A2. Crude associations between hospitalizations for asthma and bronchiolitis in children aged less than 5 years of age, living in a 7.5 km buffer around the industrial complex and daily exposure variables at lag 0

Pollutants	n day (n case days)	Daily mean		Hourly max	
		IQR (h/day)	OR (95%CI)	IQR	OR (95%CI)
Percentage of hours downwind[a]	4328 (990)	27% (6.48)	1.11 (1.02-1.22)	N/A	N/A
SO_2 (ppb)[b]	4174 (958)	9.22	1.03 (0.98-1.08)	47.00	1.06 (0.99 -1.14)
$PM_{2.5}$ ($\mu g/m^3$)[c]	1860 (426)	4.29	0.94 (0.84-1.05)	10.00	0.96 (0.86-1.07)

[a] 16 cases (i.e., 16 case days + 72 control days) were excluded due to missing wind data in case-crossover strata

[b] 48 cases (i.e., 48 case days + 204 control days) were excluded due to missing SO_2 concentrations data in case-crossover strata

[c] 580 cases (i.e., 580 case days + 2519 control days) were excluded due to missing $PM_{2.5}$ concentrations data in case-crossover strata

Table A3. Adjusted[a] associations between hospitalizations for asthma and bronchiolitis in children aged less than 5 years of age, living in a 2.5 km buffer around the industrial complex and daily exposure variables at lag 0

Pollutants	n day (n case days)	Daily mean		Hourly max	
		IQR (h/day)	OR (95%CI)	IQR	OR (95%CI)
Percentage of hours downwind[b]	351 (80)	27.24% (6.48)	0.88 (0.59-1.32)	N/A	N/A
SO_2 (ppb)[c]	327 (75)	9.22	1.11 (0.91-1.35)	47.00	1.18 (0.89-1.58)
$PM_{2.5}$ ($\mu g/m^3$)[d]	203 (46)	4.29	1.13 (0.75-1.70)	10.00	0.97 (0.67-1.42)

[a] Associations are adjusted for relative humidity, wind speed and temperature

[b] 11 cases (i.e., 11case days + 45 control days) were excluded due to missing wind data in case-crossover strata

[c] 16 cases (i.e., 16case days + 66 control days) were excluded due to missing SO_2 concentrations data in case-crossover strata

[d] 45 cases (i.e., 45case days + 192 control days) were excluded due to missing $PM_{2.5}$ concentrations data in case-crossover strata

Table A4. Adjusted[a] associations between hospitalizations for asthma and bronchiolitis in children aged 2 to 4 years old, living in a 7.5 km buffer around the industrial complex and daily exposure variables at lag 0

Pollutants	n day (n case days)	Daily mean		Hourly max	
		IQR (h/day)	OR (95%CI)	IQR	OR (95%CI)
Percentage of hours downwind[b]	1099 (251)	27.24 % (6.48)	1.12 (0.93-1.35)	N/A	N/A
SO_2 (ppb)[c]	1060 (242)	9.22	1.10 (1.00-1.20)	47.00	1.18 (1.02-1.37)
$PM_{2.5}$ ($\mu g/m^3$)[d]	489 (115)	4.29	0.89 (0.72-1.10)	10.00	0.93 (0.77-1.13)

[a] Associations are adjusted for relative humidity, wind speed and temperature

[b] 5 cases (i.e., 5 case days + 22 control days) were excluded due to missing wind data in case-crossover strata

[c] 12 cases (i.e., 12 case days + 50 control days) were excluded due to missing SO_2 concentrations data in case-crossover strata

[d] 145 cases (i.e., 145 case days + 630 control days) were excluded due to missing $PM_{2.5}$ concentrations data in case-crossover strata

Table A5. Adjusted[a] associations between hospitalizations for all respiratory symptoms in children aged less than 5 years of age, living in a 7.5 km buffer around the industrial complex and daily exposure variables at lag 0

Pollutants	n day (n case days)	Daily mean		Hourly max	
		IQR (h/day)	OR (95%CI)	IQR	OR (95%CI)
Percentage of hours downwind[b]	12077 (2771)	27.24% (6.48)	1.01 (0.96-1.07)	N/A	N/A
SO_2 (ppb)[c]	11551 (2668)	9.22	1.02 (0.99-1.05)	47.00	1.06 (1.01-1.11)
$PM_{2.5}$ ($\mu g/m^3$)[d]	5057 (1170)	4.29	0.95 (0.89-1.03)	13.00	0.97 (0.91-1.05)

[a] Associations are adjusted for relative humidity, wind speed and temperature

[b] 41 cases (i.e., 41 case days + 184 control days) were excluded due to missing wind data in case-crossover strata

[c] 131 cases (i.e., 131 case days + 552 control days) were excluded due to missing SO_2 concentrations data in case-crossover strata

[d] 1700 cases (i.e., 1700 case days + 7411 control days) were excluded due to missing $PM_{2.5}$ concentrations data in case-crossover strata

Table A6. Adjusted[a] associations between hospitalizations for asthma and bronchiolitis in children aged less than 5 years of age, living in a 7.5 km buffer around the point source B and daily exposure variables at lag 0

Pollutants	n day (n case days)	Daily mean		Hourly max	
		IQR (h/day)	OR (95%CI)	IQR	OR (95%CI)
Percentage of hours downwind[b]	4133 (952)	27.24% (6.48)	1.11 (1.01 – 1.21)	N/A	N/A
SO_2 (ppb)[c]	3945 (916)	9.22	1.03 (0.98-1.08)	47.00	1.06 (0.98-1.15)
$PM_{2.5}$ ($\mu g/m^3$)[d]	1767 (413)	4.29	0.94 (0.84 – 1.06)	10.00	0.97 (0.86 – 1.09)

[a] Associations are adjusted for relative humidity, wind speed and temperature

[b] 16 cases (i.e., 16 case days + 72 control days) were excluded due to missing wind data in case-crossover strata

[c] 48 cases (i.e., 48 case days + 204 control days) were excluded due to missing SO_2 concentrations data in case-crossover strata

[d] 577 cases (i.e., 577 case days + 2505 control days) were excluded due to missing $PM_{2.5}$ concentrations data in case-crossover strata

Table A7. Adjusted[a] associations between hospitalizations for asthma and bronchiolitis in children aged less than 5 years of age, living in a 7.5 km buffer around the point source A and daily exposure variables at lag 0

Pollutants	n day	Daily mean		Hourly max	
	(n case days)	IQR (h/day)	OR (95%CI)	IQR	OR (95%CI)
Percentage of hours downwind[b]	4127 (951)	27.24% (6.48)	1.09 (0.99 – 1.20)	N/A	N/A
SO_2 (ppb)[c]	3937 (914)	9.22	1.03 (0.98-1.08)	47.00	1.07 (0.99-1.16)
$PM_{2.5}$ ($\mu g/m^3$)[d]	1741 (407)	4.29	0.94 (0.83 – 1.06)	10.00	0.97 (0.86 – 1.09)

[a] Associations are adjusted for relative humidity, wind speed and temperature

[b] 15 cases (i.e., 15 case days + 68 control days) were excluded due to missing wind data in case-crossover strata

[c] 48 cases (i.e., 48 case days + 204 control days) were excluded due to missing SO_2 concentrations data in case-crossover strata

[d] 582 cases (i.e., 582 case days + 2526 control days) were excluded due to missing $PM_{2.5}$ concentrations data in case-crossover strata

Table A8. Adjusted[a] associations between hospitalizations for asthma and bronchiolitis in children aged less than 5 years of age, living in a 7.5 km buffer around the point source C and daily exposure variables at lag 0

Pollutants	n day	Daily mean		Hourly max	
	(n case days)	IQR (h/day)	OR (95%CI)	IQR	OR (95%CI)
Percentage of hours downwind[b]	4190 (966)	27.24% (6.48)	1.09 (0.99 – 1.19)	N/A	N/A
SO_2 (ppb)[c]	3998 (927)	9.22	1.02 (0.97-1.07)	47.00	1.04 (0.97-1.13)
$PM_{2.5}$ ($\mu g/m^3$)[d]	1802 (420)	4.29	0.95 (0.84 – 1.07)	10.00	1.00 (0.89 – 1.13)

[a] Associations are adjusted for relative humidity, wind speed and temperature

[b] 18 cases (i.e., 18 case days + 80 control days) were excluded due to missing wind data in case-crossover strata

[c] 51 cases (i.e., 51 case days + 206 control days) were excluded due to missing SO_2 concentrations data in case-crossover strata

[d] 586 cases (i.e., 582 case days + 2537 control days) were excluded due to missing $PM_{2.5}$ concentrations data in case-crossover strata

11

Concentration, Size Distribution and Electrostatic Charge of Laying Hen House Particulate Matter

Emad A. Almuhanna[1]

[1] College of Agricultural Scneces and Food, King Faisal University, Hofuf, Saudi Arabia

Correspondence: Emad Ali Almuhanna, Department of Agricultural Systems Engineering, College of Agricultural Scneces and Food, King Faisal University, Hofuf, Saudi Arabia.

Abstract

Published data on indoor air contaminants in livestock buildings in Saudi Arabia are relatively scarce. The main objective of this study was to determine the airborne concentrations of particles and electrostatic charge acquired by airborne particles in a multiple tier housing system (Manure Belt Cage System) under the climatic conditions of Saudi Arabia. In this house, the mean of total suspended particle (TSP) concentration was 0.99 mg/m^3, the PM$_{10}$ concentration (particulate matter with a diameter less than or equal to 10 μm) was 0.47 mg/m^3, and the PM$_{2.5}$ concentration (particulate matter with a diameter less than or equal to 2.5 μm) was 0.05 mg/m^3. The particle size distribution results obtained from the layer house revealed that the GMD (geometric mean diameter) was 6.95 μm, based on the mass concentration of the particles. Alternatively, based on the number concentration of particles, the GMD was 0.82 μm. The cumulative percentage of mass concentration for the particles ranging from 0.3–10 μm showed that the major fraction of the particles was larger than 2.5 μm (>85%). The net charge-to-mass ratio (q$_N$) of airborne particles was -0.86 mC/kg (s.d.=0.27). The measured q$_N$ value of airborne particles in the layer house varied due to the nature of the particles in addition to the environmental conditions and high concentration of airborne particles inside the poultry housing. In general, PM concentrations did not exceed the recommended values and those cited in literature.

Keywords: layer, particulate matter (dust), size distribution, electrostatic charge

1. Introduction

Various poultry housing systems have been employed in Saudi Arabia by poultry industry. In the 375 eggs production projects producing 3473 million eggs in Saudi Arabia, the caged systems are usually used for layer chickens and egg production (MOA, 2011). Whereas, broiler chickens is typically raised within enclosed structures, where the floor is covered with absorbent bedding material. Environmental concerns and nuisance issues related to poultry housing air emissions are an important issue currently affecting the poultry industry (Ritz et al., 2006). For laying hens, caged systems offer opportunities for better management and reduce production costs. Important welfare considerations also include environmental conditions including air quality and hen health. Unfortunately, these parameters are not well documented for different laying hen housing systems (Green et al., 2007).

Almuhanna (2011) conducted a study to characterize air contaminants (particle size distribution and concentration of airborne particles and toxic gases) inside floor raised poultry houses equipped with natural and mechanical ventilation systems, and to determine the effect of the ventilation system on air contaminants within both types of houses. In both housing systems, the industry is faced with air quality challenges including the emission of particulate matter (TSP, PM$_{10}$ and PM$_{2.5}$), ammonia, and other toxic gases. The total suspended particulate matter concentration (TSP) can be defined as the amount of particulate matter (PM) captured on a filter with a particle size of approximately 100 μm or less. PM$_{10}$ includes particles with an aerodynamic diameter of 10 μm or less also defined as the PM that passes through a size-selective inlet with a 50% cut-off at 10 μm aerodynamic equivalent diameter. Lastly, PM$_{2.5}$ is particulate matter with an aerodynamic diameter equal to or less than 2.5 μm, it is also refers to the PM that passes through a size-selective inlet with a 50% cut-off at 2.5 μm aerodynamic equivalent diameter (EPA, 1999). This scale classification is used mostly in studies of ambient air quality in the U.S. and

Saudi Arabia. For European studies they used to report PM as respirable and inhalable particles for occupational health parameters, but increasingly use TSP, PM_{10} and $PM_{2.5}$ for outside air quality related studies. Inhalable particles refer to those smaller than 100 µm, and respirable particles are those smaller than 4 µm (Li et al., 2011). Similar scale is used by US occupational health professionals (e.g., American Conference of Industrial Hygienists, ACGIH). Furthermore, thoracic particles refers to particles with an aerodynamic diameter of 10 µm or less and comparable to PM_{10}. In Saudi Arabia, ambient particulate matter and other pollutants are regulated by the Presidency of Meteorology and Environment (PME-KSA), which establishes the General Environmental Law, including Environmental Protection Standards (PME, 2013). Unfortunately, these standards do not cover indoor air for livestock buildings. Similar standards have been developed in the United States of America (USA) and European Union (EU), where ambient air contaminates was regulated by the U.S. Environmental Protection Agency (EPA), and European Environmental Agency (EEA).

Airborne particulate matter (dust) is one of the primary means by which disease-causing organisms are spread throughout poultry housing. Reductions in airborne dust levels are associated with significant reductions in airborne bacteria (Mitchell et al., 2004). The results of a recent study (Almuhanna et al., 2011) suggested that the increase of air contaminants and gases negatively affect the general productive performance and immune response under commercial conditions of poultry industry. The characteristics of PM (e.g., concentration, number, and mass) inside livestock housing vary according to the type of animal, building, and environmental conditions. Previous studies (Li et al., 2011; Green et al., 2007; Lim et al., 2007; Heber et al., 2005; Vucemilo et al., 2007; Shuhai et al., 2009) showed that the mean PM concentration measured inside layer housed (cage system) were 1.96, 0.33, 0.032 mg/m³ for TSP, PM_{10}, and $PM_{2.5}$, respectively. The ranges of these concentrations were 0.75-4, 0.03-0.56, and 0.03-0.04 mg/m³ for TSP, PM_{10}, and $PM_{2.5}$, respectively. It is widely acknowledged that stress imposed on the poultry by environmental, nutritional, pathological and other factors can decrease production. These factors, either individually or synergistically, are likely to greatly affect growth rate, production, reproduction, behavior and, ultimately, profit.

The suggested threshold values for indoor air contaminants in livestock housing are provided in Table 1.

Table 1. Suggested threshold values for indoor air contaminants in livestock buildings

Air contaminant	Humans	Animals	Reference
Inhalable (Total) dust, mg/m³	2.40*	3.70	(Donham & Cumro, 1999a)
	-	3.40	(Wathes, 1994)
Respirable dust, mg/m³	0.16*	-	(Donham & Cumro, 2002)
	0.23	0.23	(Donham & Cumro, 1999b)
	-	1.70	(Wathes, 1994)

* Specific threshold concentrations are defined as mixed exposures between NH_3 and PM in poultry CAFOs (Donham et al., 2000).

The majority of studies on the environment in poultry houses have analyzed the concentration of air contaminants (e.g., ammonia, carbon dioxide, dust, airborne microorganisms, and toxins). However, particulate matter is one of the primary aerial pollutants in poultry housing facilities (Visser et al., 2006; Liu et al., 2006; Roumeliotis et al., 2007). Moreover, ammonia gas is produced in the housing environment from the decomposition of uric acid, which is excreted by the birds. PM and ammonia have been identified by the US-EPA as the important hazardous air pollutants to be emitted from concentrated animal feeding operations (CAFOs), such as poultry (Bunton et al., 2007). PM (TSP, PM_{10}, $PM_{2.5}$) emissions from CAFOs can consist of feed materials, various body parts (e.g., dead skin, feathers), dried feces, various microorganisms and endotoxins (Chai et al., 2009).

Particle size is one of the most important parameters for characterizing atmospheric particles in terms of future impact. The size of the particles directly relates to their role in causing health problems because it affects their deposition rate, which determines their location within the respiratory tract (Zihan et al., 2009). The PSD (particle size distribution) of PM is a very important physical property governing particle behavior; moreover, very limited investigations characterizing PSD in poultry operations have been conducted and reported in the literature, specifically in Saudi Arabia.

The behavior of airborne particles is governed primarily by their size and shape. It is also greatly affected by their electrostatic charge. Brown (1997) indicated that simultaneous measurement of size and charge is necessary if the properties of particles are to be understood and their behavior controlled. In controlling indoor air quality in air spaces (e.g., industrial workplaces, livestock housing, etc.), knowledge of electrostatic charge of airborne particles is essential in designing effective particle control devices.

Toljic et al. (2010) stated that the value of charge is a critical parameter that needs to be determined in order to accurately predict behavior of a charged particle. Furthermore, electrostatic charging of particles is an important phenomenon that involves various applications, including electrostatic precipitation. Electrostatic charge can be beneficial as in the control of dust by the use of electrically charged filters or charged water droplets (Almuhanna et al., 2008, 2009). Similarly, charge effects on pulmonary deposition may make it useful in medication or a complication so far as hazardous dust is concerned. Almuhanna et al. (2008) carried out an experiment under controlled laboratory conditions and concluded that spraying with charged water improves the efficiency to remove PM. They also found that the removal efficiency is significantly greater during longer charged water spray durations (4 and 6 min) than during shorter duration (2 min), while the spraying method and the charge polarity did not significantly influence particle removal efficiency.

Almuhanna et al. (2009) developed a prototype electrostatically assisted particulate wet scrubber (EPWS) for controlling particulate matter (dust) in livestock buildings and tested under laboratory and field conditions. Under laboratory conditions, the EPWS with the negatively charged water spray had significantly higher particle removal efficiency (79%) than either the uncharged wet scrubber (58%) or the control i.e., only the fan was operated (21%). Field tests in a swine building proved that the EPWS was effective in removing airborne particulate matter. Moreover, PM levels could be significantly reduced if ventilation systems with electrostatic dust collection systems are used (Mitchell et al., 2002). Cambra-López et al. (2009) evaluated the performance of an ionization system in a broiler house and concluded that the ionization system effectively reduced the total PM_{10} and $PM_{2.5}$ mass emissions by 36% and 10%, respectively. Understanding PM characteristics will ultimately lead to the development of best suitable methods for dust control (Almuhanna et al., 2008, 2009; Brown, 1997).

Published data on indoor air contaminants in livestock buildings in Saudi Arabia are relatively scarce. More efforts are needed to study the effect of the elevated concentrations on the in house concentration of PM where it is a typical problems facing arid and semiarid countries like Saudi Arabia and the effect of emitted pollutants from these houses on the surrounding and vice versa.

The main objective of the present study was to determine the airborne concentrations of PM (TSP, PM_{10}, and $PM_{2.5}$) and electrostatic charge acquired by airborne particles in a multiple tier housing system (Manure Belt Cage System) under the climatic conditions of Saudi Arabia. Whereas, the specific objectives were to do the following: (a) measure the particle size distribution (mass and number based); (b) determine the concentration of PM; (c) measure the electrostatic charge acquired by airborne particles; (d) compare the measured concentrations to the recommended values.

2. Materials and Methods

The layer house field measurements were conducted in the Poultry Unit at the Experimental and Training Station of King Faisal University, Al-Ahsa, Saudi Arabia.

2.1 Poultry Housing Facilities

The layer house possessed a total width, length, and height of 12 m, 16 m, and 3.6 m, respectively. The surface area of the floor and volume of the building were 196 m^2 and 762 m^3, respectively, as shown in Figure 1. The mechanically ventilated layer house was oriented in an east-west direction. The side walls were made of 20-cm thick concrete bricks, and the ceiling was made of insulated reinforced concrete. The longitudinal side walls (north and south) were equipped with an evaporative cooling fan-pad system that served as the ventilation air inlet, with a total area of 17 m^2 of cooling pads. The house was equipped with 6 exhaust fans that were 45 cm in diameter (Model DVN 183, Windy, Dongkun Industrial Co., Ltd, S. Korea), which were installed on the east side walls of the building and giving maximum flow rate of 110 m^3/min. In total, 640 cages (L=55cm × W=63 cm × H=50 cm) were arranged in four batteries (Manure Belt Cage System) with three central alleys. Each cage row was serviced by feeding systems and nipple water dispensers. The layer house accommodated a total of 2560 hens, and 4 birds were housed in each cage.

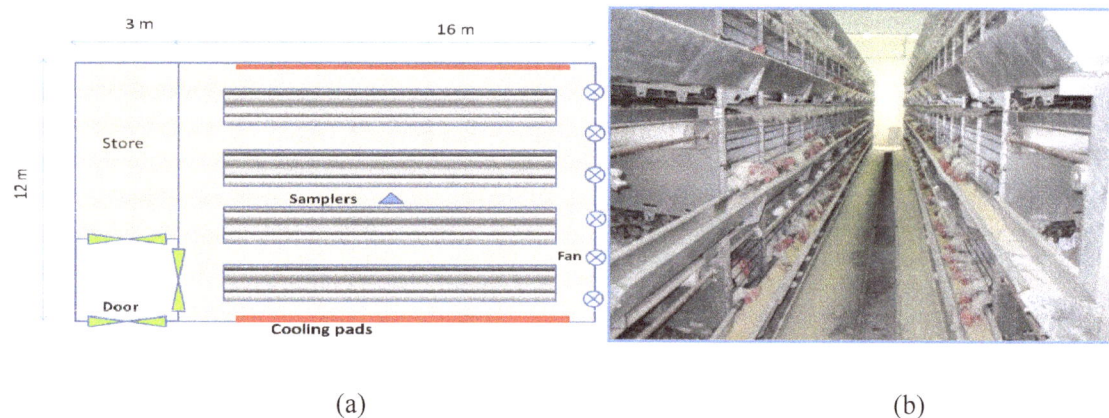

(a)   (b)

Figure 1. (a) Schematic depiction of the layer house equipped with mechanical ventilation [plan view - not drawn to scale], (b) Photo of the same house

Figure 2. Removal of manure by belt system using self-cleaning rubber belt conveyer

2.2 Measurement of Environmental Parameters

During this study, the used birds were adults (12 months old) and at their production stage. The house was ventilated according to the temperature and humidity, and the average ventilation rate during the sampling time (winter season) was 381 m^3/min. Using the ventilation control board, the indoor air temperature inside the house was adjusted according to the age of the birds and environmental parameters.

The air temperatures and relative humidity at various locations inside the building were measured using a HOBO® U12 Logger with accuracy of ±0.35 °C; ambient temperature and relative humidity were recorded for the length of study with time interval for data recording equal to 60 minutes with data acquisition every one minute for integrated measurements. Meteorological data outside the building, including air temperature, wind speed, direction and the relative humidity of the air were measured by a meteorological station (HOBO U30-NRC Weather Station, Onset Computer, USA) which was located 50 m away from the building.

2.3 Measurement of Size Distribution, Mass and Number Concentration

The size distribution and number concentration of airborne particles were monitored using a particle counter (Model GW3016A, GrayWolf Sensing Solutions, Advanced Environmental Measurements). The spectrometer measured particles with aerodynamic diameters ranging from 0.3 to 10 μm at an air sampling rate of 2.83 LPM (0.1 CFM). Moreover, six channels were used, and a counting efficiency of 100% was employed for particles with diameters of >0.45 μm. The spectrometer displayed the particle count and mass concentration readings in μg/m^3.

Figure 3. Instruments used to characterize airborne particles inside the poultry house

A fixed station (Topas, Turnkey Optical Particle Analysis System) monitor was designed to continuously record environmental TSP, PM_{10}, $PM_{2.5}$, and PM_1 particles was used to monitor PM concentrations. It uses the principle of light scattering of the single particle. The sample air was heated to avoid the relative humidity and can record concentrations up to 6500 μg/m³, with a ±0.1μg/m³ measuring accuracy with a sampling resolution of 1 min. Samples were collected with an average of one hour readings and sampling height was approximately 2 m above the ground.

The particle size distributions (number and mass) were analyzed by calculating the following statistics (Hinds, 1999):

a. Mean diameter

$$\bar{d}p = \frac{\sum n_i d_i}{N} \tag{1}$$

b. Standard deviation (SD)

$$\sigma = \left(\frac{\sum n_i \left(d_i - \bar{d}p\right)^2}{N-1} \right)^{0.5} \tag{2}$$

c. Geometric mean diameter (d_g or GMD)

$$d_g = exp\left(\frac{\sum n_i \left(ln\, d_i\right)}{N} \right) \tag{3}$$

d. Geometric standard deviation (σ_g or GSD)

$$\sigma_g = exp\left(\frac{\Sigma n_i \left(ln\, d_i - ln\, d_g\right)^2}{N-1} \right)^{0.5} \tag{4}$$

where:

d_i–Diameter of specific particles size (i), μm

d_p–Mean diameter, μm

d_g–Geometric mean diameter by mass of sample, μm (GMD)

σ_g–Geometric standard deviation (GSD)

n_i–Number of particles of specific size (i)

N–Total number of particles

2.4 Measurement of Airborne Particulate Matter Charge

For the charge measuring of airborne particles inside layer house, the CMD device was developed as reported by Almuhanna (2010) and was used to measure the net charge-to-mass ratio of airborne particles. The device consisted of two conducting enclosures, one enclosed and insulated from the other. The inner enclosure had two small openings, one for air inlet and the other for air outlet; the openings were kept small to reduce leakage of external field into the cup. It was electrically connected to the electrometer input which had a particle collection filter with a back-up metal screen. The device insulated from the outer enclosure by a rigid, high resistance insulator (Poly-tetra-fluoroethylene PTFE). The outer enclosure was connected to a grounded base and served as a shield for the inner enclosure from the external fields that could affect measurement.

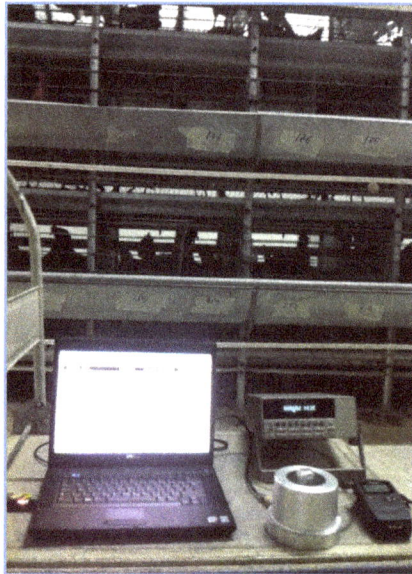

Figure 4. Components of the Charge Measuring Device (CMD)

The CMD was connected to a low-volume sampling pump that draws air and particles into the device and collects the particles onto a filter (type AE, SKC, Eighty Four, PA). The mass of collected particles on the filter was measured by weighing the filter before and after sampling in an electronic analytical microbalance (Model AWD-120D, Shimadzu Corporation, Kyoto Japan, with a accuracy of ±0.01 mg). The device was electrically connected to an electrometer (Model 6514, Keithley Instruments, Inc., Cleveland, OH), which was controlled by a computer. The electrometer had a very high sensitivity of the order of 10^{-15} A and high input impedance. Data from the electrometer were collected and managed by ExceLINX® software (Keithley Instruments, Inc., Cleveland, OH).

The net charge-to-mass ratio (mC/kg) of particles was calculated using the following equation:

$$q_N = \frac{q - q_b}{m_p} \tag{5}$$

Where, q_N is the net charge-to-mass ratio, q is the measured charge, q_b is the device background charge, and m_p is the mass of particles collected on the filter.

Data values were statistically analyzed using PROC GLM of SAS (Version 9.1, SAS Institute, Inc., Cary, N.C., 2001). Means were compared using a Duncan's multiple range test at a significance level of 5%.

3. Results and discussion

3.1 Environmental Conditions

The environmental conditions inside the layer house were compared to the external climatic conditions. Fluctuations in the air temperature surrounding the birds plays an important role in their growth rate, development, and productivity whereas, RH, are considered as an important factor that affects PM generation (CIGR, 1994). The air temperature inside the layer house varies between 18.6 °C and 25.1 °C with an average of 22.7 °C (SD = 2.5 °C). The relative humidity inside the layer house ranges from 36.2% to 51.1% with an average of 42.2% (SD = 5.9%). On the other hand, the outside air temperature ranges from 15.2 °C to 29.9 °C, and the average temperature was 22.5 °C (SD = 7.5 °C). Lastly, the relative humidity of the outside air ranges from 20.7% to 49.4%, and the average humidity is 33.1% (SD = 12%).

3.2 Particle Mass Concentration

The average concentration of TSP inside the layer house is summarized in Table 2. The average TSP inside the layer house has a significantly (P<0.05) greater mean value (0.99 mg/m^3) than the outside ambient air value (0.39 mg/m^3) during the experimental period. However, the average concentration of TSP does not exceed the acceptable range of the threshold for indoor air contaminants in livestock houses (3.4–3.7 mg/m^3) proposed by Wathes et al. (1994) and Donham and Cumro (1999b). The average PM$_{10}$ concentration inside the layer house has a significantly (P<0.05) greater mean value (0.47 mg/m^3) than the outside ambient air value (0.23 mg/m^3) during the experimental period. The weekly average PM$_{2.5}$ concentration mean value (0.05 mg/m^3) inside the layer house did not significantly (P<0.05) differ from outside ambient air value (0.05 mg/m^3) during the experimental period. Both PM$_{10}$ and PM$_{2.5}$ values do not exceed the threshold for indoor air contaminants in livestock houses (0.23 mg/m^3) proposed by Donham and Cumro (1999b). Table 2 shows the mean, standard deviation, and range of values for TSP, PM$_{10}$, and PM$_{2.5}$ during the experimental period.

Table 2. Mean, standard deviation, and range of values (mg/m^3) for TSP, PM$_{10}$, and PM$_{2.5}$ during the experimental period

House	TSP			PM$_{10}$			PM$_{2.5}$		
	Mean[*]	SD	Range[**]	Mean[*]	SD	Range[**]	Mean[*]	SD	Range[**]
Layer	0.99 a	0.2	0.7-1.11	0.47 a	0.56	0.41-0.53	0.05 a	0.01	0.04-0.05
Outside	0.39 b	0.1	0.23-0.6	0.23 b	0.05	0.18-0.54	0.05 a	0.02	0.02-0.16

[*] Column means followed by the same letter are not significantly different at a 5% level of significance.

[**] Maximum values were observed in at the layer house.

Li et al. (2011) listed (Table 3) the experimental conditions and measurements of some previous studies on PM concentrations in laying-hen houses. These findings were compared with the results of the present study.

Table 3. Production and measurement[a] conditions of studies on PM concentrations in laying-hen caged houses compared to this study

Reference	Location of Study	Bird Age and Weight	Vent. Mode	PM Mea.	Measurement Frequency[b]	TSP (mg/m^3)	PM$_{10}$ (mg/m^3)	PM$_{2.5}$ (mg/m^3)
Takai et al. (1998)[c]	Northern Europe	-	-	GF	Intermittent (note 2)	0.75-1.64	0.03-0.27	-
Wathes et al. (1997)[c]	U.K.	18-69 wk, 1.94-2.18 kg	MV	GF	Intermittent (note 3)	1.70	0.31	-
Lim et al. (2007)	Ohio	1.65 kg	MV	TEOM	Continuous	2.37	0.565	-
Martensson & Pehrson (1997)	Sweden	3-16 wk	MV	GF	Intermittent (note 5)	2.5 (1.0-3.6)	1.0 (0.3-1.3)	-
Guarino et al. (1999)[c]	Italy	34-42 wk	MV	GF	Intermittent (note 6)	1.58	0.32	-
Davis & Morishita (2005)	Ohio	-	MV	Optical	Intermittent (note 7)	-	<2.00	-
Vucemilo et al. (2007)	Croatia	-	MV	GF	Intermittent (note 8)	1.6-2.8	-	-
Li et al. (2011)	Iowa	50-90 wk	MV	TEOM	Continuous	-	0.393	0.044
This study	Saudi Arabia	50-60 wk	MV	Optical	Continuous	0.99	0.47	0.05

[a] GF = gravimetric filtration, TEOM = tapered element oscillating microbalance, MV = mechanical ventilation, and NV = natural ventilation.

[b] The following notes apply: 1 = 22 buildings surveyed, each measured over a summer day and winter day; 2 = 26 buildings surveyed, each measured over a summer day and winter day; 3 = four buildings surveyed, each measured over a summer day and a winter day; 4 = measured one day per week; 5 = measured three to four times a day; 6 = measured three times a day, five days each week, one week each month; 7 = measured once a week; and 8 = measured 15 times a day;

[c] Inhalable and respirable fractions of PM were reported.

Figure 5 shows the change in the airborne PM (dust) concentration inside the layer house over 6 days period.

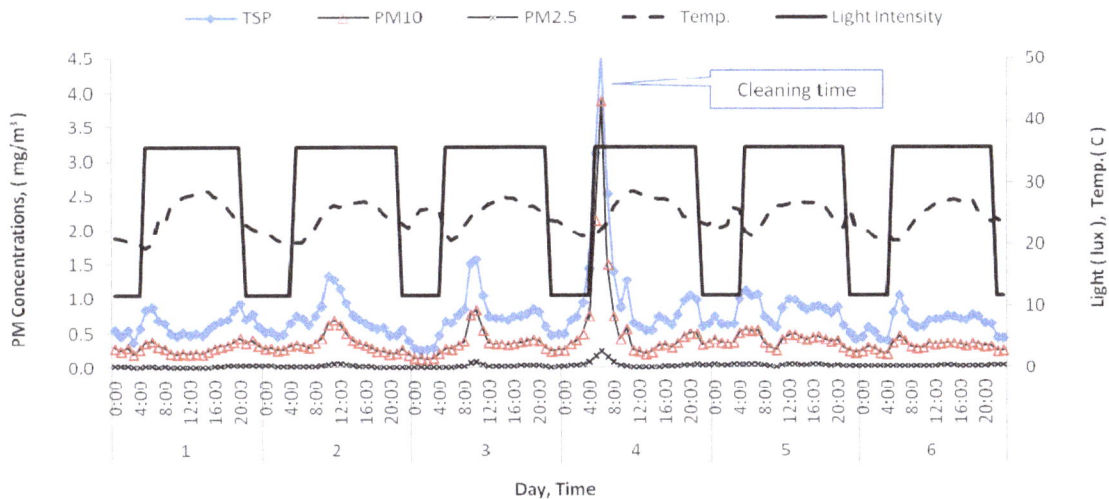

Figure 5. Cyclic changes in airborne particulate matter (dust) concentrations inside the layer house during 6 days of sampling period

There was 18 hours of light starting from 4:00 to 20:00 during this period it is obvious that there is an increse in air tempreture in addition to an increase in dust concentrations of different sizes, this is could be related to the increase of outside air tempreture and birds activities. Cleaning process could cause an elevation of dust concentration inside the building as shown in day 4 of sampling period (Figure 5).

3.3 Particle Size Distribution

Particle size distribution (PSD) of particulate matter is a very important physical property governing particle behavior; moreover, very limited investigations characterizing PSD in poultry operations have been conducted and reported in the literature, specifically in Saudi Arabia. The particle size distribution inside the layer house is summarized in Table 4.

Table 4. Particle statistics of the layer house on a number and mass basis

Parameter	Number Distribution			Mass Distribution		
	Mean	SD	Range	Mean	SD	Range
Mean Diameter (μm)	0.94	0.16	0.8-1.11	6.04	0.26	5.74-6.22
Standard deviation	1.32	0.21	1.12-1.53	2.31	0.19	2.19-2.52
Geometric Mean Diameter (μm)	0.82	0.09	0.73-0.92	6.95	0.60	6.26-7.32
Geometric Standard Deviation	2.14	0.17	2.0-2.33	2.11	0.20	1.94-2.33

The geometric mean diameter (GMD) based on the number distribution inside the layer house is 0.82 μm, and the geometric standard deviation (GSD) is 2.14. Based on the mass distribution inside the layer house, the geometric mean diameter (GMD) is 6.95 μm, whereas the geometric standard deviation (GSD) is 2.11. Figure 6 shows the particle size distribution inside the house based on number and mass concentration. It is clear that the layer barn particulate concentrations are lower in number and higher in mass concentrations than outside values.

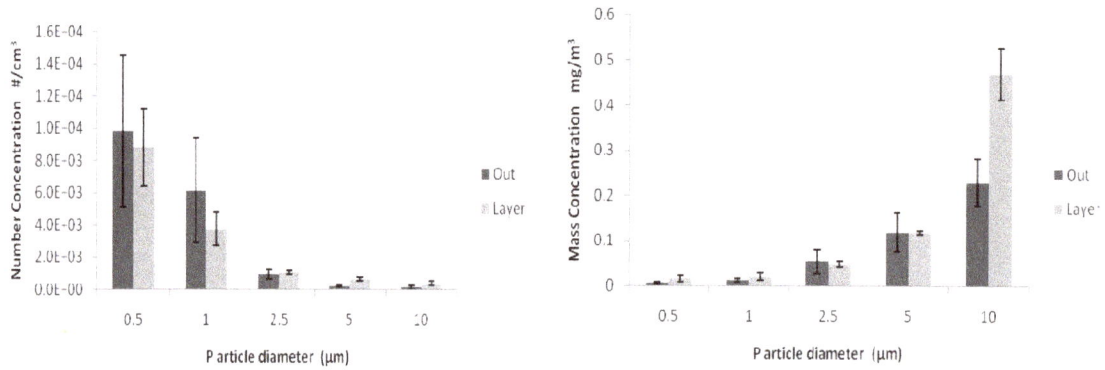

Figure 6. Particle size distribution inside the layer house compared with outside air based on number and mass concentration of particles, Error bars represent standard deviation

The cumulative percentage of mass concentration for the particles range from 0.3-10 μm (Figure 7a) shows that a major fraction of the particles are larger than 2.5 μm in diameter (>85%). The cumulative percentage of number concentration for the particles range from 0.3-10 μm (Figure 7b) shows that the major fraction of the particles is smaller than 2.5 μm (> 95%). This indicates that a greater part of the PM (dust) mass has a high probability of settling out of the air or being collected in the nasal and pharyngeal regions if inhaled. Consequently, only a small proportion will penetrate into the more sensitive lower respiratory regions where greater damage can occur.

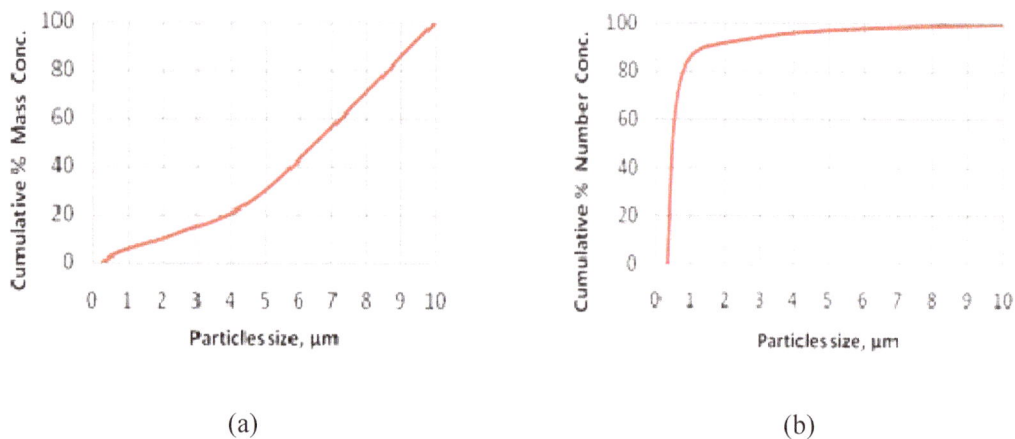

(a)   (b)

Figure 7. Measured cumulative percentage of particle size distribution based on number and mass concentration

3.4 Airborne PM Charge

In order to have a wider idea about the behavior airborne particles that is governed primarily by their size and shape and also greatly affected by their electrostatic charge. Knowledge of electrostatic charge of airborne particles is essential in designing effective particle control devices. Inside the layer house, the measured mean charge to mass ratio q_N of airborne particles was -0.86 (s.d.=0.27) mC/kg. The measured q_N value of airborne particles in the laying hen house varies due to the nature of the particles in addition to the environmental conditions and high concentration of airborne particles inside the housing.

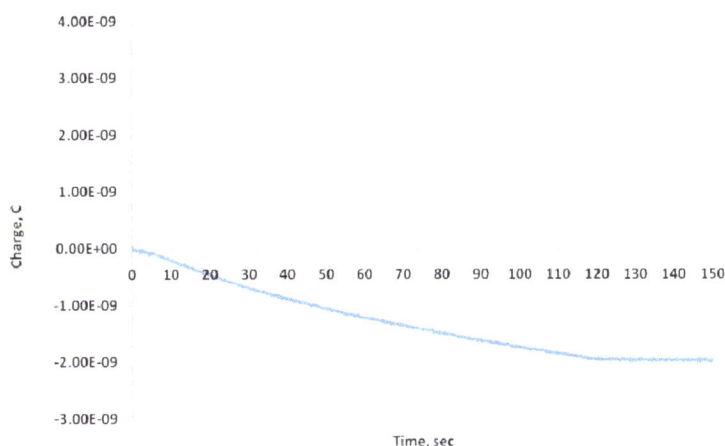

Figure 8. Electrostatic charges of airborne particles as measured by the charge measuring device

Figure 8 shows a typical plot of the measured charge. Before sampling the pump was turned on, q_b (i.e. background charge) was first measured for about 20 s. When the sampling pump was turned on, q started to increase due to the accumulation of particles on the collection filter. When the pump was turned off, q stabilized, and at this point the measured charge was used to calculate q_N of the collected particles.

Understanding PM characteristics will ultimately lead to the development of best suitable methods for dust control. Almuhanna et al. (2009) developed a prototype electrostatically assisted particulate wet scrubber (EPWS) for controlling particulate matter (dust) in livestock buildings and tested under laboratory and field conditions. Under laboratory conditions, the EPWS was effective in removing airborne particulate matter. Cambra-López et al. (2009) evaluated the performance of an ionization system in a broiler house and concluded that the ionization system effectively reduced the total PM_{10} and $PM_{2.5}$ mass emissions by 36% and 10%, respectively. The negative value of airborne particles suggesting the use of positively charged collection electrodes or objects to achieve the highest collection efficiency by dust removal techniques.

4. Conclusions

Particle concentrations, size distribution and electrostatic charge of airborne particles in a layer poultry operation in Al-Ahsa, Saudi Arabia were measured and analyzed. From the results of the present study, the following conclusions could be drawn:

The weekly average concentration of TSP in the layer house, which is equivalent to the inhalable particulate matter (dust) content, was 0.99 mg/m^3. The weekly average PM_{10} concentration in the layer house was 0.47 mg/m^3. The $PM_{2.5}$ weekly average concentration in the layer house was 0.05 mg/m^3. The $PM_{2.5}$ values for the house were lower than the suggested threshold for indoor air contaminants in livestock housing.

The literature values for TSP concentrations ranged from 0.75–3.6 mg/m^3, as compared to 0.99 mg/m^3 in the present study. PM_{10} concentrations ranged from 0.03–2 mg/m^3, as compared to 0.47 mg/m^3 in the current study. $PM_{2.5}$ concentration of previous study (Li et al., 2011) was 0.044 mg/m^3, as compared to 0.05 mg/m^3 in the current study.

The particle size distribution results obtained from the layer house revealed that the GMD was 6.95 μm, based on the mass concentration of the particles. Alternatively, based on the number concentration of particles, the GMD was 0.82 μm.

The cumulative percentage of mass concentration for the particles ranged from 0.3-10 μm showed that the major fraction of the particles was larger than 2.5 μm (>85%). This indicates that a greater part of the PM (dust) mass has a high probability of settling out of the air or being collected in the nasal and pharyngeal regions if inhaled. Consequently, only a small proportion will penetrate into the more sensitive lower respiratory regions where greater damage can occur.

Inside the layer house, the measured mean charge to mass ratio of airborne particles was -0.86 (s.d.=0.27) mC/kg.

The negative value of airborne particles suggesting the use of positively charged collection electrodes or objects to achieve the highest collection efficiency by dust removal techniques.

Acknowledgements

This work was financially supported by the Deanship of Scientific Research of King Faisal University, Hofuf, Saudi Arabia. Author is very grateful to this help and support.

References

Almuhanna, E. A. (2011). Characteristics of air contaminants in naturally and mechanically ventilated poultry houses in Al-Ahsa of Saudi Arabia. *Transactions of the ASABE, 54*(4), 1433-1443.

Almuhanna, E. A., & Maghirang, R. G. (2010). Measuring the electrostatic charge of airborne particles. *Journal of Food, Agriculture & Environment, 8*(3&4), 1033-1036.

Almuhanna, E. A., Ahmed, A. S., & Al-Yousif, Y. M. (2011). Effect of air contaminants on poultry immunological and production performance. *Int. J. Poult. Sci., 10*(6), 461-470. http://dx.doi.org/10.3923/ijps.2011.461.470

Almuhanna, E. A., Maghirang, R. G., Murphy, J. P., & Erickson, L. E. (2008). Effectiveness of electrostatically charged water spray in reducing dust concentration in enclosed spaces. *Transactions of the ASABE, 51*(1), 279-286.

Almuhanna, E. A., Maghirang, R. G., Murphy, J. P., & Erickson, L. E. (2009). Laboratory-scale electrostatically assisted wet scrubber for controlling dust in livestock buildings. *Applied Eng. in Agric., 25*(5), 745-750.

Brown, R. C. (1997). Tutorial review: Simultaneous measurement of particle size and particle charge. *J. Aerosol Sci., 28*, 1373-1391. http://dx.doi.org/10.1016/S0021-8502(97)00034-7

Bunton, B., Shaughnessy, P. O., Fitzsimmons, S., Gering, J., Hoff, S., Lyngbye, M., ... Werner, M. (2007). Monitoring and modeling of emissions from concentrated animal feeding operations: overview of methods. *Environ. Health Persp.*, 303-307.

Cambra-López, M., Winkel, A., van Harn, J., Ogink, N. W. M., & Aarnink, A. J. A. (2009). Ionization for reducing particulate matter emissions from poultry houses. *Transactions of the ASABE, 52*, 1757-1771.

Chai M., Lu, M., & Keener, T. (2009). Using an improved electrostatic precipitator for poultry dust removal. *Journal of Electrostatics, 67*(6), 870-875. http://dx.doi.org/10.1016/j.elstat.2009.07.006

CIGR. (1994). *Aerial environment in animal housing: Concentration in and emissions from farm buildings*. Working Group Report Series No 94.1. International Commission of Agricultural Engineering. Retrieved from www.cigr.org/documents/ CIGR-workinggroupreport1994.pdf

Cox, N. A. (2002). Reducing airborne pathogens, dust and Salmonella transmission in experimental hatching cabinets using an electrostatic space charge system. *Poultry Science, 81*, 49-55.

Davis, M., & Morishita, T. Y. (2005). Relative ammonia concentrations, dust concentrations, and presence of *Salmonella* species and *Escherichia coli* inside and outside commercial layer facilities. *Avian Diseases 49*(1), 30-35. http://dx.doi.org/10.1637/0005-2086(2005)49[30:RACDCA]2.0.CO;2

Donham, K. J., & Cumro, D. (1999a). Setting maximum dust exposure levels for people and animals in livestock facilities. In *Proc. Intl. Symposium on Dust Control in Animal Production Facilities*, 93-109. Horsens, Denmark: Danish Institute of Agricultural Sciences.

Donham, K. J., & Cumro, D. (1999b). Synergistic health effects of ammonia and dust exposure. In *Proc. Intl. Symposium on Dust Control in Animal Production Facilities*, 166. Horsens, Denmark: Danish Institute of Agricultural Sciences (DIAS), Research Centre Bygholm.

Donham, K. J., Cumro, D., Reynolds, S. J., & Merchant, J. A. (2000). Dose-response relationships between occupational aerosol exposures and cross-shift declines of lung function in poultry workers: Recommendations for exposure limits. *J. Occup. Environ. Med., 42*(3), 260-269. http://dx.doi.org/10.1097/00043764-200003000-00006

EPA. (1999). Sampling of ambient air for total suspended particulate matter (SPM) and using high volume (HV) sampler. Cincinnati, Ohio: U.S. Environmental Protection Agency, Center for Environmental Research Information.

Green A. R., Wesley, I., Trampel, D. W., & Xin, H. (2007). Air quality and hen health status in three types of commercial laying hen houses. *2007 ASABE Annual International Meeting*. Minneapolis, Minnesota 17–20

June 2007.

Guarino, M., Caroli, A., & Navarotto, P. (1999). Dust concentration and mortality distribution in an enclosed laying house. *Transactions of the ASABE, 42*(4), 1127-1133.

Heber, A. J., Ni, J. Q., Lim, T. T., Chervil, R., Tao, P. C., Jacobson, L. D., ... Sweeten, J. M. (2005). Air pollutant emissions from two high-rise layer barns in Indiana. *Proceeding of the annual conference and exhibition of the air and waste management association*, Pittsburgh, PA: AWMA.

Hinds, W. C. (1999). *Aerosol Technology*. New York, N.Y.: John Wiley and Sons.

Li, S., Li, H., Xin, H., & Burns, R. T. (2011). Particulate matter concentrations and emissions of a high-rise layer house in Iowa. *Transactions of the ASABE, 54*(3), 1093-1101.

Lim, T. T., Sun, H., Ni, J. Q., Zhao, L., Diehl, C. A., Heber, A. J., & Hanni, S. M. (2007). Field tests of a particulate impaction curtain on emissions from a high-rise layer barn. *Transactions of the ASABE, 50*(5), 1795-1805.

Liu, Z., Wang, L., & Beasley, D. B. (2006). *A review of emission models of ammonia released from broiler houses*. ASABE Paper No. 064101. St. Joseph, Mich.: ASABE.

Martensson, L., & Pehrson, C. (1997). Air quality in a multiple tier rearing system for layer type pullets. *J. Agric. Safety and Health, 3*(4), 217-228.

Mitchell, B. W., Richardson, L. J., Wilson, J. L., & Hofacre, C. L. (2004). Application of an electrostatic space charge system for dust, ammonia, and pathogen reduction in broiler breeder house. *Applied Eng. in Agric., 20*(1), 87-93.

MOA. (2011). *Nineteenth annual agricultural statistical book*, Ministry of Agriculture, KSA. Retrieved from http://moa.gov.sa/status/st19_table/index3.htm

PME. (2013). General environmental law and rules for implementation. Riyadh, Saudi Arabia: Presidency of Meteorology and Environment. Retrieved from www.pme.gov.sa/en/env_law.asp. Accessed 20/4/2013

Ritz, C. W., Mitchell, B. W., Fairchild, B. D., Czarick III, M., & Worley, J. W. (2006). Improving in-house air quality in broiler production facilities using an electrostatic space charge system. *J. Appl. Poultry Res., 15*(2), 333-340.

Roumeliotis, T., & Van Heyst, B. (2007). Size-fractionated particulate matter emissions from a broiler house in southern Ontario, Canada. *Sci. Total Environ., 383*(1-3), 174-182. http://dx.doi.org/10.1016/j.scitotenv.2007.05.003

Shuhai, L., Li, H., Xin, H., & Burns, R. (2009). Particulate matter emissions from a high-rise layer house in Iowa. *2009 ASABE Annual International Meeting*. Reno, Nevada. June 21–June 24. Paper Number: 095951.

Takai, H., Pedersen, S., Johnsen, J. O., Metz, J. H. M., Koerkamp, P. W. G. G., Uenk, G. H., ... Wathes, C. M. (1998). Concentrations and emissions of airborne dust in livestock buildings in northern Europe. *J. Agric. Eng. Res., 70*(1), 59-77. http://dx.doi.org/10.1006/jaer.1997.0280

Toljic, N., Castle, G. S. P., & Adamiak, K. (2010). Charge to radius dependency for conductive particles charged by induction. *Journal of Electrostatics, 68*, 57-63. http://dx.doi.org/10.1016/j.elstat.2009.10.002

Visser, M., Fairchild, B., Czarick, M., Lacy, M., Worley, J., Thompson, S., ... Naeher, L. (2006). Fine particulate matter measurements inside and outside tunnel-ventilated broiler houses. *J. Appl. Poultry Res., 15*(3), 394-405.

Vucemilo M., Katkovic, K., Vinkovic, B., Jaksic, S., Granic, K., & Mas, N. (2007). The effect of animal age on air pollutant concentration in a broiler house. *Czech J. Anim. Sci., 52*(6), 170-174.

Wathes, C. M. (1994). Air and surface hygiene. In C. M. Wathes, & D. R. Charles (Eds.), *Livestock Housing* (pp. 123-148).Wallingford, U.K.: CAB International Press.

Wathes, C. M., Holden, M. R., Sneath, R. W., White, R. P., & Phillips, V. P. (1997). Concentrations and emissions rates of aerial ammonia, nitrous oxide, methane, carbon dioxide, dust, and endotoxin in U.K. broiler and layer houses. *British Poultry Sci., 38*(1), 14-28. http://dx.doi.org/10.1080/00071669708417936

Zihan C., Wang, L., Liu, Z., Li, Q., & Beasley, D. B. (2009). Particle size distribution of particulate matter emitted from a layer operation in southeast U.S. 2009 ASABE Annual International Meeting. Reno, Nevada, USA.

12

Measurement of Monocyclic Aromatic Amines in an Urban Air

Inkyu Han[1,2], Je-Seung Lee[3] & Soo-Mi Eo[3]

[1] Division of Epidemiology, Human Genetics, and Environmental Sciences, University of Texas Health Science Center at Houston School of Public Health, Houston, TX, USA

[2] Southwest Center for Occupational and Environmental Health, University of Texas Health Science Center at Houston, Houston, TX, USA

[3] Seoul Metropolitan Government Research Institute of Public Health and Environment, Seoul, Korea

Correspondence: Inkyu Han, Division of Epidemiology, Human Genetics, and Environmental Sciences, University of Texas Health Science Center at Houston School of Public Health, 1200 Pressler Street, Houston, TX 77030, USA. E-mail: Inkyu.Han@uth.tmc.edu

Abstract

Personal air exposure to monocyclic aromatic amines (MAA) is a growing concern, in large part, due to their ubiquitous presence in the general environment and their potential health risk for bladder cancer. It is unclear what other sources of airborne MAA are for general population, due to low concentrations in the air. Detecting "trace" levels of MAAs requires a sensitive analytical method and field evaluation. In this study, an analytical method was developed to detect 2,3-dimethylaniline [2,3-DMA]; 3,5-DMA; and 3-ethylaniline [3-EA] in general air environment. During a 12-hr sampling periods, the estimated limit of quantifications (LOQs) were less than 4.13 ng/m^3 for 2,3-DMA; 3,5-DMA; and 3-EA. Desorption efficiencies (recovery rates) were at least 89% with 1 ng of each 2,3-DMA; 3,5-DMA; and 3-EA per tube. The storage effect for three MAAs showed that all three MAAs remained above 60% on the sorbent tubes and filters over 10 days. A field study was conducted in Seoul, Korea to validate sampling method in a real-world busy street with traffic, an office near the same street, and a residential home away from the busy street. Gas-phase 2,3-DMA was detected only in the indoor home sample (3.26±0.60 ng/m^3), and 3,5-DMA was not detected in all samples. Particle-bound 3-EA was detected in the street (10.92±4.73 ng/m^3), office (9.47±6.11 ng/m^3), and residential home (7.53±4.17 ng/m^3). The results suggested that the proposed analytical and field sampling methods can useful for environmental exposure assessment of these MAAs.

Keywords: monocyclic aromatic amine, analytical method, field evaluation, urban air, exposure assessment

1. Introduction

The health effects of 2,3-dimethylaniline (2,3-DMA); 3,5-dimethylaniline (3,5-DMA); and 3-ethylaniline (3-EA), a class of aromatic amines (AAs), are a growing concern due to their similar chemical structures to carcinogens such as aniline and 2,6-dimethylaniline (Gan et al., 2004; Skipper et al., 2006). Exposure to these compounds can be ubiquitous because these AAs are widely used in work places as well as general environments as dyes, pharmaceuticals, cosmetics, agricultural pesticides, antioxidants in polymers, and motor fuels (Kutting et al., 2009; Skipper, Kim, Sun, Wogan, & Tannenbaum, 2010; Angerer, Ewers, & Wilhelm 2007; Scherer, 2005).

Personal exposure assessment to airborne AAs has been conducted in work places in the past 20 years (Ward et al., 1996; Talaska & Al-Zoughool, 2003). Occupational exposure studies found that aniline and benzidine, a class of AAs, were associated with increase risk for bladder cancer in dyers, painters, and hairdressers (Naito et al., 1995; Sathiakumar & Delzell, 2000; Talaska, 2003). In the general environment, previous exposure studies found that smoking contributed to increase in indoor concentrations of some AAs such as aniline and 4-aminobiphenyl (Van Hemelrijck, Michaud, Connolly, & Kabir, 2009; Hammond et al., 1993). Recent exposure studies, however, suggested that a set of specific AAs including 2,3-DMA; 3,5-DMA; and 3-EA may not be attributed to tobacco smoke (Gan et al., 2004; Skipper et al., 2010). Unlike aniline and 4-aminobiphenyl, the environmental sources of personal air exposure to these compounds are still unknown and very little is understood about the airborne concentrations that potentially put an individual at risk for bladder cancer, because of lack of robust methods for quantifying AAs in air.

Few studies have tried to measure airborne 2,3-DMA; 3,5-DMA; and 3-EA in general environments worldwide (Palmiotto, Pieraccini, Moneti, & Dolara, 2001; Zhu & Aikawa, 2004; Ning et al., 2005). Furthermore, few studies have successfully detected airborne concentrations of these chemicals due to the absence of a sensitive method that can detect trace levels in the air. In addition, while these compounds can exist in the air as gas-phase and particle-phase, only one study has examined both gas-phase and particle-phase of these compounds (Akyuz, 2008). It is clear that a significantly more sensitive, rapid, and simple method is needed to allow for robust measurements of these compounds for the general population.

In this study we developed an analytical method to detect low levels of airborne gas and particle phase of 2,3-DMA; 3,5-DMA; and 3-EA. We also evaluated the feasibility of detecting ambient AAs in a real-world urban environment.

2. Methods

2.1 Sampling Locations

Sampling locations in Seoul Korea were a street vendor booth on a street with busy traffic (Gangnam St.), an office located on the same street (first floor), and a residential home located 5 km east from the Gangnam site. Gangnam Street is an 8-lane road which supports all types of vehicles including passenger cars, taxis, as well as diesel-powered buses and trucks. Sampling was conducted from September 23 to October 1, 2010. Street vendor booth samples were collected for 7 days, office samples were collected for 3 days, and residential home samples were collected for 6 days. Street vendor booths are ubiquitous in this area, and the booths are typically located about 1 m from the car lane. The samplers were deployed on the outside wall of the booth in sequential 4-hr intervals from 07:00 to 19:00 (12 hrs). The concentrations of AAs for the street vendor booth were reported as a 12-hr integrated sample by combining the three samples for each day. For the office measurement, a sampler was deployed on a desk between 08:00-18:00 (10 hrs) for 3 days. For indoor home sampling, a sampler was deployed in a living room at 06:00 and removed 20:00 (14 hrs) for 6 days. The inlet of all samplers in each location was placed approximately 1.5 m above ground level.

2.2 Sampling Equipment

To collect particle bound AAs, we used 37 mm glass fiber filters treated with 0.26 N sulfuric acid in polystyrene cassettes assembled by the manufacturer (SKC Inc., Eighty Four, PA). Gas phase AAs were collected using XAD-7 sorbent tubes coated with 10% phosphoric acid (SKC Inc., Eighty Four, PA). The cassette and sorbent tube were attached in series to a low flow sampling pump (SKC Inc, Pocket Pump 210 series, Eighty Four, PA). The sampling flow rate was calibrated at 200 mL/min prior to field sampling and then measured after field sampling. At the end of each sampling period samplers were capped and stored at ambient temperature until transport to the Seoul Metropolitan Government Research Institute of Public Health and Environment (SMG-RIPHE). A field blank was used at each location for each sampling day.

2.3 Reagents

A standard stock solution was prepared by adding 1 mL of 2,3-DMA (99%); 3,5-DMA (98%); and 3-EA (98%), separately, into a 10 mL volumetric flask. Reagent grade methanol (ACS reagent grade (GC) > 99.8%, Sigma-Aldrich, St. Louis, MO) was added into the volumetric flask to make 10 mL of stock solution with 100 mg/mL for each compound. 2,3-DMA; 3,5-DMA; and 3-EA were purchased from Sigma-Aldrich (St. Louis, MO). The concentrations of standard solution ranged from 0.025 ng/μL to 4.9 ng/μL. Internal standard (aniline-2,3,4,5-d$_5$ solution, Supelco, Park Bellefonte, PA) was separately prepared and added into all standard solutions to a concentration of 0.5 ng/μL. Desorbing solution was prepared by adding ammonium hydroxide (5.0 N in water, Sigma-Aldrich, St. Louis, MO) to methanol at 0.2 N concentrations. Internal standard was added into the desorbing solution.

2.4 Sample Extraction

To assess ambient concentrations of AAs, we used a modification of the National Institute of Occupational Safety and Health (NIOSH) 2017 method for particle bound AAs and Occupational Safety and Health Administration (OSHA) PV2079 method for gas phase AAs. Glass fiber filters taken from the polystyrene cassette were transferred to 4 mL amber vials. Sorbent tubes were opened and the front and back section of each tube were placed in separate 4 mL vials. Each amber vial was desorbed with 2 mL of the desorbing solution. The vial was sealed immediately and sonicated in a water bath (Branson, Model 2510, Danbury, CT) for 1 hour to extract AAs. After cooling the vials to room temperature, the aliquots were filtered through 0.25μm polypropylene syringe filters (Whatman, Puradisc 13, Clifton, NJ) into 2 mL amber vials. The filtered aliquots were analyzed by Gas Chromatography (GC).

2.5 Analytical Conditions

An Agilent 7890A GC coupled with 5975C XL MSD (Agilent Technologies, Palo Alto, CA) was used for AAs analyses. AAs were separated on a 30 m × 0.25 mm i.d. × 250 nm film thickness Agilent DB-5 MS Column (Palo Alto, CA) with the following oven temperature program: initial temperature was set at 50 °C for 2 min, increased to 110 °C at 3 °C/min and finally increased to 280 °C at 30 °C/min holding at 280 °C for 6 min. Temperatures of the sample inlet and MS source were 280 °C and 290 °C, respectively. Flow rate on the column was 1.1 mL/min. Selected Ion Monitoring (SIM) mode was used for MS operations.

3. Results

3.1 Validation of Analytical Method

Instrument detection limit (IDL) was determined with a standard solution of 0.025 ng/μL for each alkylaniline in 7 replicate analyses (Table 1). The IDL was determined by multiplying three times the standard deviation (SD) of the 7 replicates. IDL ranged from 0.02 to 0.18 ng/μL for all three AAs. The limit of quantification (LOQ) was estimated by $(3.3 \times \text{IDL})/0.144 \text{ m}^3$ of air assuming a 12-hr sample at 200 mL/min. The estimated LOQs ranged between 0.46-2.06 ng/m^3 for 3-EA; 3.21-4.13 ng/m^3 for 3,5-DMA; and 1.83-2.25 ng/m^3 for 2,3-DMA. Desorption efficiency was tested for the three AAs by spiking 1 ng per tube and filter. The desorption efficiency ranged from 89 to 118%. To determine precision, both sorbent tubes and glass fiber filters were separately spiked with a 1μl of standard solution containing 0.5 ng/μL of each AA. Precision was reported as percent relative standard deviation (% RSD). Precision for each AA was less than 10% except for gas-phase 3,5-DMA (12%).

Table 1. Summary of analytical condition, instrument detection limit (IDL), estimated limit of quantification (LOQ) and precision

Compound	RT (min)	M.W	Q1	Q2	Media	IDL (ng/μl)	LOQ (ng/m^3)	Precision (% RSD)
3-EA	18.248	106	121	77	XAD-7	0.02	0.46	6.6
					Filter	0.09	2.06	8.8
3,5-DMA	18.474	121	106	77	XAD-7	0.18	4.13	12.2
					Filter	0.14	3.21	4.7
2,3-DMA	19.446	121	106	77	XAD-7	0.11	2.52	7.1
					Filter	0.08	1.83	3.1

3.2 Storage Effect

The stability of all three AAs on XAD-7 sorbent tubes and filters was tested at the UTSPH laboratory, to examine when field samples should be analyzed after sample collection. Sorbent tubes and filters were spiked with a 1μl of standard solution containing 1 ng/μL of each AA, and stored in amber jars for 1, 3, 5, and 10 days at room temperature prior to being analyzed. The percent amount remaining of AA per storage days is shown in Figure 1. The spiked AAs were retained at 75-97 % on sorbent tubes from day 1 to day 10 (Figure 1a). NIOSH method PV 2079 reported that XAD-7 sorbent tubes spiked with 570 μg (5 ppm) of aniline retained 81-87 % after 7 days of ambient temperature storage. The three AAs spiked on glass fiber filters decreased to 72-78 % at day 1, 62-68 % at day 5 and 59-72 % by day 10 (Figure 1b). Given our results, all field samples were analyzed within 5 days of sampling.

3.3 Measurement of Alkylanilines in Urban Air

A total of 32 samples (16 particle phase and 16 gaseous phase) were analyzed for 2,3-DMA; 3,5-DMA; and 3-EA   at three locations.   Gas phase 2,3-DMA was only detected and quantified in 5 out of the 6 sampling days in a residential home (Table 2). The average concentration of 2,3-DMA in the home was 3.78 ± 0.83 ng/m^3. Gas phase 2,3-DMA was not detected in either Gangnam Street and the office. Only particle bound 3-EA was detected and quantified by GC/MS at all locations (Table 3). Average concentrations of particle bound 3-EA were 10.92 ± 4.73 ng/m^3 at Gangnam Street followed by the office (9.47 ± 6.11 ng/m^3) and home (7.53 ± 4.17 ng/m^3). 3,5-DMA was not detected in either of the particle bound or gas phase samples for all locations.

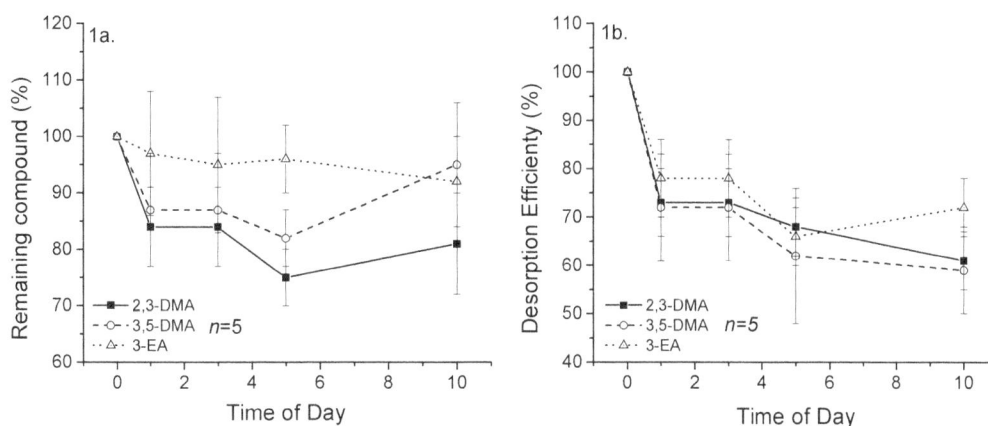

Figure 1. Storage effect of 2,3-dimethylaniline; 3,5-dimethylaniline and 3-ethylaniline. Percentages of three MAAs remaining in the XAD-7 tubes (Figure 1a) and glass fiber filters (Figure 1b) after different storage time at room temperature

Table 2. Descriptive results of gas phase 2,3-dimethylaniline

Gaseous 2,3-DMA	9/23 (Thr)	9/24 (Fri)	9/25 (Sat)	9/26 (Sun)	9/27 (Mon)	9/28 (Tue)	9/29 (Wed)	9/30 (Thr)	10/1 (Fri)
Street (n=7)	ND	ND	ND	(NS)	ND	ND	ND	ND	(NS)
Home (n=6)	(NS)	**4.46**	**4.65**	(NS)	ND	**2.95**	**2.87**	**3.95**	(NS)
Office (n=3)	(NS)	(NS)	(NS)	(NS)	(NS)	(NS)	ND	ND	ND

Note: Not detected (ND); no sampling (NS).

Table 3. Descriptive results of particle-bound 3-ethylaniline

Particle 3-EA	9/23 (Thr)	9/24 (Fri)	9/25 (Sat)	9/26 (Sun)	9/27 (Mon)	9/28 (Tue)	9/29 (Wed)	9/30 (Thr)	10/1 (Fri)
Street (n=7)	ND	ND	ND	(NS)	ND	**5.46**	**13.86**	**13.44**	(NS)
Home (n=6)	(NS)	ND	ND	(NS)	ND	**4.44**	**12.27**	**5.87**	(NS)
Office (n=3)	(NS)	(NS)	(NS)	(NS)	(NS)	(NS)	**5.89**	**6.00**	**16.52**

Note: Not detected (ND); no sampling (NS).

4. Discussion

The estimated LOQs in this study ranged from 0.46 ng/m^3 to 4.13 ng/m^3 for three MAAs. The results in this study demonstrate improved sensitivity to detect these aromatic amines using a GC/MS compared to previous studies. For example, Research by Zhu and Aikawa (2004) reported LOQ values for 2-EA and 4-EA of 10 ng/m^3 and 20 ng/m^3, respectively (Zhu & Aikawa, 2004), Another study by Zhu, Newhook, Marro and Chan (2005) reported a method detection limit of 1.2 µg/m^3 for 3,5,-DMA (Zhu, Newhook, Marro, & Chan, 2005). The desorption efficiency in this study was similar to other studies. For instance, NIOSH PV2079 method reported average desorption efficiency of 92% for aniline and other environmental monitoring studies also reported desorption efficiency ranges between 92 and 114% for some of methyl- and ethyl anilines (Akyuz, 2008; Zhu & Aikawa, 2004; Zhu et al., 2005). The storage effects of the filters and sorbent tubes were similar to previous studies. In a Canadian study, Zhu and Aikawa (2004) tested storage effects of desorption tubes spiked with 1 ng of 2-EA and 4-EA (Zhu & Aikawa, 2004) and reported a retention of 60-80% after 5 days of storage.

In our field feasibility study, the detection of gas phase 2,3-DMA only in the indoor home samples suggest that the source of 2,3-DMA is related to cooking or other domestic activities. A housewife reported that she used vegetable cooking oils to fry food for meals at least once a day between 9/24-9/30. Other potential sources of

2,3-DMA in indoor homes can be the use of paints, cleaning agents, pesticides, and smoking. However, none of these potential sources were observed during the sampling period in the home.

Sources of airborne AAs can be emissions from industrial activities (Ge, Wexler, & Clegg, 2011), motor vehicle exhaust, cooking, and biomass burning (He et al., 2010; Mohr et al., 2009; Sun et al., 2011). It is also possible that ambient air AAs can be formed from photochemical reactions with volatile organic compounds (Sun et al., 2011). Particle phase 3-EA in the non-smoking office and the home without smokers is likely derived from infiltration of PM from ambient air. In this study the average ratios of indoor/ambient (Gangnam St.) of particle bound 3-EA in home and office were less than 1. The results are consistent with a previous study reported the indoor/outdoor ratio of aniline (0.93) in Canada (Zhu & Aikawa, 2004).

The analytical condition and field sampling method showed that the three AAs are present as both gas and particle phase in ambient air and indoor air. This study provides the substantial evidence that general populations are exposed to monocyclic aromatic amines not attributed to smoking. In future health studies, this exposure assessment method can be useful to identify environmental exposure sources of MAA and to quantify the relationship between exposure and bladder cancer risk.

5. Conclusion

We have developed a sensitive analytical method to measure airborne 3-EA; 2,3-DMA; and 3,5-DMA and evaluated the feasibility of measuring these AA compounds in a real-world urban environment. Given the results from 8 days of intensive gas and particle phase sampling at three different locations, we reach the following conclusions: (1) limit of quantifications (LOQs) ranged 0.46-4.13 ng/m^3 while those in previous studies ranged 10-20 ng/m^3. (2) Desorption efficiencies ranged from 89-118% after spiking with 1 ng of each 3-EA; 2,3-DMA; and 3,5-DMA per tube. (3) Low concentrations of AAs spiked on sorbent tubes and filters stored at room temperature remained above 60% after 10 days. (4) Field investigation showed that particle bound 3-EA was detected in indoor and outdoor ambient air. (5) Gas phase 2,3-DMA was detected only in the indoor home air and its emission source was potentially cooking activities. In conclusion, the established analytical method and field sampling strategy can be used for exposure assessment to these aromatic amines in general environment.

Acknowledgements

The study is in part supported by CDC/NIOSH T42OH008421 and the Office of the Dean at the University of Texas School of Public Health. Although the research described in this manuscript has been funded wholly or in part by CDC/NIOSH T42OH008421 to the UTSPH, it has not been subjected to the Agency's required peer and policy review and therefore does not necessarily reflect the views of the Agency and no official endorsement should be inferred.

References

Akyuz, M. (2008). Simultaneous determination of aliphatic and aromatic amines in ambient air and airborne particulate matters by gas chromatography-mass spectrometry. *Atmospheric Environment, 42*, 3809-3819. http://dx.doi.org/10.1016/j.atmosenv.2007.12.057

Angerer, J., Ewers, U., & Wilhelm, M. (2007). Human biomonitoring: state of the art. *International Journal of Hygiene and Environmental Health, 210*, 201-28. http://dx.doi.org/ 10.1016/j.ijheh.2007.01.024

Gan, J. P., Skipper, P. L., Gago-Dominguez, M., Arakawa, K., Ross, R. K., Yu, M. C., & Tannenbaum, S. R. (2004). Alkylaniline-hemoglobin adducts and risk of non-smoking-related bladder cancer. *Journal of the National Cancer Institute, 96*, 1425-1431. http://dx.doi.org/ 10.1093/jnci/djh274

Ge, X. L., Wexler, A. S., & Clegg, S. L. (2011). Atmospheric amines - Part I. A review. *Atmospheric Environment, 45*, 524-546. http://dx.doi.org/10.1016/j.atmosenv.2010.10.012

Hammond, S. K., Coghlin, J., Gann, P. H., Paul, K., Taghizadeh, K., Skipper, P. L., & Tannenbaum, S. R. (1993). Relationship between Environmental Tobacco-Smoke Exposure and Carcinogen Hemoglobin Adduct Levels in Nonsmokers. *Journal of the National Cancer Institute, 85*, 474-478.

He, L. Y., Lin, U., Huang, X. F., Guo, S., Xue, L., Su, Q., ... Zhang, Y. H. (2010). Characterization of high-resolution aerosol mass spectra of primary organic aerosol emissions from Chinese cooking and biomass burning. *Atmospheric Chemistry and Physics, 10*, 11535-11543. http://dx.doi.org/10.5194/acp-10-11535-2010

Kutting, B., Goen, T., Schwegler, U., Fromme, H., Uter, W., Angerer, J., & Drexler, H. (2009). Monoarylamines in the general population - A cross-sectional population-based study including 1004 Bavarian subjects.

International Journal of Hygiene and Environmental Health, 212, 298-309. http://dx.doi.org/10.1016/j.ijheh.2008.07.004

Mohr, C., Huffman, J. A., Cubison, M. J., Aiken, A. C., Docherty, K. S., Kimmel, J. R., ... Jimenez, J. L. (2009). Characterization of Primary Organic Aerosol Emissions from Meat Cooking, Trash Burning, and Motor Vehicles with High-Resolution Aerosol Mass Spectrometry and Comparison with Ambient and Chamber Observations. *Environmental Science & Technology, 43*, 2443-2449. http://dx.doi.org/10.1021/es8011518

Naito, S., Tanaka, K., Koga, H., Kotoh, S., Hirohata, T., & Kumazawa, J. (1995). Cancer Occurrence among Dyestuff Workers Exposed to Aromatic-Amines - a Long-Term Follow-up-Study. *Cancer, 76*, 1445-1452. http://dx.doi.org/10.1002/1097-0142(19951015)76:8<1445::AID-CNCR2820760823>3.0.CO;2-R

Ning, Z. W., Zhao, S. T., Wang, J., Hu, B., Zhu, Z. G., Huang, Y. D., & Wang, D. (2005). Determination of aliphatic amines and heterocyclic amines in indoor air by a modified GC and GC/MS methods. *Indoor Air 2005: Proceedings of the 10th International Conference on Indoor Air Quality and Climate, 1-5*, 2048-2052.

Palmiotto, G., Pieraccini, G., Moneti, G., & Dolara, P. (2001). Determination of the levels of aromatic amines in indoor and outdoor air in Italy. *Chemosphere, 43*, 355-361. http://dx.doi.org/10.1016/S0045-6535(00)00109-0

Sathiakumar, N., & Delzell, E. (2000). An updated mortality study of workers at a dye and resin manufacturing plant. *Journal of Occupational and Environmental Medicine, 42*, 762-771. http://dx.doi.org/10.1097/00043764-200007000-00012

Scherer, G. (2005). Biomonitoring of inhaled complex mixtures--ambient air, diesel exhaust and cigarette smoke. *Experimental and Toxicolgic Pathology, 57*(Suppl 1), 75-110. http://dx.doi.org/10.1016/j.etp.2005.05.007

Skipper, P. L., Kim, M. Y., Sun, H. L. P., Wogan, G. N., & Tannenbaum, S. R. (2010). Monocyclic aromatic amines as potential human carcinogens: old is new again. *Carcinogenesis, 31*, 50-58. http://dx.doi.org/10.1093/carcin/bgp267

Skipper, P. L., Trudel, L. J., Kensler, T. W., Groopman, J. D., Egner, P. A., Liberman, R. G., ... Tannenbaum, S. R. (2006). DNA adduct formation by 2,6-dimethyl-, 3,5-dimethyl-, and 3-ethylaniline in vivo in mice. *Chemical Research in Toxicology, 19*, 1086-1090. http://dx.doi.org/10.1021/tx060082q

Sun, Y. L., Zhang, Q., Schwab, J. J., Demerjian, K. L., Chen, W. N., Bae, M. S., ... Lin, Y. C. (2011). Characterization of the sources and processess of organic and inorganic aerosols in New York city with a high-resolution time-of-flight aerosol mass spectrometer. *Atmospheric Chemistry and Physics, 11*, 1581-1602. http://dx.doi.org/10.5194/acp-11-1581-2011

Talaska, G. (2003). Aromatic amines and human urinary bladder cancer: Exposure sources and epidemiology. *Journal of Environmental Science and Health Part C-Environmental Carcinogenesis & Ecotoxicology Reviews, 21*, 29-43. http://dx.doi.org/10.1081/GNC-120021372

Talaska, G., & Al-Zoughool, M. (2003). Aromatic amines and biomarkers of human exposure. *Journal of Environmental Science and Health Part C-Environmental Carcinogenesis & Ecotoxicology Reviews, 21*, 133-164. http://dx.doi.org/10.1081/GNC-120026234

Van Hemelrijck, M. J. J., Michaud, D. S., Connolly, G. N., & Kabir, Z. (2009). Secondhand Smoking, 4-Aminobiphenyl, and Bladder Cancer: Two Meta-analyses. *Cancer Epidemiology Biomarkers & Prevention, 18*, 1312-1320. http://dx.doi.org/10.1158/1055-9965.EPI-08-0613

Ward, E. M., Sabbioni, G., DeBord, D. G., Teass, A. W., Brown, K. K., Talaska, G. G., ... Streicher, R. P. (1996). Monitoring of aromatic amine exposures in workers at a chemical plant with a known bladder cancer excess. *Journal of the National Cancer Institute, 88*, 1046-1052. http://dx.doi.org/10.1093/jnci/88.15.1046

Zhu, J. P., & Aikawa, B. (2004). Determination of aniline and related mono-aromatic amines in indoor air in selected Canadian residences by a modified thermal desorption GC/MS method. *Environment International, 30*, 135-143. http://dx.doi.org/10.1016/S0160-4120(03)00168-5

Zhu, J. P., Newhook, R., Marro, L., & Chan, C. C. (2005). Selected volatile organic compounds in residential air in the city of Ottawa, Canada. *Environmental Science & Technology, 39*, 3964-3971. http://dx.doi.org/10.1021/es050173u

13

Would Use of Contaminated Water for Irrigation Lead to More Accumulation of Nitrate in Crops?

Guangwei Huang[1]

[1] Graduate School of Global Environmental Studies, Sophia University, Tokyo, Japan

Correspondence: Guangwei Huang, Graduate School of Global Environmental Studies, Sophia University, Tokyo, Japan. E-mail: huang@genv.sophia.ac.jp

Abstract

Nitrate content in agricultural crops is of interest to governments and the general public owing to the possible implications for health. There are two pathways for nitrate to enter into human body: food and drinking water. The dietary intake of nitrate is usually much larger than that from drinking water. This work investigated nitrate content in various crops and its relevance to the nitrate concentration of irrigation water in an arid region, Northwest China. It revealed that irrigation water with high concentration of nitrate tends to produce crops with high build-up of nitrate. Nevertheless, all sampled crops adjust their nitrate distributions in a way that the edible parts contain much less nitrate except lettuce. The nitrate content in the edible root of lettuce reached up to 5900 mg/kg exceeding the limit set by the European Commission. The findings, although preliminary, supplemented the existing literature and raised questions for further study.

Keywords: nitrate, distribution in crops, wastewater, arid region

1. Introduction

Water use has been growing globally putting unprecedented pressure on renewable, but finite water resources, especially in water-stressed regions. An increasing number of rivers now run dry before reaching the sea for substantial periods of the year. In many areas, groundwater is being pumped at rates that exceed replenishment, depleting aquifers and the base flows of rivers (Postel, 2000). In the future, climate change and bio-energy demands are expected to amplify the already complex relationship between world development and water demand. According to the WHO reports that within the next 50 years, more than 40% of the world's population will live in countries facing water stress or water scarcity (WHO, 2006). At the same time, the production and discharge of domestic wastewater is rapidly increasing in developing countries due to population growth, urbanization, and economic development. The produced municipal wastewater in China increased from 23.03 (10^9 m^3/yr) in 2002 to 35.48 (10^9 m^3/yr) in 2009. Despite billions invested in improving wastewater management worldwide, Ujang and Henze (2006) pointed out that 95 per cent of wastewater generated enters the environment without proper treatment. In developing countries, wastewater treatment plants represent one of the major investments. However, restricted local budgets, lack of local expertise, and lack of funding resulted in inadequate operation of wastewater treatment plants in developing countries (Paraskevas et al., 2002).

Due to the increasing trend in both water demand and wastewater discharge and the high cost of treatment facilities, there is an increasing use of partially treated and untreated wastewater in irrigated agriculture in developing countries. Indeed, the use of domestic wastewater for crop production has been practiced for several centuries in one form or another in countries such as China, India, Australia, Germany and UK. However, the importance of wastewater reuse in relation to sustainable development was recognized in recent decades. The question about from where the extra water for further development is to come, has led to a scrutiny of present water use strategies. It appears that reuse of wastewater is a key toward rational and efficient use of available water.

Benefits of wastewater use in agriculture may include: reliable and low cost source of water that is available to farmers whenever they need it (unlike canal irrigation) even in the dry season; nutrients present in wastewater may replace chemical fertilizers saving a lot of money to farmers and increasing crop yields as well. Pescod (1992) estimated that typical wastewater effluent from domestic sources could supply all of the nitrogen and much of the phosphorus and potassium that are normally required for agricultural crop production. Therefore, it can be considered as a means of reducing poverty for a large portion of people in developing countries. Applying wastewater to land also limits the pollution of rivers, canals and other surface-water resources, which would otherwise be used as disposal outlets. From the environmental engineering point of view it is of interest that agricultural fields can function as a tertiary treatment step through using effluents in irrigated agriculture. Besides, using wastewater means that there is less demand for fresh water for irrigation.

On the other hand, use of wastewater irrigation may result in serious environmental problems, contaminating groundwater with nitrates and heavy metals (especially if the wastewater contains industrial wastes). Such pollutants may build up over time in the soil and be taken up by food crops. Therefore, policymakers need to develop better understanding on the consequence of using wastewater and comprehensive strategies for managing wastewater tailored to local socioeconomic and environmental conditions. It must be assured that short-term benefits of wastewater irrigation are not offset by health and environmental costs. In this work, the food safety issue in relation to the use of wastewater was examined from the angle of nitrate accumulation in crops.

Nitrate (NO_3^-) is an integral part of the nitrogen cycle in the environment. It is oxidized form of nitrogen and highly soluble in water. It is invisible, odorless and tasteless. Nitrate is commonly found in soil and water and is the most effective form by which plants obtain their nitrogen. However, it is also most readily lost from soil by leaching in drainage water, which may be considered as one of the reasons for over-application of nitrogen fertilizers (Addiscott, 2005; Razowska-Jaworek & Sadurski, 2004). A particular concern of nitrogen use in agricultural practices is the nitrate contamination in waters and accumulation in agricultural products. The primary reason for the concern is the threat it poses to human health. The toxicity of nitrate to humans is mainly attributable to its reduction to nitrite. The major biological effect of nitrite in humans is its involvement in the oxidation of normal Hemoglobin (Hb) to met-Hemoglobin (met-Hb), which is unable to transport oxygen to the tissues. The Hb of young infants is more susceptible to met-Hb formation than that of older children and adults. Drinking water with elevated nitrate levels has been highlighted as the cause of the so-called "blue baby syndrome" (Majumdar, 2003). It is also associated with respiratory and reproductive system illness, some cancers, and thyroid problems (Dutt & Lim, 1987; Moore, 2003). Study by Parslowa et al. (1997) indicated a positive relationship between the incidence of childhood-onset insulin-dependent diabetes mellitus and levels of nitrate in drinking water. On the other hand, studies have shown that vegetables eaten by people contribute about 72-94% of the total daily intake of nitrate (Dich et al., 1996; Eichholzer & Gutzwiller, 1998). De Martin and Restani (2003) showed that leafy green vegetables accumulate high amounts of nitrates, concentrations reaching up to 6000 mg/kg. Petersen and Stoltze (1999) surveyed the contents of nitrate and nitrite in lettuce, leek, potato, beetroot, Chinese cabbage and white cabbage on the Danish market. The highest content of nitrate was found in lettuce followed by beetroot and Chinese cabbage. Although previous studies have greatly improved the understanding on the possible range of nitrate contents in various agricultural products, the knowledge of nitrate distribution within crops is still limited. Santamaria et al. (2001) showed that the nitrate content in vegetable organs can be listed in the decreasing order from petiole, leaf, stem, root to seed. Nevertheless, it is not clear if the decreasing order from stem, root to seed also holds for tall crop plant such as corn. Factors such as plant species, soil, nitrogen fertilization, light, climatic conditions and water availability affect nitrate accumulation.

The present study examined the nitrate distribution in various crops. The objective is to further improve the understanding on the health risk of nitrate contamination of crops by differentiating its concentration in edible parts from non-edible parts and highlighting its relevance to irrigation water. It is hypothesized that irrigation water quality is another fact influencing crop nitrate accumulation. The rationale is that different irrigation water resources may contain different levels of nutrients and have different ranges of water temperature which might affect the nitrate intake of crops and nitrate translocation inside crops as well. By choosing survey sites within the same climate zone having similar soil condition, and farming methods, the difference in nitrate contents of crops among survey sites, if detected, may be attributed to the difference in irrigation water quality.

2. Method

The middle reaches of the Heihe River was chosen as the study area. Heihe River is the second largest inland rivers in China. Its main stream, with a length of 821 km, originates from the Qilian Mountains of Qinghai Province, flows through the Zhangye Basin, which is part of the ancient Silk Road, and ends up in the Inner Mongolia Autonomous Region. The catchment of the middle reaches, or the Zhangye Basin, covers an area of 1.08×10^4 km^2 extending from $38°30'$ to $39°50'$ N and $99°10'$ to $100°52'$ E. Along the main stream, the middle reaches starts from the Yingluo Gorge and ends at the Zhengyi Gorge (Figure 1).

Zhangye Basin is characterized by a dry continental climate, with mean annual precipitation of about 200 mm, and annual evaporation of 2000 mm or more. The renewable water resource per capita in Zhangye is 1250 m^3/yr, just 5% of the world average. Since 2001, about 50~60% of the total river discharge have been diverted annually from the middle to the lower reaches of the river following the regulation imposed by the Ministry of Water Resources of China (MWR). Consequently, groundwater abstraction and sewage water reuse for agricultural purpose has been increasing gradually to compensate for the reduced river water supply.

Grain crops, particularly seed corn, are densely cultivated in the Zhangye Basin, sustained by the continuous application of chemical nitrogen fertilizers. In 2005, the total amount of nitrogen fertilizers applied on the corn fields was more than 300 kg ha^{-1} year^{-1} and, more recently, was more than 450 kg ha^{-1} year^{-1} (Su et al., 2007).

The study by Yang and Liu (2010) showed that 32.4% of the groundwater well samples in Zhangye had NO_3^- - N concentrations greater than the allowed value set by the WHO. Fang and Ding (2010) reported the NO_3^- concentration of groundwater as high as 84.1 mg/L in the urbanized area of the Zhangye Basin and about 67% of irrigation land area was associated with medium (13-30 mg/L) NO_3^- concentration. Zhang et al. (2004) reported nitrate concentrations in groundwater ranging from 45 to 150 mg/L in irrigated area. Field survey by Qin et al. (2011) also showed that groundwater with high concentration of nitrate (NO_3^- > 45 mg/L) appeared in the center of the irrigated area.

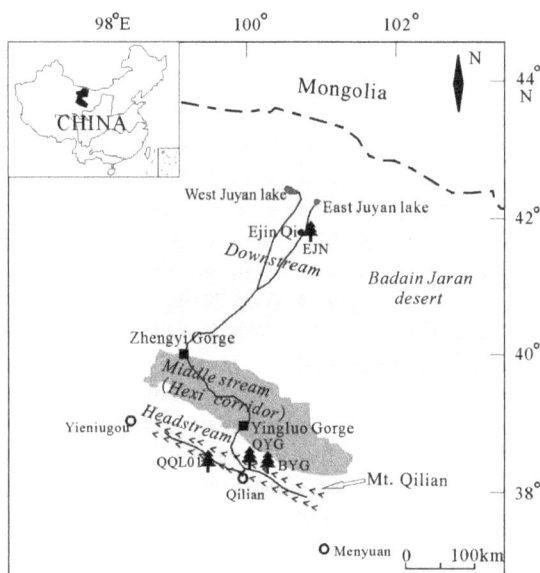

Figure 1. Watershed of the Heihe River

In the Zhangye Basin, there are three sources of irrigation water: river water; groundwater and the mixture of river water and sewage water. The use of sewage water for irrigation is due to its water scarcity. In this study, three types of farmland using different source of water for irrigation were selected to see how crops are affected by the type of irrigation water. The soil type is mainly sandy loam at all sites surveyed in this study. The sampling sites are depicted in Figure 2.

Gaozai village and the Genming village are located in the north and northeast part of the Zhangye City, where the irrigation water is the sewage mixed with river water from a tributary of the Heihe River. Guojiabao village and Erzha village are relatively close to the urbanized area of the city, where the source of irrigation water is groundwater. Yangjiazha village is in the southern part of the city and uses water from the Heihe River for irrigation. Crop samples were taken from these fields and on-site measurements of nitrate concentration were conducted. The method is to extract out plant juice of a volume between 0.3 ml and 2 ml from crop sample by a squeezer, and measure its nitrate concentration with the Horiba compact nitrate ion meter (LAQUAtwin-B741), which is based on the colorimetric method of Gilbert (1926). The sensor was calibrated by the two-calibration mode with 300 mg/kg and 5000 mg/kg standard solutions on a daily base. For seed-corn plant, the measurements were conducted every 10 cm along the stalk from the brace roots to the top flag leaf. Moreover, the nitrate concentration of corn niblets was also measured. For comparison, both mature and immature corn plants were measured. For vegetables, leaf and root were measured separately. Besides, the nitrate and ammonium (NH_4^+) concentrations of irrigation water entering into the farmland were measured with the Horiba compact nitrate ion meter (LAQUAtwin-B743) for water and HACH Portable Meter (DR/890), respectively. Furthermore nitrate concentrations in drinking well water were also tested where the nitrate concentrations in crops were measured. The field investigation was conducted from August 8 to August 18, 2012 and all measurements were done under fine weather during daytime. A conventional approach to study nitrate accumulation in crops is to conduct split plot experiments under well-controlled conditions. Such kind of study is useful for better understanding the fundamental mechanisms related to nitrate accumulation in crops. However, this approach does not fit the present study because the objective of this work was to grasp the actual situation about nitrate contamination of market ready crops in the target area. Besides, agricultural products sold in markets were not used in this study because the storage process may alter the nitrate contents and its relevance to irrigation water could not be discussed.

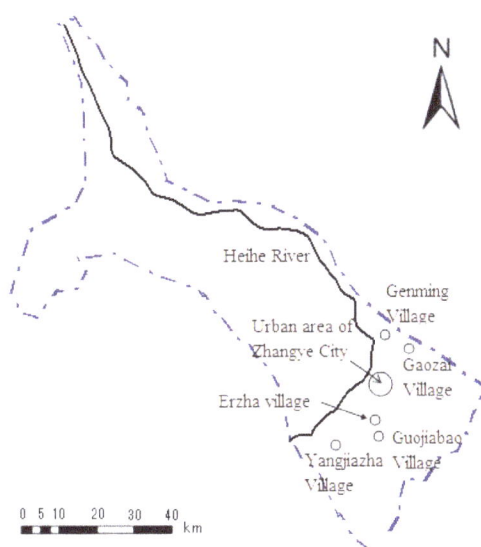

Figure 2. Zhangye Basin and sampling sites

3. Results and Discussions

3.1 Wastewater-Fed Field

At the two sites where sewage-dominated water was used for irrigation, the concentrations of nitrate in water entering into corn fields were found to be 20 mg/L in the Gaozai village and 110 mg/L in the Genming village, respectively at the time of sampling. And the concentrations of ammonium (NH_4^+) in the water were also tested at the same time and they were 6 mg/L and 0.2 mg/L, respectively. In the Gaozai village, we randomly selected two corn plants; one mature and another immature. The maturity is judged by the size and color of corn ears. The mature corn ear was 20 cm long and golden color while the immature was 10 cm long and pale green color. However, in the Genming village, immature corn was not spotted at the time of survey.

Figure 3 shows the nitrate concentration distributions along the corn stalks. The distributions along the mature corns from both sites showed very similar pattern. The nitrate concentration was as high as 3700 mg/kg at the brace roots but decayed quickly along the stalk. In the immature plant, the nitrate concentration at the bottom was not significantly higher than the top. The maximum concentration occurred somehow in the middle of the plant where a single ear grew. The nitrate concentrations in niblets of both the mature and immature corns were 110 mg/kg and 210 mg/kg, respectively. This comparison suggests that the corn plant adjust its nitrate build-up through its growing period in a way that the edible part contains the least nitrate at the end. In other words, the final distribution of nitrate in corn plant is in favor of human beings. Besides, the relatively low concentration of nitrate at the brace roots of immature corn plant implies that the plant accumulated more nitrate or assimilated less nitrate at its late growing stage. This provides a hint for better nitrogen fertilizer application. The nitrate concentrations in the drinking well in Gaozai village was also tested and found to be 68 mg/kg, exceeding the WHO standard but not Chinese standard. Although evidence is not available at this stage of investigation to assert that the drinking water (groundwater) contamination of nitrate is related to the use of wastewater in irrigation, it is a concern that should be addressed and clarified.

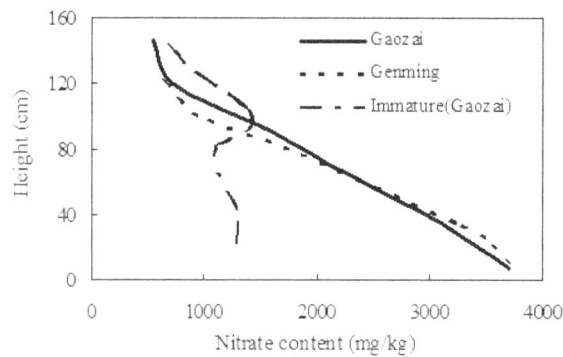

Figure 3. Nitrate distributions in wastewater-irrigated corn stalks

3.2 Groundwater-Fed Field

In the Guojiabo and Erzha villages, the groundwater is used to irrigate vegetable farmlands. The two villages are about 1.5 km apart and their irrigation water was taken from wells about 30 m deep. Cauliflower, broccoli, lettuce, beet, tomato, Chinese cabbage, green pepper, red pepper, lantern chilli, cucumber were sampled for nitrate concentration. The measurement results are compiled in Figure 4.

The nitrate concentration in broccoli leaf was as high as 5300 mg/kg. However, the nitrate concentrations in broccoli thick stalk and flower head were 2600 mg/kg and 860 mg/kg, respectively. Cauliflower and Chinese cabbage contained high amount of nitrate in leaf but much less in edible part. Among all measured, lettuce was the only one having concentrations higher than 3000 mg/kg in both leaf and root. Beet leaf and root had quite the same but moderate amount of nitrate. It should be noted that both the leaf and root of lettuce are consumed as food by humans. For broccoli, fleshy flower head and the top part of thick stalk are edible. Beet is now usually used for making sugar. The distinctive patterns of nitrate distribution within Lettuce and Beet revealed in this study may serve as a call for further investigation on nitrate accumulation mechanisms in crops. Besides, water quality test showed that the nitrate concentrations in the drinking well water was 78 mg/kg in the Guojiabo village and as high as 160 mg/kg in the Erzha village. This suggested that nitrate contamination of groundwater has progressed in the vegetable-growing villages.

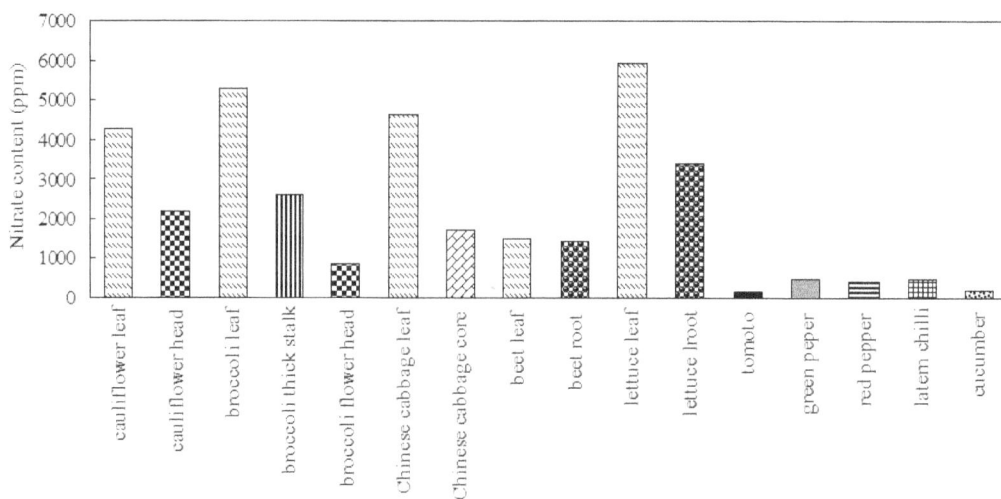

Figure 4. Nitrate contents in various agricultural products

3.3 River Water

In the Yangjiazha village where the water from the Heihe River is used for irrigation, the nitrate concentration in the irrigation water was tested to be 15 mg/l. Figure 5 shows the nitrate distribution along a corn plant sampled from this village. The corn plant in this river-water-irrigated area shared similar characteristics with that in the wastewater-irrigated area in terms of the vertical decreasing pattern. However, the nitrate concentration in the river-water-fed plant decreased vertically much faster than the wastewater-fed plant. Therefore, the total build-up of nitrate is much less when river water is used for irrigation. In the Yangjiazha village, the nitrate concentration of the drinking well water was 34 mg/kg, below the WHO standard.

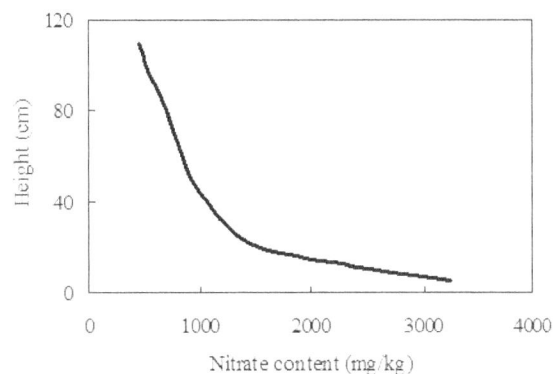

Figure 5. Nitrate distributions in river-water- irrigated corn stalk

3.4 Discussion

Prior to the work of Santamaria et al. (2001), other studies (Maynard et al., 1976; Ostrem & Collins, 1983) also reported that leaf blades have a lower nitrate content than stems and petioles, and young leaves show a lower nitrate concentration than older leaves. The present study improved the understanding on nitrate distribution in vegetables by showing that there could be a difference between edible and non-edible parts. Examiningthe nitrate distributionin such a manner is important from the food safety perspective. Such a classification and corresponding knowledge may help develop guideline for food processing. Besides, the vertical profiles along corn stalks obtained from this study supplemented the literature with regard to nitrate distribution in high plants.

The reason for more accumulation of nitrate in corn stalks irrigated by contaminated water could not be explained by the available data from the current study. However, a hypothesis is that it might be influenced by water temperature because rates of nutrient uptake and internal transport mechanisms could be considered as temperature-dependent. Since the river water originated from snow-melting, the preliminary field data found that the river water temperature was 2~3 °C lower than the sewage-dominated irrigation water.

Plants have the ability to maintain an internal environment with a composition different from that of their surroundings. The internal environment (chemical contents) of the plant body remains more or less constant whereas the outside environment is highly variable. Whenever an ion moves into or out of a cell unbalanced by a counter ion of opposite charge, it creates a voltage difference across the membrane called electrogenic pump. A fundamental expression for characterizing such a system is the Nernst potential E as below:

$$E=(RT/zF)\ln(C_o/C_i) \tag{1}$$

Where R is the gas constant, T is the absolute temperature, z is the charge of ion, and F is the Faraday constant. C_o and C_i are the ion concentrations outside and inside the cell, respectively.

Therefore, nutrient uptake by plant is temperature-affected in general although the sensitivity might be species-dependent.

In river water-irrigated fields, fertilizer is applied into soil so that nutrients are uptaken by plant roots. In wastewater-fed fields, flood irrigation is the normal practice so that nutrients may also be absorbed via stalk due to the direct contact with nutrients-rich wastewater.

Because water temperature measurements at different locations were not done continuously over a sufficiently long period of time, and plant growth was not simultaneously monitored with water temperature, quantitative evaluation of water temperature effect on nutrient uptake was not possible at this stage. Nevertheless, the preliminary findings highlighted the need for further study to quantify the true difference in water temperature regime among different irrigation waters and analyze its effects on nutrient uptake and vertical transport.

3.5 Statistical Analysis

In total, 10 mature corn plants were sampled in the Zhangye Basin in August 2012. The concentrations of root part varied from 3300 mg/kg to 3700 mg/kg with the mean and standard deviation being 3588 mg/kg and 118 mg/kg, respectively. Meanwhile, the concentrations of top part varied from 450 mg/kg to 770 mg/kg with the mean and standard deviation being 624 mg/kg and 132 mg/kg, respectively. Moreover, the concentrations of corn nib lets varied from 96 mg/kg to 210 mg/kg with the mean and standard deviation being 146 mg/kg and 41 mg/kg, respectively. To confirm that the observed difference in nitrate content connotes any real difference between wastewater-fed and river-water-fed groups, the test of significance should be conducted. For large samples, it can be done by a Z-test. Since the sample size is limited in the present study, the significance test was conducted by the so-called change-point analysis, which is a non-parametric method. It is an effective and powerful statistical tool for detecting mean shifts in a time series. The previous applications of change point detection methods include bio-informatics applications (Erdman & Emerson, 2008), network traffic analysis (Kwon et al. 2006), climatology (Reeves et al., 2007) and oceanography (Killick et al., 2010). The procedure constructed by Taylor (2000) for performing a change-point analysis iteratively uses a combination of cumulative sum charts (CUSUM) and bootstrapping without replacement. To use this technique under the context of this study, nitrate concentrations of corns measured in sewage-irrigated fields were appended by concentration data taken from river-water-irrigated fields to form sequential datasets for different parts of corn. Then, the change-point analysis was applied to datasets to detect any significant change in data. If a significant shift in mean appeared between sewage-related and river-related data, the null hypothesis of no difference in nitrate content between wastewater-fed and river-water-fed corns could be rejected. As shown in Figure 6, for the middle part of corn, there was a significant shift in mean concentration of nitrate from wastewater-fed to river-water-fed groups. However, for the root and top parts and niblets of corn, change-point analysis revealed that the null hypothesis of no significant difference in nitrate content between wastewater-fed and river-water-fed groups was retained.

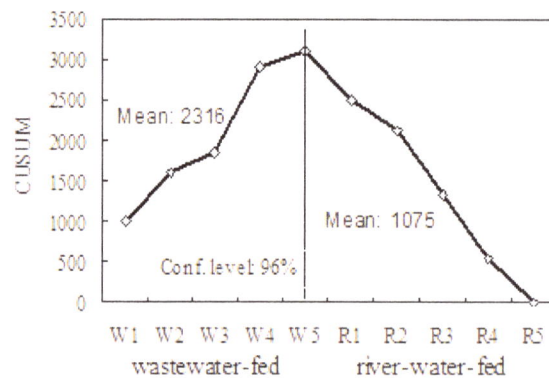

Figure 6. Test of significance by change-point analysis for the nitrate content in the middle part of corn plant

4. Conclusions

By looking into the difference in nitrate content among agricultural crops grown with different types of irrigation water, it revealed that the accumulation of nitrate in crops is affected by the type of irrigation water. The use of sewage water for irrigation due to water scarcity may cause more nitrate build-up in crops. In particular, high concentration of nitrate was found in the middle portion of corn stalk when irrigated with wastewater. Such a finding has never been reported before and it was hypothesized to be attributable to water temperature. Nevertheless, by measuring nitrate contents in different parts of various agricultural crops, the present work further revealed that even irrigated with contaminated water, the edible parts of various crops such as corn niblets, broccoli flower head are much less contaminated compared with non-edible parts such as corn stalk and Chinese cabbage leaf. The nitrate in lettuce and beet was found to have quite uniform distribution, but nitrate content in seed-corn plant can be listed in the decreasing order from brace roots, stalk to niblets. Among all sampled, lettuce is the only one having very high concentration in edible parts. Therefore, it can be stated that many crops adjust their nitrate content distributions in favor of humans.

Although, the use of low quality water for irrigation may not cause significant contamination of edible parts of crops, it can pollute groundwater to the extent that it is no longer suitable for drinking purpose. Therefore, integrated management is required to minimize potential negative effects of wastewater reuse.

Acknowledgements

Thanks should be given to Liu Huan, Li Ling, graduate students of Sophia University, for their assistance in field work. Appreciation also goes to Dr. T. Akiyama, the University of Tokyo, and Dr. Li Jia, the University of Niigata Prefecture, for their cooperation prior to and during the field investigation. Above all, the most sincere gratitude and appreciation go to Prof. Li Xin and Prof. Ma Mingguo, Cold and Arid Regions Environment and Engineering Research Institute, Chinese Academy of Sciences for their generous support and cooperation.

References

Addiscott, T. (2005). *Nitrate, Agriculture And The Environment*. CABI.

De Martin, S., & Restani, P. (2003). Determination of nitrates by a novel ion chromatographic method: Occurrence in leafy vegetables (organic and conventional) and exposure assessment for Italian consumers. *Food Additives and Contaminants, 20*, 787-792. http://dx.doi.org/10.1080/0265203031000152415

Dich, J., Jarvinen, R., Knekt, P., & Penttila, P. L. (1996). Dietary intakes of nitrate, nitrite and NDMA in the Finnish mobile clinic health examination survey. *Food Additives and Contaminants, 13*, 541-552. http://dx.doi.org/10.1080/02652039609374439

Dutt, M. C., Lim, H. Y., & Chew, R. K. H. (1987). Nitrate consumption and the incidence of gastric cancer in Singapore. *Food Chem. Toxicology, 25*, 515-520. http://dx.doi.org/10.1016/0278-6915(87)90202-X

Eichholzer, M., & Gutzwiller, F. (1998). Dietary nitrates, nitrites, and N-Nitroso compounds and cancer risk: a review of the epidemiologic evidence. *Nutr. Rev., 56*(4), 95-105. http://dx.doi.org/10.1111/j.1753-4887.1998.tb01721.x

Erdman, C., & Emerson, J. W. (2008). A Fast Bayesian Change Point Analysis for the Segmentation of Microarray Data. *Bioinformatics, 24*(19), 2143-2148. http://dx.doi.org/10.1093/bioinformatics/btn404

Fang, J., & Ding, Y. J. (2010). Assessment of groundwater contamination by NO_3^- using geographical information system in the Zhangye Basin, Northwest China. *Environ Earth Sci., 60*, 809−816. http://dx.doi.org/10.1007/s12665-009-0218-y

Heffer, P. (2009). *Assessment of Fertilizer Use by Crop at the Global Level 2006/07–2007/08*. International Fertilizer Industry Association, Paris, France.

JECFA. (1995). Evaluation of Certain Food Additives and Contaminants. *WHO Technical Report Series, 859*, 32-35.

Killick, R., Eckley, I. A., Jonathan, P., & Ewans, K. (2010). Detection of changes in the characteristics of oceanographic time-series using statistical change point analysis. *Ocean Engineering, 37*(13), 1120-1126. http://dx.doi.org/10.1016/j.oceaneng.2010.04.009

Kwon, D. W., Ko, K., Vannucci, M., Reddy, A. L. N., & Kim, S. (2006). Wavelet methods for the detection of anomalies and their application to network traffic analysis. *Quality and Reliability Engineering International, 22*, 953-969. http://dx.doi.org/10.1002/qre.781

Majumdar, D. (2003). The Blue Baby Syndrome: Nitrate poisoning in humans. *Resonance, 8*(10), 20-30. http://dx.doi.org/10.1007/BF02840703

Moore, E., & Matalon, E. (2011). *The Human Costs of Nitrate-contaminated Drinking Water in the San Joaquin Valley*. Report of Pacific Institute.

Parslow, R. C., Law, G. R., McKinney, P. A., Staines, A., Williams, R., & Bodansky, H. J. (1997). Incidence of childhood diabetes mellitus in Yorkshire, Northern England, is associated with nitrate in drinking water: an ecological study. *Diabetologia, 40*, 550-56. http://dx.doi.org/10.1007/s001250050714

Pescod, M. B. (1992). Wastewater treatment and use in agriculture. *FAO Irrigation and Drainage Paper 47*. FAO, Rome.

Petersen, A., & Stoltze, S. (1999). Nitrate and nitrite in vegetables on the Danish market: content and intake.

Food Addit Contam., 16(7), 291-9. http://dx.doi.org/10.1080/026520399283957

Postel, S. L. (2000). Entering an era of water scarcity: the challenges ahead. *Ecol. Appl., 10*(4), 941-948. http://dx.doi.org/10.1890/1051-0761(2000)010[0941:EAEOWS]2.0.CO;2

Qin, D., Qian, Y., Han, L., Wang, Z., Li, C., & Zhao, Z. (2011). Assessing impact of irrigation water on groundwater recharge and quality in arid environment using CFCs, tritium and stable isotopes, in the Zhangye Basin, Northwest China. *Journal of Hydrology, 405*, 194-208. http://dx.doi.org/10.1016/j.jhydrol.2011.05.023

Razowska-Jaworek, L., & Sadurski, A. (2004). Nitrate in Groundwaters. *IAH Hydrogeology Selected Papers, 5*, 247-258.

Reeves, J., Chen, J., Wang, X. L., Lund, R., & Lu, Q. (2007). A review and comparison of changepoint detection techniques for climate data. *Journal of Applied Meteorology and Climatology, 6*, 900-915.

Santamaria, P., Elia, A., & Serio, F. (2001). Ways of reducing rocket salad nitrate content. *Acta Horticulturae, 548*, 529-537.

SCF. (1995). Scientific Committee for Food. Opinion on nitrates and nitrites. *Reports of the Scientific Committee for Food 38th Series*, 1-33.

Su, Y. Z., Zhang, Z. H., & Yang, R. (2007). Amount of irrigation and nitrogen application for maize grown on sandy farmland in the marginal oasis in the middle of Heihe River Basin (in Chinese). *Acta Agron Sinica, 33*(1), 2007-2015.

Taylor, W. A. (2000). *Change-Point Analysis: A Powerful New Tool For Detecting Changes*. Retrieved 20 May, 2013 from http://www.variation.com/cpa/tech/changepoint.html

Ujang, Z., & Henze, M. (Eds.) (2006). Municipal Wastewater Management in Developing Countries. *Principles and Engineering*, 352. London: IWA Publishing.

Vroomen, H. (1989). *Fertilizer Use and Price Statistics*. Resource and Technology Division, ERS, USDA, Statistics Bulletin 780.

Yang, R., & Liu, W. (2010). Nitrate contamination of groundwater in an agroecosystem in Zhangye Oasis, Northwest China. *Environ Earth Sci., 61*, 123-129. http://dx.doi.org/10.1007/s12665-009-0327-7

Zhang, C. Y., Wang, Z., & Cheng, X. X. (2004). Studies of nitrogen isotopes in sources of nitrate pollution in groundwater beneath the city of Zhangye (in Chinese). *J Arid Land Resour Environ., 18*(1), 79-85.

14

A Study on Mining Industry Pollution in Chapagaon, Nepal

Deshar Bashu Dev[1]

[1] Environmental Economic System, Rissho University, Tokyo, Japan

Correspondence: Deshar Bashu Dev, 5-13-3, Nishi Kamata, Ogawasou 202, Ota-ku, Tokyo, Japan. E-mail: bashudev2013@yahoo.com

Abstract

Stone mining industries in Chapagaon, Lalitpur area is in regular operation since more than 35 years. In this long period, the operators of mines and stone crushing have cleared up the vast area of forest for the purpose of stone mining. As a result the fragile forest ecology, biodiversity and scenery beauty of this area have been widely devastated. Environment of Chapagaon indicates that the destruction of forest area has already affected the local population in terms of declining fresh water sources; drying of wells, reduction on ground water level, livestock productivity and loss of scenic beauty of the place. The agriculture crop yield has also decreased vastly and the area is gradually converting into dry land, the top soils have been eroding and crop plants are being covered with pollutants which are effecting directly or indirectly to the local people.

In this paper, health impact of local people, environmental and economic impact of locality by mining industries are examined and evaluated considering questionnaire and available data from several sources such as government publications, related researches, websites and other references.

This paper concludes with recommendations in order to control environment pollution, to reduce the impact of gravel, sand and stone mines.

Keywords: stone mining, stone-crushing, biodiversity, scenic beauty, pollutants

1. Introduction

Chapagaon is located in the outlying area of the south part of Kathmandu Valley. It is dense traditional settlement area inhabited predominantly by the Newars. The area of Chapagaon VDC (Note 1) is 6.76 km^2 with total population of 12789, out of this 6516 male and 6273 female. It is taken one of the fastest growing VDC in Kathmandu Valley. The development prospect of proposed outer ring road of Kathmandu Valley and closeness to the city center of Kathmandu, this area has been an attractive residential location for many migrants (CBS, 2011).

In this VDC no proper considerations are made for settlement planning, mining, crushing and management of these. The mining industries of this area are taken problematic not only because of the lack of proper considerations about the environment but also because of the narrow road networks of this area. The hauling trucks of this area create the sound whole the day and pollute the air. Mining industries in Chapagaon are concentrated in Ward no. 6 which is adjoining Ward with Lele VDC. This ward is taken as the best for crushing and stone mines because of lots of open spaces and availability of stones. Streams of Chapagaon; Nakhu and Karmanasa are disturbing by stone mines and crushing industries, the volume of drinking water has been reduced gradually because of the extraction of upper layer stones and soil of spring catchment area. In the past years (2010/11) there were more than 30 stone and crushing industries but in 2012 gradually these reduced below No. 9 in Chapagaon and shifted in other neighboring VDCs like Lele, Nallu, Bhardeu etc. (VDC Profile, 2010).

2. Objective of the Study

The general objective of the study is to show environmental, social and health impacts of the disordered natural resource exploitation, the specific objectives are as follows:

i. To find out the consequences of mining industries in study area;

ii. To assess the condition of natural resources; and

iii. To recommend some of the ways of natural resources conservation to the concern agencies

3. Justification

The criteria adopted for categorizing the mines units as small, medium and large scale, differ from country to country and there exists no universal yardstick (Ghose & Roy, 2007). Recognition of the fact that small-scale mining can make a significant contribution to development objectives, which has been one of the principal motives for this persistent interest (Noetstaller, 1994).

Mining was a flourishing activity in the remote past. But it was conducted with crude technique and to a limited scale. Iron ore, copper, slate, mica, marble, lead, lignite, etc. were the chief minerals exploited and utilized in those days. Subsequently, it was gradually abandoned due to three basic reasons: (a) Exhaustion of easily accessible top deposits (b) Lack of improved techniques (c) Negligence of the then ruling classes (Shrestha, 2004).

Rapid development has been spurring the mining industry in Nepal. Until the early 1990s, there were just few mining sites producing gravel for house and road construction. During the 1990s the population in Nepal's cities grew rapidly and building construction techniques changed requiring greater quantities of gravel (CLSR, 2009) (Note 2). At the same time the road network was expanding in many districts which increased demand for gravel. As well as mining industries are taken as the main sources of building materials. Stone crushers are small scale industries in the unorganized sector. They provide basic material for road and building construction. They are highly labor intensive. The various unit operations involved in stone crushing viz., size reduction, size classification and transfer operations have the potential to emit process and fugitive dust (Sivacoumar, Jayabalu, Subrahmanyam, Jothikumar, & Swarnalatha, 2012). The three stages of mineral development, viz exploration, mining and processing, have caused different types of environmental damages, which include ecological disturbance, destruction of natural flora and fauna, pollution of air, land and water, instability of soil and rock masses, landscape degradation and radiation hazards (Aigbedion & Iyayi, 2007).

The industries in and around large and small urban areas are increasing day by day. In Nepal, The Mines and Mineral Act, 1985 A.D., its amendment, 1993 A.D. and Regulation 1999 A.D. introduced for the management of mines and mineral industries of the country. After this time the mines and mineral industries are regulated by the Department of Mines and Geology under the Ministry of Industry, Commerce and Supplies. The responsibility of regulation of established mines and mineral industries in Nepal remains on the hand of government with its several local bodies but the proper follow-up and regulation seems very weak.

The case of Chapagaon remains the same; the mining industries established are found not within the parameters of government's approval, these are out of the limitations because of which the local environmental resources are degraded on greater extent. Mainly the mining industries are responsible to decrease the scenic beauty of the place, degradation of the productivity of land, dust and smoke pollution etc. So this study attempts to find out such impacts which are related to the consequences of mining industries in this area.

4. Study Area

Chapagaon is the village which is spread in 68 km^2 with the shape of conch shell. It is very close to Mangal Bazar (palace area of Malla Kings) which is just 10 km^2 away from Chapagon. Chapagon can also be known as "Wadey", "Champapu". However the presence of Bajarnarahi temple helps to identify the village more easily. Historically, the village was full of forest with Chanp trees. At first, the village was named as Champapur, as it was developed by demolishing the forest of Chanp tree and later the name was revolted into Chapagaon, Similarly, the name Wadey was assigned from Newari Wa which means rice and Dey means state. Thus, the name Wadey means state of rice. This name was specifically selected as the village has large number of rice production with better quality as compared with other areas of Patan. In various manuscripts, this village was generally indicated with the names of Wadey, Chapagon and Champapur. However, the village is commonly familiar with the name Chapagaon in all government sectors. In order to retain its traditionalist, the municipality of this village is named as Champapur. During the period of Shivadev the Lichhavi king, the village was found to be well development and act as an important business center. For instance, business man from other areas had to pay tax for selling their fish in Chapagaon. Till now, the village has specific plans like Bhansar Tole to execute their implications. This has been evident from the script written on the statue of Jalahari situated near Brahma (Note 3) statue in the ear of Basantadev. From this, it is understood that Chapagon is a well-developed and planned village as of the ancient times. Several historical evidences indicated that Chapagon was highly developed during the period of Malla kings. The culture and religious values are higher in Chapagaon. The most important function of this village is Jatra. This function is celebrated twice a year, one on the Astami (Note 4) of Kartic and other on full Moon Day of Chaitra. At present, Jatra on Astami of Kartic is not celebrated because of the mislaid of idol in 2046 B.S. Bulu and Pyangaun has their own Jatra of Chandra Bhairav and Jatra of

Mahadev. Apart from this, there are other ceremonies also conducted such as Dipankar Walk, Samaydhyo Bwayagu, Ganesh and Saraswati Jatra of Jhyalipati (Note 5), Bhairab Jatra (Note 6), Juga Chareor Samyak Dan (Note 7), Pond Fair of Khasimar and Tika Bhairab Jatra (Note 8).

Figure 1. Area map of Chapagaon

5. Mine and Mineral Acts and Regulations in Nepal

There are two acts and regulations concerning mines and minerals in Nepal, they are as follows:

i. The Mines and Mineral Act, 2042 B.S. (1985 A.D.) and its amendment, 2050 B.S. (1993 A.D.).

ii. Mines and Mineral Regulation 2056 B. S. (1999 A. D.).

Source: Ministry of Industry, Department of Mines and Geology, 2066 B.S. (2009 A.D.).

These two mines and mineral acts and regulations encompass with policies for handling, monitoring, and administrating the mineral department of Nepal. These policies are open to all qualified aspirant. The main intention of promulgating such policies is to regulate, manage, and operate mining of all minerals excluding natural gas and petroleum. The license system for mining minerals is of two stages. Such license can be obtained by any person having technical and financial competency to perform mining operation. Two kinds of license are provided include:

•   To target mineral source, Prospecting licenses are provided wherein volume and grade of mineral has not been identified and

•   Mining licenses are provided to perform mining operation wherein amount and quality has been previously determined by corresponding department.

Mining license holders are allowed to do mineral exploration not more than 25 km^2 and not less than 0.25 km^2 in the first 10 to 30 years based on the level of minerals. This period may extend up to 1 to 10 years. Mining license can be provided by Department of Mines and Geology by identifying the mineral deposits of applicants through exploration activities (Pradhan, 2011). The expense encounter by the department in exploration activities can be either converted into share or it can be retrieved from qualified aspirants.

Similarly, prospecting license holders are not allowed to perform mining operation more than 25 km^2 and below 0.25 km^2 for the first 2 to 4 years of operation. This is with a provision of extension up to 1 to 2 years. Eligible candidate must complete the exploration activity within 2 years for ordinary nonmetallic minerals and 4 years for valuable nonmetallic and metallic minerals. Further, the Mines, and Mineral Act and Regulation of Nepal provide the subsequent provision with reference to:

•   Improve safe and secure of mine and miners' workers.

•   Proper usage of land.

•   Undertake mining activities in an environmental responsible manner.

- Get approval from corresponding department through sending sample for lab testing and exploring outside the country.

- The Government of Nepal has all rights to conduct the mining activities either directly or by selecting eligible persons to conduct activities. They also have rights to participate in the mineral developing activities directly or by any form of contribution.

The Mining Act and Regulation also provide several offers and rights for lessee while performing mining operations. It includes:

- Right to access the land and materials for mining operations.

- Right to trade and export the mining products.

- Payment renewed on the basis of minerals productions and commodities.

- Right to import equipment and machines for mining activities.

The expense for perceiving Prospecting and Mining license is very cheap in Nepal when compared with other countries. The Government provides royalty of mineral products based on the quality, its type, and volume of mineral production. Consequently, the minerals are categorized into non-metallic, fuel, metallic and construction minerals. For metallic minerals, the royalty can be fixed based on metal production. For others, it can be fixed based on the minerals production. At present, 25% income tax is amended for corporate governance. Further, extra charges are not imposed on the interest on foreign loans. The Government also provides tax deduction from corporate income. In addition to this, the Government of Nepal also regulates some legal policies to prevent double taxation on FDI during agreement or providing license for investment (MOI, 2009).

6. Methodology

This section deals briefly with the research methodology applied in the study. This is purely academic research based on social science.

6.1 Rationale for the Selection Study Area

The selection of the study area is one of the critical issues while undertaking research work. I admit that my study site is pro-urban area, considered to be one of the famous mining industries site. The rationale for the selection of study area includes:

i. The mining industries were operated in this site before 35 years.

ii. The mining industries in this site are operated in forest areas.

iii. The mining industries in this site are operated out of the limit of government's approval.

iv. The researcher is familiar with the ecology and environment of this site.

6.2 Research Design

i. The research design is based on descriptive and exploratory.

ii. It is descriptive as it is based on detailed investigation and record of the mining industries in this area.

iii. It is exploratory in the sense that analysis focused on exploring whether mining industries in this site are rationale to the environment or not. An attempt has been made to make the mining industries more responsible to the local environment.

6.3 Nature and Source of Data

Both primary and secondary data have been collected for the purpose of study.

i. Primary data are based on household survey, observation, and interview.

ii. Both published and unpublished documents, records, books and relevant materials related to the subject matters have been incorporated as secondary data.

6.4 Universe and Sampling Procedure

This study has been confined within Chapagaon VDC of Lalitpur District, Nepal. This VDC contains 12789 as total population of the VDC. Among these the 100 households of Ward no. 6 were selected for the survey and 6 mining industries were assessed for the purpose of this study. There are all together 10 mining industries in Chapagaon VDC but 6 mining industries were operating in this field visit period.

6.5 Data Collection Technique and Instruments

For the collection of primary data, the following techniques were adopted.

Household Survey: Household survey was conducted to gather more information about the impacts of mining industries. Various information regarding to the pollution and other impact was collected from structured questionnaire.

Observation: Non-participatory observation was applied during research to study the location, and concerning environmental impacts of mining industries.

Interview with Key Informants: Some knowledgeable persons such as elderly persons, members of forest-user group, community based representative personnel, teachers and local leaders were selected as key informants to carryout research. Checklist and guidelines were prepared for key informant's interview.

6.6 Method of Data Analysis

The collected data were edited, coded, classified and tabulated for data organization. The quantitative data have been presented in tabular form and suitable statistical tools like percentage, ratio, mean etc. has been adopted for data analysis. Bar-diagram and trend analysis have been presented to make figure attractive. The quantitative data have been interpreted and analyzed in descriptive way based on their numerical characteristics.

7. Results

7.1 Increased Trend of Mining Industries in Chapagaon VDC

The following figures show the increasing trend of mining industries in Chapagaon VDC in 5 years period of time (2007 to 2012):

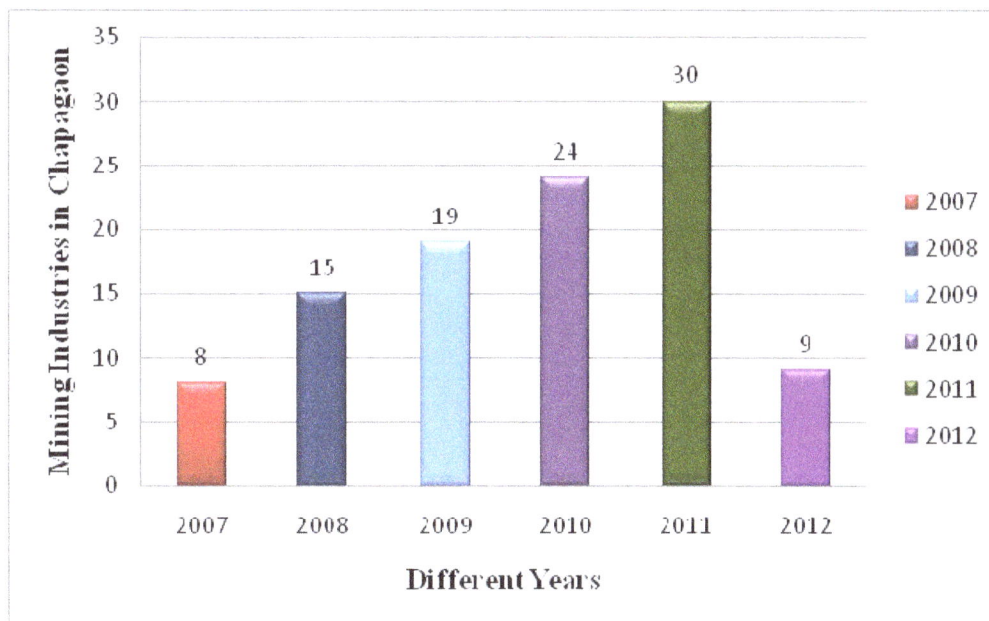

Figure 2. Yearly average income of Chapagaon VCD through mining industries (adopted from field survey, 2012)

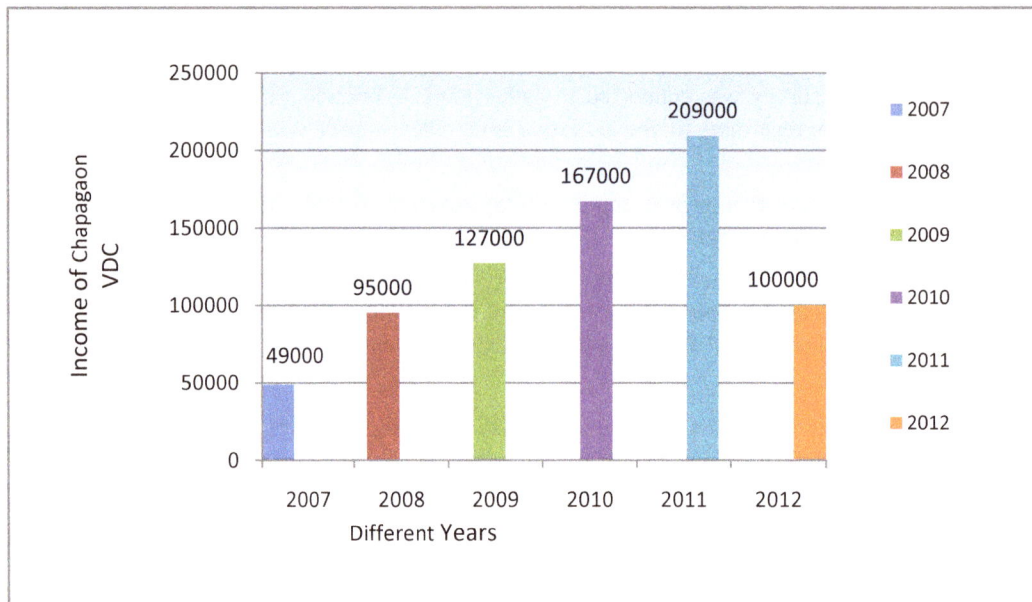

Figure 3. Increase trend of mining industries in Chapagaon (adopted from field survey, 2012)

Figure 2 and Figure 3 show that the income of Chapagaon VDC in the year 2007 by 8 mining industries is Rs. 49000 (about 500 US$-in the rate, 1 US $ = Rs. 98), in 2008 by 15 mining industries is Rs. 95000 (970 US$), in 2009 by 19 mining industries is Rs. 127000 (1296 US$), in 2010 by 24 mining industries is Rs. 167000 (1704 US$), in 2011 by 30 mining industries is Rs. 209000 (2132 US$), and in 2012 by 9 mining industries is Rs. 100000 (1020 US$). The fluctuation of income amount of the VDC symbolizes the increasing rate of tax and the system of tax pay of the VDC. Sometimes, because of the disturbances over mining industries the tax amount were not paid by the mine owners.

But Figure 2 doesn't represent the regulation of mining industries whole the year, this only represents the establishment of mining industries in Chapagaon in year basis. Some of the industries established there but production activities remained very less because of the disturbances of workers and local people too, so the income is not as equal to the number of mining industries.

7.2 Positive and Negative Impact Assessment of Mining Industries in Chapagaon

In Chapagaon 100 households of mining area (Ward No. 6) were survived, the responses of them about negative and positive impacts of mining industries are mentioned in the table below:

Table 1. Positive and negative impact of mining industries in Chapagagon

No. of Respondents	Positive Impacts of Mining Industries	Negative Impacts of Mining Industries	Positive Percentage	Negative Percentage
100	41	59	43	57

Table 1 shows that majority of the respondents of Chapagaon respond negative impacts of Mining industries in their area. Out of 100 respondents 41 percent only respond the positive impacts of these industries rest of these respond the negative impacts. They were also asked about the aspects of negative impacts too, most of them were agreed on dust, smoke, vehicular congestion, sound pollution and loss of scenic beauty of the place. At the time of survey, about 50 percent respondents were unknown about the tax payment system of mining industries in this area.

7.2.1 Positive Impacts

-      Income generation for the VDC.

-      Employment generation for the local people.

- Frequent maintenance of local roads by mining industries owners.

- Construction materials on their site on cheap price.

- Use of local resources/mobilization of local resources.

7.2.2 Negative Impacts

- Disturbance: Sound pollution, water pollution, land pollution.

- Degradation of scenic beauty of the area and degradation of tourism resources.

- Extinction of flora and fauna (plants and animals).

- Extinction of aquatic diversity (water species).

- Gradual drying of drinking water bodies causes the scarcity of drinking water.

- Loss of grazing land for the cattle.

7.3 Demographic and Land Use Features of Chapagaon

Table 2. Demographic and land use features of Chapagaon

Male/Female	No of Male/Female	In percentage (Male/Female)
Male	6516	50.95%
Female	6273	49.05%
Total	12789	100.00%

Ward-wise Households and Population			Area Coverage		Land Use	
Ward No.	No. of Households	Population	Ward-wise Area Coverage		Land Use Coverage	
			Ward	Area in km^2	Use in	Area in km^2
1	189	960	1	0.23	Settlement	0.12
2	211	1114	2	0.76	Bushes	1.22
3	258	1242	3	0.64	Cultivable Land	4.76
4	301	1565	4	0.5	Forest	0.99
5	252	1219	5	0.31	Useless Land	0.02
6	481	2580	6	1.77	Sandy Land	0.09
7	294	1442	7	1.44	Land Cover by Water	0.02
8	219	1073	8	0.96	Total	7.22
9	326	1594	9	0.61		
Total		12789	Total	7.22		

Source: CBS (2011).

Table 2 shows the total population and land area of Chapagaon VDC according to the preliminary results of census survey, 2011. As per this report this VDC contains 12789 total populations and 7.22 km^2 total land area. Out of this settlement area covers 0.12 km^2 area, bushes, 1.22 km^2, cultivable land 4.76 km^2, forest area 0.99 km^2, useless land 0.02 km^2, sandy land 0.09 km^2 and land covered by water 0.02 km^2 area. This further indicates that the large amount of land area in this VDC is cultivable. The mining industries which have been established on the useless land (means out of human use) are not disturbed by the local people but which have been established on cultivable land, bushes area and pasture land are being disturbed frequently by the local people.

7.4 Concentration of Mining Industries in Chapagaon and Their Production

Mining industries in Chapagaon are concentrated in Ward no. 6 which is adjoining Ward with Lele VDC (Village Development Committee). This ward is taken as the best for crushing and stone mines because of lots of open spaces and availability of stones. Streams of Chapagaon; Nakhu and Karmanasa are disturbing by stone mines and crushing industries, the volume of drinking water has been reduced gradually because of the extraction of upper layer stones and soil of spring catchment area. In the past years (2010/011) there were more than 30 stone and crushing industries but in 2012 gradually these reduced in Chapagaon and shifted in other neighboring VDCs like Lele, Nallu, Bhardeu etc. (Field Survey, 2012).

The following table shows the Crushing and Stone Mines currently existed in Chapagaon VDC with production capacity of these:

Table 3. Crushing and stone mines existed in Chapagaon VDC

Name of Crushing/Stone Industry	Location	Total Worker	Production Type	Production Capacity
Bajrabarahi Roda Dhunga Udyog	Ward No. 6	5	Gravel and stones	69 Mini Truck/day
Bhanjyang Dhunga Khani	Ward No. 6	6	Gravel and stones	71 Mini Truck/day
Bhuwaneshwor Roda Dhunga Udyog	Ward No. 6	7	Gravel and stones	75 Mini Truck/day
Champapur Dhunga Roda Udyog	Ward No. 6	6	Gravel and stones	72 Mini Truck/day
Excel Stone Crusher	Ward No.6	5	Gravel and stones	67 Mini Truck/day
Lalit Concrete P.vt.Ltd.	Ward No. 6	7	Sand, gravel, and stones	73 Mini Truck/day
Nepal Roda Dhunga Udyog	Ward No. 6	6	Gravel and stones	71 Mini Truck/day
Purna Dhunga Khani	Ward No.6	6	Gravel and stones	69 Mini Truck/day
Santi Roda Dhunga Udyog	Ward No.6	5	Sand, gravel and stones	65 Mini Truck/day

Source: Field Survey, 2012.

These Mining industries of Chapagaon produce mainly 5 types of stones which use in buildings and road construction such as a large stone for the basement of building, small stones for the flooring, and gravel for the road construction, next product has been used for the wall painting dust.

7.5 Tax Payment by Mining Industries in Chapagaon

The mining industries of Chapagaon pay the tax to the VDC as according to their volume of export of mine products, the tax has been approved by DDC and VDC has taken the tax on the following manner:

i. Large Truck per trip Rs. 400 (about 4.08US$-in the rate, 1 US$=Rs. 98).

ii. Mini Truck per trip Rs. 150 (about 1.53US$-in the rate, 1 US$=Rs. 98).

Above tax payment system further symbolizes the contribution of mining industries for the development of this VDC. Chapagaon VDC's record shows that yearly in average 100,000 taxes has been collected, this tax has mainly used for the maintenance of road and environmental cleanliness.

7.6 Pollution from Mining Industries in Chapagaon

Mining and mineral activities generally affect the outside environment. In the process of making products pollution and waste are produced which ultimately threaten the human health and the surrounding environment. The similar cases are found in Chapagaon area especially from the production and transport of gravel, sand and stones. Different types of impacts of these productions and transport are analyzed below:

7.6.1 Health Impact to Local People

Exposure to heavy dust concentration from stone crushers may produce several diseases, chief among them being pneumoconiosis (Zenz et al., 1994). Silicosis, caused by inhalation of dust containing silica, is an important form of this disease. The impact caused by gravel, sands and stone mines in in Chapagaon area is air pollution and its associated health impacts to the local people those located in nearby mine area. In open areas of

Chapagaon, the impact of such mines on human health is not likely to be significant. But in the residential areas, when one truckloads of sand and gravel from its excavation and starts transport to its destination mostly in Lalitpur Sub-Metropolitan City and Kathmandu Metropolitan City it fills dirt, smoke, and sound pollution. The following are the main impacts caused by sand, gravel and stone mines.

- Increased air pollution which directly impacts the health of surrounding people
- Dissemination of dust and fumes from gravel and sand at the mining area
- Dispersion of dust due to lack of proper monitoring and lethargic operations
- Fleeting of dust and fumes from exposed or opened dump trucks
- Contamination of ground water due to mixing of waste water from mines.

Each of the impacts listed above produces greater impact to the human health but these are hard to measure. In earlier times, minimal populations and establishment of few mines in Chapagaon made these impacts less noticeable. But now these are observed further serious most of the local people who are on the side of narrow black-topped road area like Pyanggaon, Chapakot, Bajrabarahi etc. are facing several health problems such as headache, dry nose, eye dimming problem, asthmas, respiratory diseases and lungs problems. The field survey has revealed that about 30 people per month directly or indirectly are affected by the cause of pollution from stone and stone-crushing industries in this area. In a day 1264 times the hauling trucks inter and exit from this site which produce the large amount of pollution not only smoke and dust but also sound.

7.6.2 Pollution on Agricultural Land

The stone and sand extraction process entails the removal of large amounts of waste too, which becomes pollution for the agricultural land. The deposits and wastes from the stone and stone-crushing mines disturb the general flow of streams and rivers that causes river and stream bank cutting in the rainy season. Most of the agricultural lands especially paddy fields are on the side of Nakhu and Karmanasha streams in Chapagaon which are under the threat of stone mine and crushing industries. The field survey of this area has recorded that about 5 Ropani paddy field of this VDC has been destroyed in average per year especially in rainy season. Moreover, the local people are agree on the fact that the vegetables farmed on about 11 Ropani of land of this VDC are also destroying per month by the dropping sand, gravel and stones on the side of road because of the unsafe coverage of hauling trucks.

7.6.3 Impact on Soil Quality

In Chapagaon, soils over large area are destroyed by mining activities. Moreover, agricultural lands near mining and crushing site are particularly affected. The fugitive dust has created significant impact over the agriculture land of Chapakot, Pyanggaon and Bajrabarahi area. Erosion of exposed soils, wind-blown dust, dropped pieces of sand, gravel and stone are usually posing the greatest risk over the soil quality.

7.6.4 Water Pollution

In most mines, the soil potential and sedimentation get contaminate and thereby affecting quality of water. As the mining activities occupy large area of land and hence large amount of ground materials exposed at site generate soil erosion, which is the major concern at hard-rocking mining sites. Due to erosion, considerable loading of sediments to nearby water bodies, especially during severe storm events and rainy season. In rainy season, excess of contaminated particles are mixed with rain water and drained in rills, natural channels or gullies.

In Chapagaon VDC, two streams namely Nakhu and Karmanasa are polluted through the flow of mine waste. The main factors influencing water bodies pollution includes the volume and velocity of runoff from precipitation events. The stone mines are located on upper slope area of Karmanasa stream and the mine depositions of dry season slip down with the volume of wind and runoff. The heavy rainfall in rainy season sweeps down all deposited items of mines, it causes several floods to the stream and full of sedimentation on Karmanasa stream. The case of Nakhu stream is different than that of Karmanasa. Nakhu stream is affected by the quarrying practices on the sides of it. Due to the cause of heavy quarrying, the paddy field of both sides of this stream has converted as a water flowing area of stream, this stream is widening year by year. The water of both of these streams is polluted so the people of adjoining areas are facing the problem of drinking water and the use for secondary purposes.

Some cumulative impacts of mining in Chapagaon area on water bodies and water species are mentioned below:

- Lost access of locals to the clean water;
- Deposition of mining waste on the water bodies;

- Extinction of water diversity i.e. fishes, frogs, snakes, leaches, worms etc.
- Lost access of locals to the secondary use of water such as irrigating, swimming, washing, fishing etc.

7.6.5 Erosion/Sediment of Mining Industries

Major sources of erosion/sediment loading at mining sites can include open pit areas, heap and dump leaches, waste rock and overburden piles etc. A further concern is that exposed materials from mining operations may contribute sediments with pollutants, principally heavy deposits of gravel. The types of impacts associated with erosion and sedimentation are numerous, typically producing both short-term and long-term impacts. In surface water the erosion and waste rock of mines fills up the depth of water level which causes toxic effects in fish. Sometimes the waste materials especially from mines flow to the stream with huge chunks and these make the heavy sediment and the fishes displace from their original dwells. With these chunks of rock and other materials flow the topsoil chemicals as used by miners.

Sediments deposited in layers in flood plains or terrestrial ecosystems can produce many impacts associated with surface waters, ground water, and terrestrial ecosystems. In Chapagaon area erosion/sediment from the upper slope areas of Nakhu and Karmanasa Streams causes the several impacts on water aquatic fish and other species. Field survey has recorded that due to the cause of erosion/sediment the streams of this area are now out of indigenous fish species and other water species, but in past years several fish species were available such as before 3 years in Karmanasha and 15 years in Nakhu Stream.

7.6.6 Socio-Cultural Impact

Gravel, sand and stone mining in Chapagaon area has increased traffic congestion and safety hazards. When operating these mines, several trucks run via Chapagaon for more than 10 hours each day. As a result of this, there has been increase in air pollution due to dust, diesel fumes and so on. In Ward no. 6 of Chapagaon VDC, where many crushing and stone mines are located, so heavy traffic hazards like trucks and other heavy vehicles are transported several times per day. In spite of this, the aesthetic degradation due to stone mining and crushing altered land mass of Chapagaon area and vanished green vegetation. Public nuisance is another important impact created by these mines. The state has not been formulated essential regulations for the operation of stone, sand and gravel mines in this area. The conversion of open spaces into built form has degraded the balance of built and non-built spaces that existed in traditional settlement planning and made the place more congested and traffic jam for the longer time. The single lane road of Chapagaon area has been carried 1262 times of transactions of hauling tracks per day. Cumulative impacts of ripping, drilling, blasting Overall, the local residents are highly affected by mining, blasting, transport, drilling, grinding and ripping. All types of above mentioned equipment produced vibration but vibration from blasting drastically affect building structure, people of local residents in a large manner. In Chapagaon area the traditional buildings are gradually disappearing because of the easy access of construction materials of new buildings. And the traditional buildings which are existed at present time are covered with thick dust flying from hauling trucks on the narrow road existed almost at the center of traditional settlements. The traditional identity of Chapagaon area and the great influence of Bajrabarahi Temple has found now on crisis.

7.6.7 Economic Impact

The social impact of large scale mining is controversial and complex to describe. Though the mining operation can create employment, roads, and schools, the profit from such operation cannot be uniformly distributed among people. The Chapagaon Village Development Committee fines only 200 rupees per month from a mining industry and takes Rs. 400/trip (about 4.08US$-in the rate, 1 US$=Rs. 98) from large truck and 150 Rs. (about 1.53US$-in the rate, 1 US$=Rs. 98) from mini truck but in reality it has not found paid by each and every. The mining entrepreneurs of Chapagaon VDC are found stronger than locals, the response and social demands of locals are not found fulfilled by industry owners. The perception of inhabitants of this area is that the community has not got any social contributions from mines. But at the time of observation, only 2 mine owners were not paid the monthly tax to the VDC office. Three mines were not operated because of the disturbance of local people. In conclusion, mining industries are not fully negative, these are contributed Rs.100, 000 (about 1,020.40 US$-in the rate, 1 US$=Rs. 98) per month. Respondents (57%) insisted that the amount paid by them is not sufficient only for the maintenance of road and environment, so the mines are not beneficial for them. Local people have taken mines as the major causes of pollution. Economic contribution of mining industries to the local has found not visible, Village Development Committee fines and takes the money from them but allocates as the total income of VDC, so the contribution of mining industries has found contradictory.

7.6.8 Impacts on Flora and Fauna

Generally, the term Wildlife refers to non-domesticated vertebrates, but in broader terms it refers to all plants and animals. By degrading green vegetation and top soil, dissemination of pollution, dislodgment of fauna, the mining operation disrupted the life of flora and fauna. Some of the cumulative impacts of mining on wildlife in Chapagaon area are mentioned below:

7.6.8.1 Loss of Habitats

The living pattern of wildlife species depends on conditions such as local weather, soil, altitude, and other local features. The existence of wildlife is directly or indirectly affected by mining operation. Influences of mining to wildlife are primarily from dislocating animals from its place of origin due to pilling of mining wastage. Moreover, the alteration created by land distribution disturbed the living pattern of wildlife and thereby reduced the survival of such species. In Chapagaon area wildlife species like bird species, reptiles, small mammals, amphibians etc. are found disappeared.

7.6.8.2 Habitat Fragmentation

Habitat fragmentation has found in Chapagaon area wherein the habitats are dispersed in to smaller groups which resulted in the increased isolation of habitat patches. Due to this, majority of the species are found disappearing from its native places.

8. Discussion

Gravel, sand, and stone mines are common across Chapagaon. Although these mines are not regulated under the Mines and Minerals Acts and Regulations of Nepal, they are registered with District Development Committee and some on Village Development Committee only. The primary environmental impact from gravel, sand and stone mines in Chapagaon area are degraded air quality from blowing dust particles, smoke, and dropping sands. Deposition of mine on the side of Karmanas and NakahLake contaminate the surface water quality of Chapagaon. Other impacts of mining operation on environment includes aesthetic degradation, gravel deposition, increased traffics on roads creates high level dust, diesel fumes which impacts the quality life of local residents.

Moreover, current environmental laws in Chapagaon area also not so effective in regulating gravel, stone crushing and mining operations. As compared with smaller minim, larger mines are considered as minor determinant for air pollution and hence these mines are allowed to work with minimal quantity of emission. This may create nuisances to local communities. However the state government did not consider the influence of these impacts. Prevailing regulations failed to consider the location of these mines near residential areas.

Modifications in existing rules and regulations may reduce the impact of gravel, sand and stone minim on environment.

9. Conclusion

Mining operations are considered one of the main sources of environmental degradation. Depletion of available land due to mining, waste from industries, conversion of land to industry and pollution of land, water and air by industrial wastes, are environmental side effects of the use of these non-renewable resources. The environmental damage has in turn resulted in waste of arable land, as well as economic crops and trees.

The number of mining industries in Chapagaon has found decreased in 2012, but the environmental impacts have found rather increased. Chapagaon is an adjoining VDC of Lele, Bhardeu and Nallu VDCs, especially the gravel, sand and stone productions of Lele VDC are to be passed through the way of Chapagaon, so the environment of Chapagaon has degraded even after shifting the mining industries from Chapagaon. The income of Chapagaon VDC in the year 2012has found Rs. 100000 (about 1,020.40US$-in the rate, 1 US$=Rs. 98); this symbolizes the reduction of income of Chapagaon VDC. The tax from gravel, sand and stones in Chapagaon remains less while the productions are not existed in Chapagaon. The productions of other VDCs just pay the tax to Chapagaon on large and small truck basis. The impacts regarding to environment of Chapagaon are related to the degradation over the scenic beauty of Chapagaon, loss of soil quality, reduction on agricultural production, air pollution, drying the source of drinking water, soil erosion, sedimentation on local streams, habitat loss and fragmentation of wild life, health impacts on local people etc.

Economic impacts of mining industries in Chapagaon have found some positive too. The collected taxes from mining industries in Chapagaon VDC have utilized for the maintenance of local roads and local environment, the tax has also been utilized for the infrastructures development too. In gist, it can be said that mining industries of Chapagaon are detrimental to the environment but beneficial to the local people. But the perception of local people has found just partially positive.

10. Recommendation

In order to control environment pollution, the following recommendation will be useful to reduce the impact of gravel, sand and stone mines.

a.   Refuse to provide permission to start new mines or reject permission to re-open the mines. Permission should be given only if the required materials are not existed in given area. This would be appropriate where the damage has already occurred and prevention of incoherent and random accumulation of sand, gravel and stone mines is required.

b.   Enforcing existed emission permits strongly and consistently. To obtain this, the state would recruit more competent inspectors to take more appropriate actions against mining operations.

c.   Refuse to give permission for operating mining in unsuitable locations. It should be ensured that the permission should not for mines to be operated in historical area of Chapagaon, residential area, rural communities as the mining will destroy charming of such areas.

d.   Motivate to use re-processed materials such as recycled stones, gravels etc. This would definitely reduce the beginning of new mines and aids to resolve the overloading problems in mining areas.

e.   The historical and natural sites of Chapagaon are to be protected for the prospects of tourism in this area, so the mining industries are essential to be shifted to the other places than presently existed.

f.   The areas which are far from the local area, these kinds of areas should be managed for the mining.

g.   Miners need to use environmentally friendly equipments.

h.   People awareness programme should be conducted from the Miners and from government about the impact of mining industries.

i.   Tree planting programme should be conducted from Miners and local people.

References

Aigbedion. I., & Iyayi, S. E. (2007). Environmental effect of mineral exploitation in Nigeria. *International Journal of Physical Sciences, 2*(2), 33-38.

CBS. (2011). *Preliminary Results of National Population Census.* Government of Nepal: National Planning Commission Secretariat.

Child Labour report. (2009). *Children Working in Mining Industry.* World Education and its Ngo partners, Kathmandu.

MOI. (2009). *Mines and Minerals Acts and Regulations.* Government of Nepal: Ministry of Industry, Department of Mines and Geology.

Ghose, M. K., & Roy, S. (2007). Contribution of small-scale mining to employment, development and sustainability–an Indian scenario. *Environment, Development and Sustainability, 9*(3), 283-303. http://dx.doi.org/10.1007/s10668-006-9024-9

Noetstaller, R. (1994). *Small-scale mining, practices, policies, perspectives, In: Small-scale Mining–A Global Overview* (pp. 3-10). In A. K. Ghose (Ed.). New Delhi: Oxford & IBH Publishing Co.

Shrestha, S. H. (2004). *Economic Geography of Nepal* (2nd ed.). Kathmandu: Educational Publishing House.

Sivacoumar, R., Jayabalu, R., Subrahmanyam, Y. V., Jothikumar, N., & Swarnalatha, S. (2012). *Air pollution in stone crushing industry, and associated health effects.* National Environmental Engineering Research Institute. CSIR, Taramani–600 113, INDIA.

Zenz, C., Dickerson, B., & Horvath, E. B. (1994). Occupational medicine, Mosby, St. Louis, 167–236.Gavin Hilson, Pollution prevention and cleaner production in the mining industry: an analysis of current issues. *Journal of Cleaner Production, 8*(2), 119-126.

Notes

Note 1. Village Development Committee (A VDC has a status as an autonomous institution and authority for interacting with the more centralized institutions of governance in Nepal).

Note 2. Child Labour Status Report

Note 3. Brahmā) is the Hindu god (deva) of creation and one of the Trimūrti, the others being Vishṇu and Śiva.

Note 4. Astami is the eighth day (Tithi) of Hindu lunar calendar.

Note 5. SaraswatiJatra (Puja) or Shree Panchami is the day to celebrate the birthday of Saraswati (the Goddess of Learning).

Note 6. A festival of Bhairab, Bhairava, sometimes known as Bhairo or Bhairon, is the fierce manifestation of Shiva associated with annihilation. He is one of the most important deities of Nepal, sacred to Hindus and Buddhists alike. Bhairava is invoked in prayers to destroy enemies.

Note 7. The Newari Festival of Samyak Mahadan occurs once every five years.

Note 8. A local festival of Tika Bhairab; Lele, Nepal.

15

Characteristics of Atmospheric Particulate Matter and Metals in Industrial Sites in Korea

Rupak Aryal[1], Aeri Kim[2], Byeong-Kyu Lee[2], Mohammad Kamruzzaman[1] & Simon Beecham[1]

[1] Centre for Water Management and Reuse, School of Natural and Built Environments, University of South Australia, Mawson Lakes, SA, Australia

[2] Department of Civil and Environmental Engineering, University of Ulsan, Ulsan, Republic of Korea

Correspondence: Rupak Aryal, Centre for Water Management and Reuse, School of Natural and Built Environments, University of South Australia, Mawson Lakes, SA 5095, Australia.

Abstract

The distribution of metals in atmospheric particulates less than 10 μm was studied at a petrochemical refinery site and at a non-ferrous heavy metals industrial site in the city of Ulsan, South Korea in both the summer and fall seasons. The samples were collected with a high volume sampling system equipped with a 9 stage cascade impactor, which effectively separated the particulate matter into 9 size ranges. Total PM_{10} was 59 ± 14 μg/m^3 in summer and 56 ± 18 μg/m^3 in fall at the petrochemical site whereas it was 52 ± 14 μg/m3 in summer and 88 ± 36μg/m^3 in fall at the non-ferrous heavy metals site. The particle size fractionation in less than 10 μm showed a typical bimodal distribution, with one peak corresponding to the particle size range of 1.1-4.7 μm and the other to the range of 4.7-10 μm. Five heavy metals (Ni, Cu, Zn, Pb, and Cd) were measured in the composite mixture of particulates (0.1-1.1, 1.1-4.7 and 4.7-10 μm). The heavy metals concentrations were found to be higher in the 1.1-4.7 μm fraction followed by 4.7-10 and 0.1-1.1 μm. Among the metals Pb showed particle size dependent whereas Zn was homogeneously mixed in all sizes. The obtained data are important for an estimation of level pollution with heavy metals in industrial sites.

Keywords: atmospheric particulate, heavy metals, seasonal variation

1. Introduction

Urban areas often produce atmospheric particulate that lead to numerous health hazards to both humans (Bünger, Krahl, Schröder, Schmidt, & Westphal, 2012; Dockery & Pope, 1994; Olsson & Benner, 1999) and ecosystems (Bell, Samet, & Dominici, 2004a, 2004b; Schwartz, Dockery, & Neas, 1996). Of these hazardous pollutants, particulate matter (PM) with aerodynamic diameter less than or equal to 10 μm (PM_{10}) are particularly important as they can carry many toxic elements within a complex mixture of anthropogenic and naturally occurring airborne particles. They are formed by two basic mechanisms: dispersion and condensation, including chemical reactions. The atmospheric particulates are a mixture of primary and secondary aerosols and usually have bimodal mass distributions (Bashurova, Koutzenogil, Pusep, & Shokhirev, 1991; Karanasiou, Sitaras, Siskos, & Eleftheriadis, 2007; Spurny, 1996). The last decade has seen a plethora of research documenting relations between particulate matter and respiratory diseases, cardiovascular mortality, morbidity and malignant lung diseases (Bräuner et al., 2007; Liu, Ying, Harkema, Sun, & Rajagopalan, 2013; Ostro, Broadwin, Green, Feng, & Lipsett, 2006; Risom, Møller, & Loft, 2005; Simkhovich, Kleinman, & Kloner, 2008). An increase in PM_{10} by a concentration change of 10 μg/m^3 can yield 1% increase in overall mortality and 3-6% increase in deaths associated with respiratory disease (Kan, Wong, Vichit-Vadakan, & Qian, 2010; Ostro, Chestnut, Vichit-Vadakan, & Laixuthai, 1999; Ostro, Hurley, & Lipsett, 1999).

Metals are commonly found in atmospheric particles. While they can be present in almost all sizes of atmospheric particulate, in general, fine particulate carries higher concentrations of toxic metals than coarse particulate (Fang & Huang, 2011; Hieu & Lee, 2010). Metals associated with respirable particles have been shown to increase numerous diseases (Hu et al., 2012; Pandey et al., 2013). Metals in the urban atmosphere are frequently associated with specific pollutant sources, and these are often used as tracers in order to identify the source of atmospheric particulate (Chen et al., 2013; Duan & Tan, 2013; Wang, Bi, Wu, Zhang, & Feng, 2013;

Weckwerth, 2001).

The determination of particle sizes in total PM_{10} and the metals associated with those particles may provide an insight into health hazards for different particle sizes. The coarser particles (aerodynamic diameter > 3 μm) are thought to be formed by high temperature combustion, crustal erosion and road dust resuspension. The finer fractions are believed to be principally emitted from anthropogenic sources including combustion, high-temperature industrial activities and automotive traffic (Dockery & Pope, 1994; Toscano, Moret, Gambaro, Barbante, & Capodaglio, 2011). Knowledge on the size of the particles and their relationship to metals is vital in understanding health hazards in the atmosphere and to develop efficient methods and technologies to tackle these problems.

Numerous studies have been undertaken to estimate PM_{10} and its associated heavy metals in the atmosphere surrounding industrial areas across several continents (Chen, 2007; Lim, Lee, Moon, Chung, & Kim, 2010; Ochsenkühn & Ochsenkühn-Petropoulou, 2008; Razos & Christides, 2010). The dispersion and accumulation of particulate matter in any location is mainly affected by the existing sources, meteorological conditions and local topography. The local wind circulations may change the particulate matter's dispersion and accumulation within a very short distance and this in turn affects pollution levels in a small area from time to time. The city of Ulsan is one of the most industrial regions in South Korea. Previous studies have been undertaken to understand the particulate matter in residential areas in Ulsan (Hieu & Lee, 2010; Lee & Hieu, 2011). However, atmospheric particulate dispersion in industrial zones has received little attention.

The aim of this study was to investigate the atmospheric particulate and associated heavy metals in the summer and fall seasons at a petrochemical (PC) industry and a non-ferrous industry (NFI) sites. In order to study the differences observed in the relevant size distributions, cluster analysis was performed on the aerosol data. The examination of the relationship between the different particle sizes and metals in particles will help in identifying the formation, dispersion and accumulation of particles and metals in two seasons: summer and fall.

2. Method

2.1 Sample Collection

The two sites (PC and NFI) are located in an industrial zone in Ulsan, South Korea (Figure 1). Site PC is approximately 2 km north-west of site NFI. A total of 10 samples were collected from each site for each of two seasons (summer and fall) using a 9 stage cascade impactor (Environmental Tisch, USA). The samples were collected from the roofs of the industrial buildings, 21 m above the street level during summer and fall (2007). The impactors contained 9 filters (0-0.4, 0.4-0.7, 0.7-1.1, 1.1-2.1, 2.1-3.3, 3.3-4.7, 4.7-5.8, 5.8-9.0, and 9.0-10 μm) and were exposed to the atmosphere for 24 h to retrieve daily average PM samples. The samples collected on the filters were kept in desiccators for 48 h at laboratory temperature (22 °C) to minimize the effect of moisture on the filters before and after taking air samples. A gravimetric analysis was performed using a 5-digit microbalance (Mettler, Toledo) to measure the collected mass in each fraction. Meteorological data (temperature, wind speed, wind direction, humidity, cloud cover and precipitation) were collected from the Ulsan Metropolitan Government.

Flow rates were measured in the field using a manometer which sensed the pressure drop across the 9 stages of the impactor. The manometer was calibrated in the laboratory using a positive displacement flow rate meter. In the field, the flow rate was set at an initial value of 28.3 m^3/h. At the end of the measurement (24 h), the corresponding value was found to be 28.2 m^3/h, that corresponded to a deviation of less than 0.35%.

Quantification of the particles collected in cascade impactor for the nine fractions were based on the theoretical impaction curve diagrams of each stage of the ambient cascade impactor provided by the impactor manufacturer. Detailed PM mass fraction calculation processes are mentioned in Table 1. The cascade measurement was verified by the by the β-ray attenuation method at the same sampling site, measured by Ulsan Network of Air Pollution Monitoring (UNAPM). The PM_{10} concentrations obtained in this study using the cascade impactor were significantly correlated ($R^2 = 0.88$) with the PM_{10} concentrations measured.

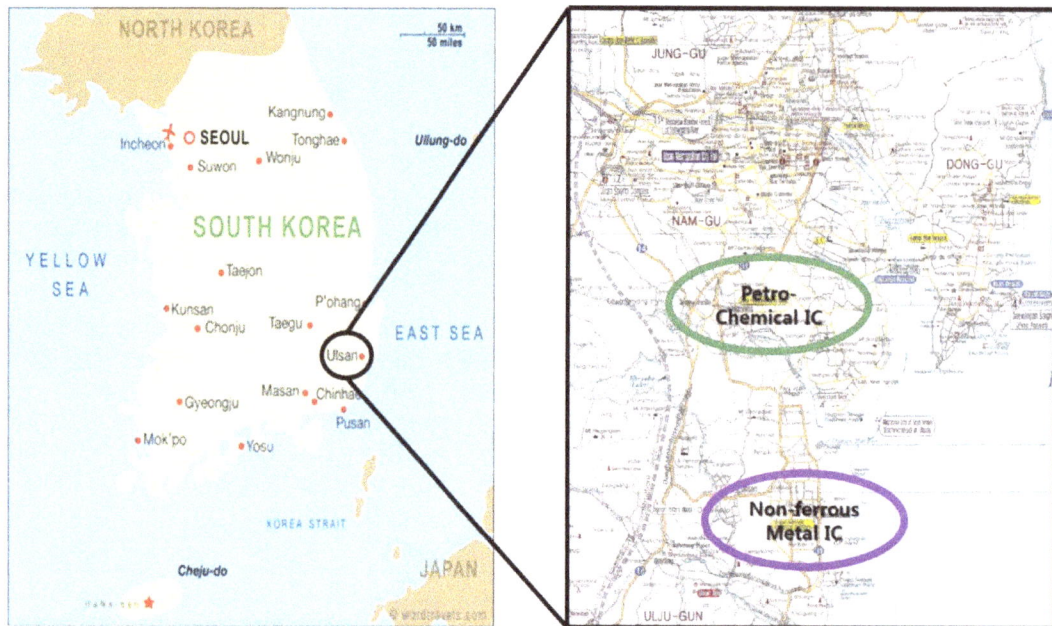

Figure 1. Ulsan map showing PC and NFI sampling sites

Table 1. PM mass calculation

Stage	Size range (µm)	Mass	Calculation formula
0	9.0-10.0	m1	$PM_{10} = m9 + m8 + m7 + m6 + m5 + m4 + m3 + m2 + 0.76 \times m1$
1	5.8-9.0	m2	
2	4.7-5.8	m3	
3	3.3-4.7	m4	$PM_4 = m9 + m8 + m7 + m6 + 0.84 \times m5 + 0.13 \times m4$
4	2.1-3.3	m5	
5	1.1-2.1	m6	
6	0.7-1.1	m7	$PM_1 = m9 + m8 + m7 + 0.44 \times m6$
7	0.4-0.7	m8	
Backup filter	<0.4	m9	

2.2 Meteorological Data Analysis

Meteorological data (wind speed and wind direction) were taken from the network station (located at the sampling site) for air quality monitoring in Ulsan. The windrose diagrams was plotted based on hourly data using Korean software for atmospheric dispersion modeling (Airmaster, version 2.0) provided by Korean Meteorological Administration.

2.3 Heavy Metals Analysis

Due to the low mass in each filter, a total of 9 samples were grouped into three composite samples (0-1.1, 1.1-4.7 and 4.7-10 µm). The composite mixture was dissolved in approximately 30 cc of 1:1 mixture of nitric acid (1.03 M) and perchloric acid (2.23 M) and extracted using an ultrasonic bath (50 kHz) for 1 hr (50% cycle). After extraction, the extract was filtered through glass fiber filters (GF75, Advantec) and diluted. The diluted extract was used to quantify heavy metals using Atomic Absorption Spectroscopy (Varian, AA 240). Five heavy metals (Ni, Cu, Zn, Pb, and Cd) were quantified.

Blank correction was done by measuring the back filters and membrane using the gravimetric method and heavy metals measurement used the method described above. The blank contribution to the total concentration was regularly under 15% for all the elements. Accuracy and precision (repeatability) were measured using Standard Urban Dust Reference Material (NIST, SRM-1648). The recovery was between 80-115%.

2.4 Statistical Analysis

Two statistical methods, Pearson correlation analysis and cluster analysis, were tested to measure the relationship among the variables (particle sizes and seasons). Pearson correlation "r" was investigated by comparing the "r" value. Pearson's "r" is invariant to scale changes where the original data is transformed to standardized form, dividing the distance from the mean by the sample standard deviation. The properties of Pearson's correlation coefficient lie between -1 to $+1$, with 0 indicates no relationship between pairs of variable sand follows "t" distribution with $n-2$ degrees of freedom, where n is the number of samples. To test the significance of correlation (r), where r differs from zero, the test statistic t_r is defined as:

$$t_r = \frac{r\sqrt{(n-2)}}{\sqrt{(1-r^2)}}$$

It was considered 1% level of significance (α) for $r \geq 0.9$ and 5% level of significance for $0.7 \leq r \leq 0.89$. We used a cumulative sum (CUSUM) method (Kamruzzaman, Beecham, & Metcalfe, 2011), to test whether there was an evidence of relative changes under the mean or not. The CUSUM at time n was calculated as:

$$C_t = \sum_{t=1}^{n} (x_i - \bar{x})$$

where x_i represent a sample size, $\bar{x}$ is the mean of the sample of length n. It is noted that if consecutive values tend to lie below the mean, then C_t will have a negative slope and if consecutive values tend to lie above the mean then C_t will have a positive slope.

Cluster analysis (CA) was performed to classify particles sizes on the basis of the similarities of their concentration. Hierarchical cluster analysis, used in this study, assisted in identifying relatively homogeneous groups of variables (particles sizes), using an algorithm that starts with each variable in a separate cluster and combines clusters until only one is left. A dendrogram was plotted to assess the distance of the clusters formed (Aryal, Lebegue, Vigneswaran, Kandasamy, & Grasmick, 2009).

3. Results

3.1 Particle Size Distribution

The total PM_{10} ranges were 59 ± 14 µg/m³in summer and 66 ± 18 µg/m³ in fall at the PC site and 52 ± 14 µg/m³ in summer and 88 ± 36 µg/m³ in fall at the NFI site. The result shows that PM_{10} concentration in fall was relatively higher than in summer. The result shows although the value did not exceed the daily average guideline 100 µg/m³ but exceeded annual average guideline 100 µg/m³ (Ministry of Environment, 2007). By the beginning of fall, the ambient temperature drops down. Increase in PM_{10} in with lowering atmospheric temperature has been reported in other cities (Aryal et al., 2008; Ho et al., 2006; Kulshrestha, Satsangi, Masih, & Taneja, 2009) and in Ulsan residential area also (Hieu & Lee, 2010; Lee & Park, 2010). Table 2 shows PM_{10} observed in other industrial sites across the world. The results show that the PM_{10} observed in this study at both sites were within the range of reported values in the literature (Fang et al., 2000; Karar, Gupta, Kumar, & Biswas, 2006).

Figure 2 shows PM_{10} distributionin both sites and seasons. The figure indicated that there was difference in PM_{10} in summer and fall on both sites not only in total but also in fractions (particle sizes). To understand the relationship among particles within sites among two seasons, we calculated Pearson Correlation "r". Table 3 shows the correlation coefficient for particles (sizes) within sites and between sites.

Table 2. PM$_{10}$ value observed in other industrial sites

Site	Features	Concentration (µg/m^3)
La Coruna, Spain (Beceiro-González, Andrade-Garda, Serrano-Velasco, & López-Mahía, 1997)	Industrial site	25
Taejon, Korea (Kim, Lee, & Jang, 2002)	Industrial site	72
Barcelona, Spain (Rodríguez et al., 2004)	Industrial site	33
Kolkatta, India (Karar & Gupta, 2006)	Industrial site	62-401
Tito Scalo, Italy (Ragosta et al., 2006)	Industrial site	24
Dunkirk, France (Alleman, Lamaison, Perdrix, Robache, & Galloo, 2010)	Industrial site	14-36
Basque country, Spain (Viana, Querol, Alastuey, Gangoiti, & Menéndez, 2003)	Urban area near to industry	34
San Joaquin valley, CA, USA (Chow et al., 1999)	Urban area near to industry	50

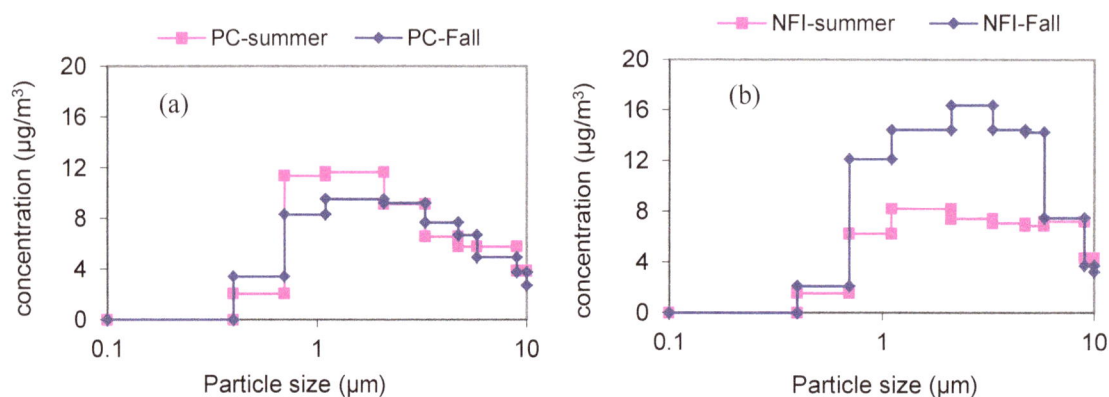

Figure 2. Particle size distributions for (a) PC and (b) NFI areas

Table 3. Relationship between particle sizes

	PC in Summer	PC in Fall	NFI in Summer	NFI in Fall
PC in Summer	1			
PC in Fall	0.91*	1		
NFI in Summer	0.77**	0.81*	1	
NFI in Fall	0.77**	0.94*	0.86*	1

*Coefficients are statistically significant at 1% level.

**Coefficients are statistically significant at 5% level.

The statistic showed that PM$_{10}$ had significant correlation in PM$_{10}$ between summer and fall (explain with what?) (r = 0.91 for PC site and 0.86 for NFI site with level of significance 1%). Among the sites (PC and NFI), the Pearson 'r' was 0.77 with level of significance were at 5%. Moreover, we observed significant mean difference (t statistics) and noticeable relative changes at 5.61 µm with p = 0.000803 for PC sites and at 4.47 µm with p = 0.002903 for NFI sites for summer and fall, when we used the cumulative sum method. Figure 3 shows relative change in particle sizes in PC and NFI in summer and fall. The solid line (vertical) shows where maximum noticeable change occurs in PM fractions in summer and fall.

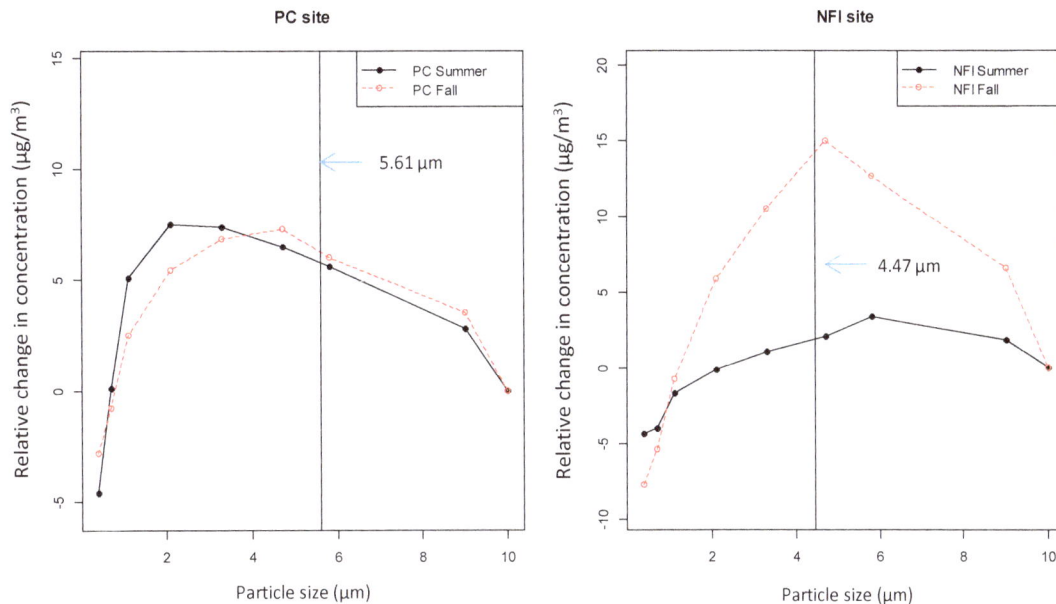

Figure 3. Particle size distributions for (a) PC and (b) NFI areas

Noticeable changes in particle size distribution for PC and for NFI in summer and fall season suggests that agglomeration of particles and/or intrusion of particles from surrounding areas may have occurred in fall. To understand the possible intrusion from surrounding sites, we plotted wind rose diagram for summer and fall (Figure 4). The wind direction in the summer is from southeast to northwest (sea to land). During fall, as the temperature drops down and the prevailing wind flows from north to south. Since NFI is at downstream of the industrial zone, two possible assumptions were thought for higher concentration of PM in NFI during fall: i) the northwesterly wind brought particulates from the upstream to NFI and ii) NFI produced more PM in fall.

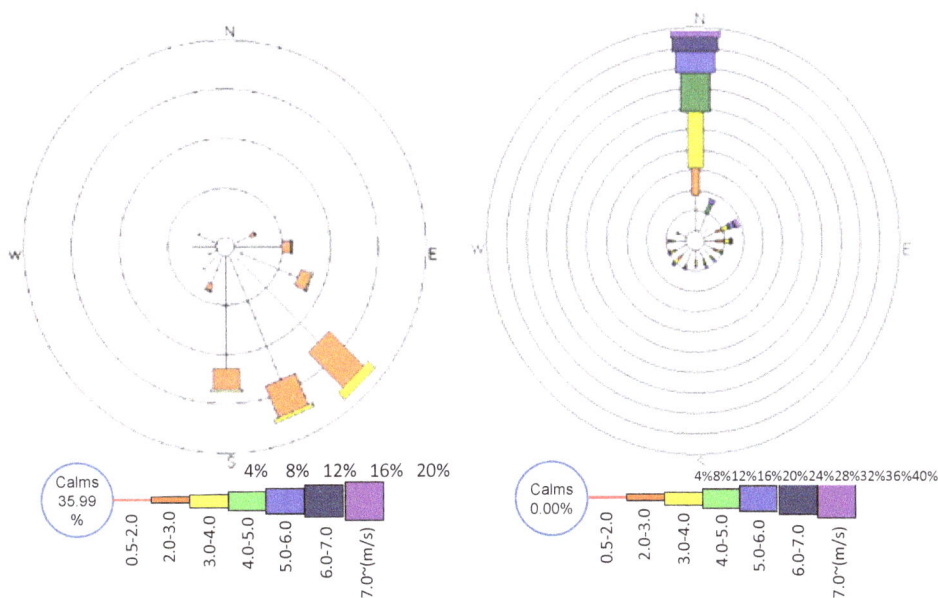

Figure 4. Windrose diagram at sampling site for summer (left) and fall (right)

To understand the particle size relationship, we plotted dendrograms using a cluster analysis. This method of analysis is described in (Aryal et al., 2009). Figure 5 shows dendrogram plots for particles collected at the PC and NFI sites in both summer and fall. The dendrograms show that the particle size distribution is mainly bimodal, with one cluster corresponding to the finer fraction 1.1-4.7 μm and the other cluster to the coarse mode 4.7-10 μm, shown by dotted boxes. Similar bimodal distributions have been reported earlier (Karanasiou et al., 2007; Toscano et al., 2011). Our observations show a close relationship between the ultrafine particles (<1.1 μm) and the coarser fraction (>5.8 μm). The relationship between ultrafine and coarse fractions suggests the possibility of similar sources and/or modes of release.

3.2 Heavy Metals Distribution in Particles

Five heavy metals (Ni, Cu, Zn, Pb, and Cd) were measured. Since the concentrations of the particles in the samples were low, composite samples were made by mixing nine different sizes to three fractions (10-4.7, 4.7-1.1 and 0.1-1.1 μm). Figure 6 shows the pseudo total heavy metals concentrations (μg·m^{-3}) in the total particulate (PM$_{10}$) calculated as mass fractions. Pseudo total heavy metal concentration refers to the sum of measurement of individual fractions. Out of total metals abundance, Zn abundance is more than 86%. It is seen that while the total particulate matter concentration increased in fall, the overall heavy metal content decreased except for Ni at NFI site. Similar and higher concentration of heavy metal contents in particulate matter in summer has been reported in literatures (Espinosa, Ternero Rodríguez, De La Rosa, & Sánchez, 2001; Gupta, Salunkhe, & Kumar, 2010; Querol et al., 2006; Razos & Christides, 2010; Shah, Shaheen, & Jaffar, 2006; Smichowski et al., 2004).

To know the possible reasons for high value of Zn and Ni in fall than summer we plotted the graph between fractional concentration and total concentration as shown in Figure 7. It can be assumed that if the sources for both seasons are similar then fractional distribution of metals with respect to total will also be similar whereas inclusion of surrounding sources changes the relationship to good or poor depending upon incursion amount. The graph showed that relationship between fractional metal concentrations with total are more linear in summer than in fall. It is to be noted that Zn concentration became slightly higher in fall than in summer (less than 2 fold) whereas Ni concentration increased almost 6 folds fold in fall than in summer in NFI. Table 3 shows the estimated fraction particle size effect and filled regression model. Due to small decrease in concentration for Zn R^2 value was relatively lower in fall than in summer. For Ni strong increase of its concentration, R^2 value changed significantly. This led us to conclude that Zn and Ni were added to the atmosphere in PC and NFI from surrounding during fall season.

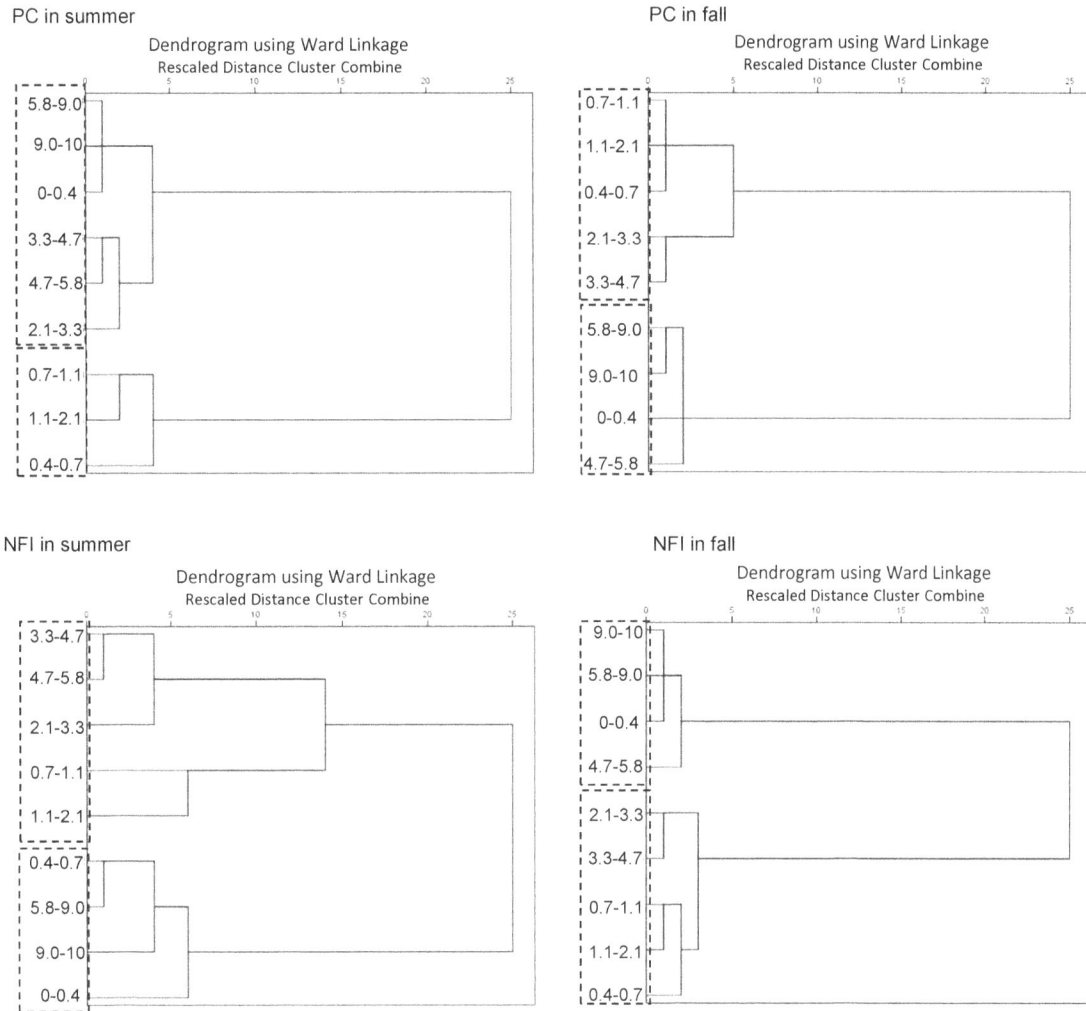

Figure 5. Particle size distributions and their relationships

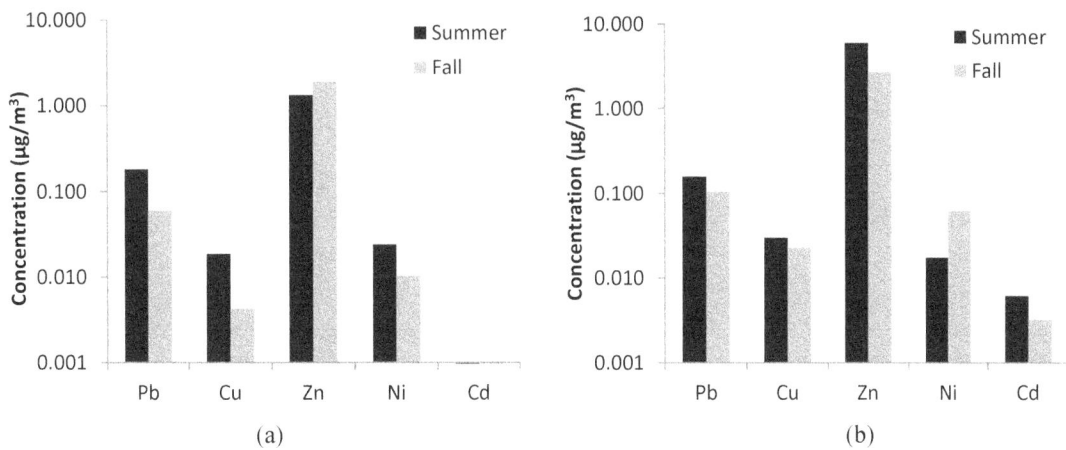

Figure 6. Pseudo total metal concentrations ($\mu g/m^3$) in total particulate matter in (a) PC and (b) NFI

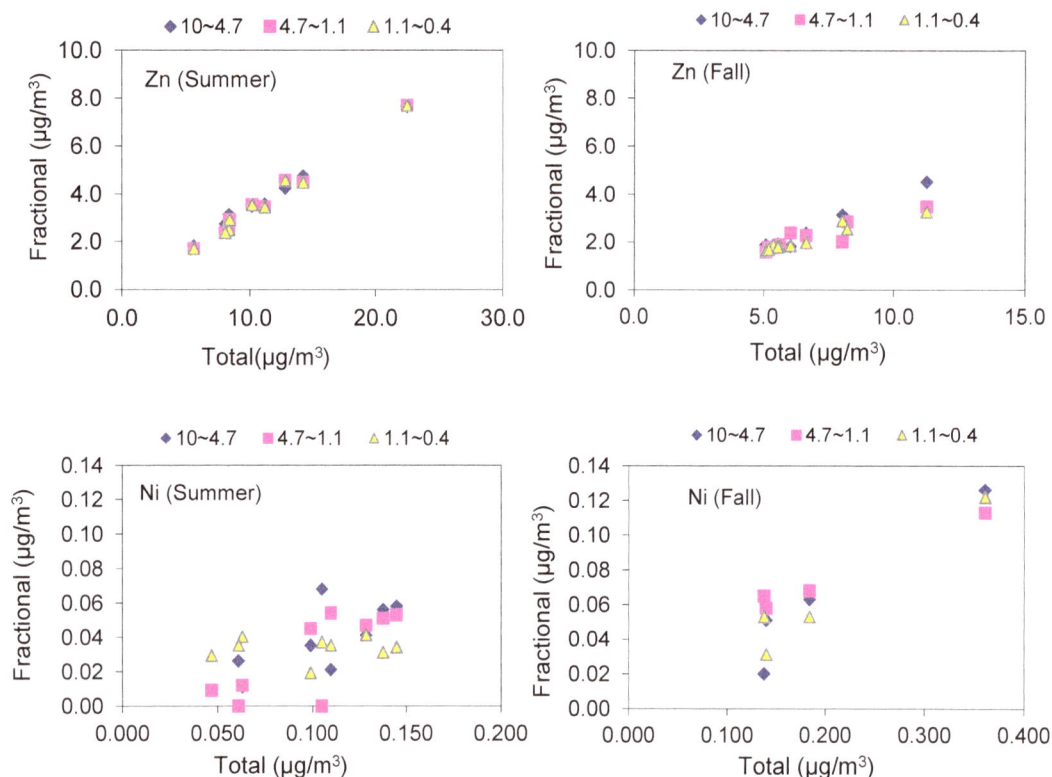

Figure 7. Relationship between fractional concentration with total for Zn (in PC) and Ni (in NFI)

4. Conclusions

Atmospheric particulate matter and associated heavy metals in less than 10 μm was investigated at two industrial sites in Ulsan, South Korea for two seasons were analyzed. The following conclusions can be drawn from this work:

Total PM_{10} ranged from 59 ± 14 μg/m^3 in summer and 56 ± 18 μg/m^3 in fall for the petrochemical (PC) site whereas for the non-ferrous heavy metals (NFI) site it was around 52 ± 14 μg/m^3 in summer and 88 ± 36 μg/m^3 in fall. An increase of PM_{10} concentration at the NFI site in fall indicated that this area possibly receives particulates from surrounding areas.

An increase of 0.4-5.8 μm particle size in fall at the NFI site indicates that the area may have received particulate matter from other areas.

The heavy metals concentration was relatively higher in the summer than in fall (except for Zn in PC and Ni in NFI) and the change in concentration was possibly due to contribution from surrounding sources.

References

Alleman, L. Y., Lamaison, L., Perdrix, E., Robache, A., & Galloo, J. C. (2010). PM_{10} metal concentrations and source identification using positive matrix factorization and wind sectoring in a French industrial zone. *Atmospheric Research, 96*(4), 612-625. http://dx.doi.org/10.1016/j.atmosres.2010.02.008

Aryal, R., Lebegue, J., Vigneswaran, S., Kandasamy, J., & Grasmick, A. (2009). Identification and characterisation of biofilm formed on membrane bio-reactor. *Separation and Purification Technology, 67*(1), 86-94. http://dx.doi.org/10.1016/j.seppur.2009.03.031

Aryal, R. K., Lee, B. K., Karki, R., Gurung, A., Kandasamy, J., Pathak, B. K., . . . Giri, N. (2008). Seasonal PM_{10} dynamics in Kathmandu Valley. *Atmospheric Environment, 42*(37), 8623-8633. http://dx.doi.org/10.1016/j.atmosenv.2008.08.016

Bünger, J., Krahl, J., Schröder, O., Schmidt, L., & Westphal, G. A. (2012). Potential hazards associated with combustion of bio-derived versus petroleum-derived diesel fuel. *Critical Reviews in Toxicology, 42*(9), 732-750. http://dx.doi.org/10.3109/10408444.2012.710194

Bashurova, V. S., Koutzenogil, K. P., Pusep, A. Y., & Shokhirev, N. V. (1991). Determination of atmospheric aerosol size distribution functions from screen diffusion battery data: Mathematical aspects. *Journal of Aerosol Science, 22*(3), 373-388. http://dx.doi.org/10.1016/S0021-8502(05)80014-X

Beceiro-González, E., Andrade-Garda, J. M., Serrano-Velasco, E., & López-Mahía, P. (1997). Metals in airborne particulate matter in La Coruna (NW Spain). *Science of the Total Environment, 196*(2), 131-139. http://dx.doi.org/10.1016/S0048-9697(96)05412-5

Bell, M. L., Samet, J. M., & Dominici, F. (2004a). *Time-series studies of particulate matter.* (Vol. 25, pp. 247-280).

Bell, M. L., Samet, J. M., & Dominici, F. (2004b). Time-Series Studies of Particulate Matter. *Annual Review of Public Health, 25*(1), 247-280. http://dx.doi.org/10.1146/annurev.publhealth.25.102802.124329

Bräuner, E. V., Forchhammer, L., Möller, P., Simonsen, J., Glasius, M., Wåhlin, P., . . . Loft, S. (2007). Exposure to ultrafine particles from ambient air and oxidative stress-induced DNA damage. *Environmental Health Perspectives, 115*(8), 1177-1182. http://dx.doi.org/10.1289/ehp.9984

Chen, B., Stein, A. F., Maldonado, P. G., Sanchez de la Campa, A. M., Gonzalez-Castanedo, Y., Castell, N., & de la Rosa, J. D. (2013). Size distribution and concentrations of heavy metals in atmospheric aerosols originating from industrial emissions as predicted by the HYSPLIT model. *Atmospheric Environment, 71*, 234-244. http://dx.doi.org/10.1016/j.atmosenv.2013.02.013

Chen, H. W. (2007). Characteristics and risk assessment of trace metals in airborne particulates from a semiconductor industrial area of Northern Taiwan. *Fresenius Environmental Bulletin, 16*(10), 1288-1294.

Chow, J. C., Watson, J. G., Lowenthal, D. H., Hackney, R., Magliano, K., Lehrman, D., & Smith, T. (1999). Temporal variations of $PM_{2.5}$, PM_{10}, and gaseous precursors during the 1995 integrated monitoring study in Central California. *Journal of the Air and Waste Management Association, 49*(SPEC. ISS.), 16-24. http://dx.doi.org/10.1080/10473289.1999.10463909

Dockery, D. W., & Pope, C. A. (1994). Acute Respiratory Effects of Particulate Air Pollution. *Annual Review of Public Health, 15*(1), 107-132. http://dx.doi.org/10.1146/annurev.pu.15.050194.000543

Duan, J., & Tan, J. (2013). Atmospheric heavy metals and Arsenic in China: Situation, sources and control policies. *Atmospheric Environment, 74*, 93-101. http://dx.doi.org/10.1016/j.atmosenv.2013.03.031

Espinosa, A. J. F., Ternero Rodriguez, M., De La Rosa, F. J. B., & Sánchez, J. C. J. (2001). Size distribution of metals in urban aerosols in Seville (Spain). *Atmospheric Environment, 35*(14), 2595-2601. http://dx.doi.org/10.1016/S1352-2310(00)00403-9

Fang, G. C., Chang, C. N., Wu, Y. S., Wang, V., Fu, P. P. C., Yang, D. G., . . . Chu, C. C. (2000). The study of fine and coarse particles, and metallic elements for the daytime and night-time in a suburban area of central Taiwan, Taichung. *Chemosphere, 41*(5), 639-644. http://dx.doi.org/10.1016/S0045-6535(99)00507-X

Fang, G. C., & Huang, C. S. (2011). Atmospheric particulate and metallic elements (Zn, Ni, Cu, Cd and Pb) size distribution at three characteristic sampling sites. *Environmental Forensics, 12*(3), 191-199. http://dx.doi.org/10.1080/15275922.2011.595052

Gupta, I., Salunkhe, A., & Kumar, R. (2010). Modelling 10-year trends of PM_{10} and related toxic heavy metal concentrations in four cities in India. *Journal of Hazardous Materials, 179*(1-3), 1084-1095. http://dx.doi.org/10.1016/j.jhazmat.2010.03.117

Hieu, N. T., & Lee, B. K. (2010). Characteristics of particulate matter and metals in the ambient air from a residential area in the largest industrial city in Korea. *Atmospheric Research, 98*(2-4), 526-537. http://dx.doi.org/10.1016/j.atmosres.2010.08.019

Ho, K. F., Lee, S. C., Cao, J. J., Chow, J. C., Watson, J. G., & Chan, C. K. (2006). Seasonal variations and mass closure analysis of particulate matter in Hong Kong. *Science of the Total Environment, 355*(1-3), 276-287. http://dx.doi.org/10.1016/j.scitotenv.2005.03.013

Hu, X., Zhang, Y., Ding, Z., Wang, T., Lian, H., Sun, Y., & Wu, J. (2012). Bioaccessibility and health risk of arsenic and heavy metals (Cd, Co, Cr, Cu, Ni, Pb, Zn and Mn) in TSP and PM2.5 in Nanjing, China. *Atmospheric Environment, 57*, 146-152. http://dx.doi.org/10.1016/j.atmosenv.2012.04.056

Kamruzzaman, M., Beecham, S., & Metcalfe, A. V. (2011). Non-stationarity in rainfall and temperature in the Murray Darling Basin. *Hydrological Processes, 25*(10), 1659-1675. http://dx.doi.org/10.1002/hyp.7928

Kan, H., Wong, C. M., Vichit-Vadakan, N., & Qian, Z. (2010). Short-term association between sulfur dioxide and daily mortality: The Public Health and Air Pollution in Asia (PAPA) study. *Environmental Research, 110*(3), 258-264. http://dx.doi.org/10.1016/j.envres.2010.01.006

Karanasiou, A. A., Sitaras, I. E., Siskos, P. A., & Eleftheriadis, K. (2007). Size distribution and sources of trace metals and n-alkanes in the Athens urban aerosol during summer. *Atmospheric Environment, 41*(11), 2368-2381. http://dx.doi.org/10.1016/j.atmosenv.2006.11.006

Karar, K., & Gupta, A. K. (2006). Seasonal variations and chemical characterization of ambient PM_{10} at residential and industrial sites of an urban region of Kolkata (Calcutta), India. *Atmospheric Research, 81*(1), 36-53. http://dx.doi.org/10.1016/j.atmosres.2005.11.003

Karar, K., Gupta, A. K., Kumar, A., & Biswas, A. K. (2006). Seasonal variations of PM_{10} and TSP in residential and industrial sites in an urban area of Kolkata, India. *Environmental Monitoring and Assessment, 118*(1-3), 369-381. http://dx.doi.org/10.1007/s10661-006-1503-9

Kim, K. H., Lee, J. H., & Jang, M. S. (2002). Metals in airborne particulate matter from the first and second industrial complex area of Taejon city, Korea. *Environmental Pollution, 118*(1), 41-51. http://dx.doi.org/10.1016/S0269-7491(01)00279-2

Kulshrestha, A., Satsangi, P. G., Masih, J., & Taneja, A. (2009). Metal concentration of PM2.5 and PM_{10} particles and seasonal variations in urban and rural environment of Agra, India. *Science of the Total Environment, 407*(24), 6196-6204. http://dx.doi.org/10.1016/j.scitotenv.2009.08.050

Lee, B. K., & Hieu, N. T. (2011). Seasonal Variation and Sources of Heavy Metals in Atmospheric Aerosols in a esidential Area of Ulsan, Korea. *Aerosol and Air Quality Research, 11*(6), 679-688.

Lee, B. K., & Park, G. H. (2010). Characteristics of heavy metals in airborne particulate matter on misty and clear days. *Journal of Hazardous Materials, 184*(1-3), 406-416. http://dx.doi.org/10.1016/j.jhazmat.2010.08.050

Lim, J. M., Lee, J. H., Moon, J. H., Chung, Y. S., & Kim, K. H. (2010). Airborne PM_{10} and metals from multifarious sources in an industrial complex area. *Atmospheric Research, 96*(1), 53-64. http://dx.doi.org/10.1016/j.atmosres.2009.11.013

Liu, C., Ying, Z., Harkema, J., Sun, Q., & Rajagopalan, S. (2013). Epidemiological and experimental links between air pollution and type 2 diabetes. *Toxicologic Pathology, 41*(2), 361-373. http://dx.doi.org/10.1177/0192623312464531

Ministry of Environment, R. o. K. (2007). *Air Korea.* Retrieved from http://eng.me.go.kr/content.do?method=moveContent&menuCode=pol_cha_air_sta_standards

Ochsenkühn, K. M., & Ochsenkühn-Petropoulou, M. (2008). Heavy metals in airborne particulate matter of an industrial area in Attica, Greece and their possible sources. *Fresenius Environmental Bulletin, 17*(4), 455-462.

Olsson, P. Q., & Benner, R. L. (1999). Atmospheric Chemistry and Physics: From Air Pollution to Climate Change By John H. Seinfeld (California Institute of Technology) and Spyros N. Pandis (Carnegie Mellon University). Wiley-VCH: New York. 1997. $89.95. xxvii + 1326 pp. ISBN 0-471-17815-2. *Journal of the American Chemical Society, 121*(6), 1423-1423. http://dx.doi.org/10.1021/ja985605y

Ostro, B., Broadwin, R., Green, S., Feng, W. Y., & Lipsett, M. (2006). Fine particulate air pollution and mortality in nine California counties: Results from CALFINE. *Environmental Health Perspectives, 114*(1), 29-33. http://dx.doi.org/10.1289/ehp.8335

Ostro, B., Chestnut, L., Vichit-Vadakan, N., & Laixuthai, A. (1999). The impact of particulate matter on daily mortality in Bangkok, Thailand. *Journal of the Air and Waste Management Association, 49*(SPEC. ISS.), 100-107. http://dx.doi.org/10.1080/10473289.1999.10463875

Ostro, B. D., Hurley, S., & Lipsett, M. J. (1999). Air pollution and daily mortality in the Coachella Valley, California: a study of PM_{10} dominated by coarse particles. *Environmental Research, 81*(3), 231-238. http://dx.doi.org/10.1006/enrs.1999.3978

Pandey, P., Patel, D. K., Khan, A. H., Barman, S. C., Murthy, R. C., & Kisku, G. C. (2013). Temporal distribution of fine particulates (PM 2.5, PM 10), potentially toxic metals, PAHs and Metal-bound carcinogenic risk in the population of Lucknow City, India. *Journal of Environmental Science and Health - Part A Toxic/Hazardous Substances and Environmental Engineering, 48*(7), 730-745. http://dx.doi.org/10.1080/10934529.2013.744613

Querol, X., Zhuang, X., Alastuey, A., Viana, M., Lv, W., Wang, Y., . . . Xu, S. (2006). Speciation and sources of atmospheric aerosols in a highly industrialised emerging mega-city in Central China. *Journal of Environmental Monitoring, 8*(10), 1049-1059. http://dx.doi.org/10.1039/b608768j

Ragosta, M., Caggiano, R., D'Emilio, M., Sabia, S., Trippetta, S., & Macchiato, M. (2006). PM_{10} and heavy metal measurements in an industrial area of southern Italy. *Atmospheric Research, 81*(4), 304-319. http://dx.doi.org/10.1016/j.atmosres.2006.01.006

Razos, P., & Christides, A. (2010). An investigation on heavy metals in an industrial area in Greece. *International Journal of Environmental Research, 4*(4), 785-794.

Risom, L., Møller, P., & Loft, S. (2005). Oxidative stress-induced DNA damage by particulate air pollution. *Mutation Research - Fundamental and Molecular Mechanisms of Mutagenesis, 592*(1-2), 119-137. http://dx.doi.org/10.1016/j.mrfmmm.2005.06.012

Rodríguez, S., Querol, X., Alastuey, A., Viana, M. M., Alarcón, M., Mantilla, E., & Ruiz, C. R. (2004). Comparative PM_{10}-$PM_{2.5}$ source contribution study at rural, urban and industrial sites during PM episodes in Eastern Spain. *Science of the Total Environment, 328*(1-3), 95-113. http://dx.doi.org/10.1016/S0048-9697(03)00411-X

Schwartz, J., Dockery, D. W., & Neas, L. M. (1996). Is daily mortality associated specifically with fine particles? *Journal of the Air and Waste Management Association, 46*(10), 927-939. http://dx.doi.org/10.1080/10473289.1996.10467528

Shah, M. H., Shaheen, N., & Jaffar, M. (2006). Characterization, source identification and aportionment of selected metals in TSP in an urban atmosphere. *Environmental Monitoring and Assessment, 114*(1-3), 573-587. http://dx.doi.org/10.1007/s10661-006-4940-6

Simkhovich, B. Z., Kleinman, M. T., & Kloner, R. A. (2008). Air Pollution and Cardiovascular Injury. Epidemiology, Toxicology, and Mechanisms. *Journal of the American College of Cardiology, 52*(9), 719-726. http://dx.doi.org/10.1016/j.jacc.2008.05.029

Smichowski, P., Gómez, D. R., Dawidowski, L. E., Giné, M. F., Bellato, A. C. S., & Reich, S. L. (2004). Monitoring trace metals in urban aerosols from Buenos Aires city. Determination by plasma-based techniques. *Journal of Environmental Monitoring, 6*(4), 286-294. http://dx.doi.org/10.1039/b312446k

Spurny, K. R. (1996). Aerosol air pollution its chemistry and size dependent health effects. *Journal of Aerosol Science, 27, Supplement 1*(0), S473-S474. http://dx.doi.org/10.1016/0021-8502(96)00309-6

Toscano, G., Moret, I., Gambaro, A., Barbante, C., & Capodaglio, G. (2011). Distribution and seasonal variability of trace elements in atmospheric particulate in the Venice Lagoon. *Chemosphere, 85*(9), 1518-1524. http://dx.doi.org/10.1016/j.chemosphere.2011.09.045

Viana, M., Querol, X., Alastuey, A., Gangoiti, G., & Menéndez, M. (2003). PM levels in the Basque Country (Northern Spain): Analysis of a 5-year data record and interpretation of seasonal variations. *Atmospheric Environment, 37*(21), 2879-2891. http://dx.doi.org/10.1016/S1352-2310(03)00292-9

Wang, Q., Bi, X. H., Wu, J. H., Zhang, Y. F., & Feng, Y. C. (2013). Heavy metals in urban ambient PM_{10} and soil background in eight cities around China. *Environmental Monitoring and Assessment, 185*(2), 1473-1482. http://dx.doi.org/10.1007/s10661-012-2646-5

Weckwerth, G. (2001). Verification of traffic emitted aerosol components in the ambient air of Cologne (Germany). *Atmospheric Environment, 35*(32), 5525-5536. http://dx.doi.org/10.1016/S1352-2310(01)00234-5

16

Estimation of Spatial Variations in Urban Noise Levels with a Land Use Regression Model

Sophie Goudreau[1], Céline Plante[1], Michel Fournier[1], Allan Brand[2], Yann Roche[3] & Audrey Smargiassi[2,4]

[1] Direction de santé publique de Montréal, Montréal, Canada

[2] Institut National de Santé Publique du Québec, Montréal, Canada

[3] Département de Géographie, Université du Québec à Montréal, Montréal, Canada

[4] Département de santé environnementale et de santé au travail, Université de Montréal, Montréal, Canada

Correspondence: Audrey Smargiassi, Département de santé environnementale et de santé au travail, Université de Montréal, Montréal, Québec, H3C 3J7, Canada.

Abstract

Background: Outdoor noise is a source of annoyance and health problems in cities worldwide. **Objective:** We developed a land-use regression using a GAM Model to estimate the spatial variation of noise levels in Montreal.

Methods: Noise levels were measured over a two week period during the summer of 2010 at 87 sites and during the winter of 2011 at 62 sites. A land use regression model was produced for both seasons to estimate noise levels as LA_{eq24h} (resolution of 20 m). A leave one out cross-validation (LOOCV) was performed.

Results: LA_{eq24h} measured range from 53.4 to 73.7 dBA for the summer and from 54.1 to 77.7 dBA for the winter. The land use regression models explained **64** % of spatial variability for the summer and **40** % for the winter. The main predictors are the Normalized Difference Vegetation Index; the length of vehicular arteries, highways, and bus lines; and the proximity to an international airport. The Root mean square error from the LOOCV was 3.3 and 4.5 dBA for the summer and the winter respectively.

Conclusion: The model explained a large part of the variability in noise levels and the RMSE remain relatively important on the noise levels scale.

Keywords: environmental noise, GAM models, land use regression, spatial variation

1. Introduction

Prolonged exposure to high environmental noise levels has been associated with annoyance, learning difficulties in children, sleep disturbances, hypertension, and other cardiovascular outcomes (Basner et al., 2014; Sørensen et al., 2012; Huss et al., 2010; Basner et al., 2006; Jarup et al., 2005; Stansfeld et al., 2005).

Spatial variations of noise levels have been observed in urban areas (Zuo et al., 2014; Seto et al., 2007; Alberola et al., 2005). These spatial differences have been associated with main emission sources and with characteristics of the built environment (Zuo et al., 2014). Main emission sources of noise in urban environments include road, aerial, and railroad traffic; public and construction works; industries, domestic noise, and noise from leisure activities (Berglund et al., 1999). Among these sources, road traffic is the greatest contributor in urban settings (Makarewicz, 1993).

Sound waves are attenuated by their geometrical spread and their atmospheric absorption. Sound wave propagation is also influenced by its absorption and reflection off different surface types, by topography, and by meteorological parameters like wind speed, direction and temperature (Piercy et al., 1977). Dense vegetation decreases noise levels through the absorption of sound waves and by disrupting their spread (Reethof, 1973).

In order to estimate environmental noise exposure of populations, a number of approaches can be used. Personal exposure measurement is the most accurate method to assess exposure but it cannot be used to assess the exposure of large populations (Zou et al., 2009). Numerical models have frequently been used to estimate noise level over a large territory, for example in Europe to produce maps of daily noise levels for large communities, in order to meet the directive requirements of the 2002 European Union Directive (Directive 2002/49/CE). Such models estimate the propagation of sound waves from emission sources, their attenuation, and subsequently noise levels over a territory. Various software are available to estimate noise levels from emission sources, such as CadnaA®, Mithra-SIG®, FHWA, CORTN (Xie et al., 2011) and SoundPlan® (Guedes et al., 2011); however their use is somewhat limited due to cost and sometimes unrealistic assumptions about dispersion patterns (Xie et al., 2011; Jerrett et al., 2005).

Land Use Regression models (statistical models, LUR), used extensively in the field of air pollution, can overcome some of the limitations of numerical models, and can easily be developed to predict noise exposure levels over a large territory. Nonetheless, LUR model have rarely been used to estimate noise levels. In LUR models, geo-referenced predictors of noise levels, available for a large territory, are used to develop an equation to predict measured noise levels. This equation is then applied to areas where noise levels were not measured in order to generate noise maps.

The objective of this study was to develop LUR models to estimate the spatial variation of noise levels on the Island of Montreal.

2. Materials and Methods

Our study site was the city of Montreal, which is the second largest city in Canada. Its population is over 1.8 million in 2006 (Statistique Canada, 2006) and covers approximately 500 km^2. The sections below describe the noise sampling campaigns over Montreal, the variables used to predict noise levels, and the statistical models developed.

2.1 Measurements of Noise Levels

Two sampling campaigns were conducted: the first campaign over a summer two-week period in 2010 (August 11th to 24th), and the second over two periods in the winter of 2011 (February 26th to March 12th, and March 12th to April 3rd). We recorded two minute noise levels continuously in decibels using a type A filter (dBA), with the Sound Level Meter and logger Noise Sentry (Paillard, 2010). According to the manufacturer, the operating temperature range of the device is between -20 and 70°C. Noise meters were covered with small zipper storage bags to protect them from the rain. Field testing showed minimal influence of the bag, even under windy and rainy conditions (data not shown).

Sampling took place at locations used to develop traffic air pollution LUR models. Locations were selected with a population-weighted location-allocation model. This model situated samplers in areas likely to have high spatial variability in traffic intensity and in population density (see Kanaroglou et al., 2005 for model description). One hundred samplers were installed at the selected locations during the summer period and 81 during the winter sampling campaign (see map in appendix). All 81 winter sites were also sampled in the previous summer sampling period. Measurements were taken concurrently for all sites during the summer, while during winter sampling about half of the sites were measured concurrently.

The samplers were installed at a height of 2.5 m above ground and were attached to street light poles, hydro-electric poles, or parking signs, usually near the sidewalk of the closest road. The samplers were installed at least 10 cm from the poles or signs. The geographic coordinates of each sampling location were recorded using Garmin eTREx Legend Cx Global Positioning System devices.

2.2 Predictors of Monitored Noise Level

In total, 69 variables of the built environment were considered as candidate predictors of noise levels. Variables included were related to vegetation cover, roads, bus and train networks, airport, industrial, commercial and residential land use zones, density of residential units and of non-residential buildings, and divisions of the Island of Montreal. Predictors were calculated as lengths, distances, density or proportion of areas for five different buffers (50, 100, 150, 500 and 750 m) for most variable of interest (see below). These variables were created in ArcGIS 9.2 and postGIS 1.5 for PostGresSQL 9.1.

Vegetation (one variable)

The Normalized Difference Vegetation Index (NDVI) was calculated using a Landsat 5TM (Montréal: 014-028) image taken on the 27th of July 2010 with a resolution of 30 m (Jackson & Huete, 1991). NDVI was the only raster-based variable included, as all other variables were produced as vectors. Sampling locations were assigned the value of the pixel at their coordinates.

Land use zone densities (35 variables)

Seven land use density variables were calculated for the five buffers using the most recent Land Use file available for Montreal (Cartographie d'utilisation du sol, version 2012, Communauté Métropolitaine de Montréal). Variables were the total area (in meters) of each of the following land uses: industrial, office building, commercial or residential (low, medium-low, medium and high density). .

Density of residential units and of non-residential buildings (10 variables)

The number of residential units and the number of non-residential buildings were calculated for each buffer (n=5) using the Montreal Island property assessment database (Rôle de l'évaluation foncière de la Ville de Montréal, 2011).

Road network (5 variables)

The length (in meters) of arteries and highways (collecting most road traffic) was calculated for the five buffers, using the Adresses Québec 2013 (© Gouvernement du Québec, Ministère des Ressources naturelles).

Bus network (10 variables)

The length (in meters) of bus lines and the number of bus stop was calculated for the five buffers. The Montreal Transport Society (STM) bus network file (2012) was used to calculate these variables.

Train network (5 variables)

The length (in meters) of train lines was calculated for the different buffers (Base de données topographiques du Québec 1:20000).

Airport (one variable)

We geo-referenced the Noise Exposure Forecast 25 map of the Montreal airport (NEF25) available at the following web site (http://www.boeing.com/commercial/noise/montreal2011.pdf; accessed 14/02/28). The NEF25 is a complex indicator of noise levels calculated based on aircraft movements, which is used for land use planning in order to avoid high level of annoyance in the population (Transport Canada. 2010). Measurements sites were categorized as being within less than a km from the NEF 25 or not.

Port of Montreal (one variable)

Terminal accesses to the Montreal port (available at http://www.port-montreal.com/) were geo-referenced, and the distance (in meters) of each sampling site to the closest access was calculated; the maximum distances was set to 10 km based on the visual inspection of the relation between noise levels and distance to the port; at 10 km, the relation reached a plateau.

Divisions of the Island of Montreal (one variable)

The Island of Montreal was divided into the following four geographic divisions based on the municipal sectors of the Origin-Destination survey 2008 of the Montreal region (AMT, 2008): East, Center, Downtown and West. This survey is used to model road traffic by Transportation ministries in Canada (Gourvil & Joubert, 2004). Sampling sites were then attributed to the region in which they were contained to capture specific noise sources for these areas.

2.3 Statistical Analyses (Development of the Land Use Regression Model)

We calculated LA_{eq24h} (equivalent noise level) at all sites, and for each sampling period, using two minute measurements (Fahy & Walker, 1998). At each site, all two-minute noise levels that were different from the average for the sampling period by three times its standard deviation were discarded as outliers. LUR models were then developed with one LA_{eq24h} per site, separately for the winter and the summer period, due to differences in the selected noise determinants for the two periods.

The set of 69 variables described above was reduced to 17 by keeping the four following variables: NDVI, NEF25, distance to terminal access and division of the island, as well as choosing only one buffer size per buffer-based variable (i.e. land use and residential and building density, road and bus and train network layers). The selection of the best buffer per variable was done as follows. First, bivariate General Additive Models (GAM) (Hastie & Tibshirani, 1990) were developed using LA_{eq24h} and each variable (n=65) independently. Our GAM model used an iteratively reweighted least square (IWLS) and a thin plate spline basis with a smoothing parameter value of three in order to describe the non-linearity of the relations between noise levels and the variables of the built environment. We then selected a single buffer size based on leave one out cross-validation (LOOCV) using the Root Mean Square Error by dropping sites individually and re-running the model to predict noise levels at the dropped site.

Finally, to develop a multivariate LUR model, an approach similar to a backward selection was applied, integrating all determinants of the built environment previously selected (n=17). But first, the shape of the relationship between LA_{eq24h} and each variable was visually inspected using LOESS, and variables in non-linear relationships were log-transformed. For variables' relations with LA_{eq24h} which were still non-linear, spline variables, with a smoothing parameter value of three, were used instead of the log transformation. GAMs were used to minimize the number of degrees of freedom associated with the use of spline variables Spline functions (Hastie & Tibshirani, 1990) were removed if their degree of freedom in the resulting model were very close to one. Variables were removed one by one, based on the value of the RMSE obtained from a LOOCV as described above. All determinants were removed one by one from the model until the RMSE value reached a minimum. All statistical analyzes were performed with the R software (version 2.11.1, package mgcv).

2.4 Application of Model and Estimation of Noise Levels in Montreal

Noise levels were estimated for all central points of a 20m x 20m grid covering the island of Montreal (~ 1.2 million points) using the equation of the summer noise model developed (see results). Noise levels estimated outside our sampling range (52.4 to 73.7) were subsequently deleted (n=26315).

3. Results

Table 1 shows descriptive statistics for LA_{eq24h} for both sampling periods. 87 and 62 sites during the summer and the winter periods with valid data, respectively. 13 and 19 samplers were stolen or malfunctioned during the two study periods. More samplers malfunctioned during the winter period due to cold temperatures during the night (<-15°C). At Most sites, at least 10 days of continuous measurements were obtained but at 15 sites during the summer and at 28 sites during the winter, sampling took place between 4 and 9 days. LA_{eq24h} levels ranged between 53.4 and 77.7 dBA across all seasons, with winter mean LA_{eq24h} noise levels being approximately 7 dBA higher than during the summer sampling period. The correlation between LA_{eq24h} summer and winter is 0.74 (n=62).

Table 1. LA_{eq24h} (dBA) for noise sampling periods on the island of Montreal

	Number of site	Arithmetic Mean (std)	Median	Range
Summer 2010	87	62.5(5.1)	62.5	53.4-73.7
Winter 2011	62	69.1(5.4)	70.2	54.1-77.7

Table 2 presents the summary statistics of the potential determinants of noise levels at the sampling locations. All associations with buffer variables selected for the multivariate model (n=17) were in the expected direction with LA_{eq24h} in bivariate models (see Figure in Appendix). The majority of predictors had a linear relationship with noise levels for each season (i.e. Positive with length of highways and arteries, with industrial land use and population density and with the airport noise curve; negative with the NDVI).

Table 2. Summary statistics of the potential determinants of noise levels at the sampling locations (n=87)[1]

| Categories | Variables | Median Min-Max Buffer radius (m) | | | | | Value at the site | Distance to nearest |
		50	100	150	500	750		
Vegetation	Normalized Difference Vegetation Index (NDVI)	---	---	---	---	---	0.446 0.183 – 0.646	---
Residential low density land use	Area within buffer	615 0-6081	3241 0-27211	7836 0-56309	92053 0-519130	245303 0-942592	---	---
Residential medium-low density land use	Area within buffer	0 0-5992	1722 0-17323	3372 0-38225	70926 0-322081	161170 0-618408		
Residential medium density land use	Area within buffer	0 0-3091	0 0-11029	832 0-17577	23360 0-174736	51539 0-323116	---	---
Residential high density land use	Area within buffer	0 0-4403	0 0-12927	0 0-24673	15373 0-137222	39096 0-300751	---	---
Commercial land use	Area within buffer	0 0-5770	0 0-22911	106 0-43837	23079 0-283900	68862 0-580901	---	---
Office building	Area within buffer	0 0-5698	0 0-15904	0 0-29470	832 0-93271	7983 0-181325		
Industrial	Area within buffer	0 0-4576	0 0-15310	0 0-40762	2130 0-590101	19204 0-1389530	---	---
Residential Buildings	Number of units within buffer	11 0-337	65 0-503	173 0-1256	1914 0-10296	4352 0-14387	---	---
Non-residential Buildings	Number within buffer	7 0-33	34 0-234	80 0-539	49 2-195	103 5-418	---	---
Road network	Length of arteries and highways within buffer (m)	0 0-415	0 0-1584	0 0-3332	0 0-13668	0 0-23208	---	---
Bus network	Length of bus line within buffer (m)	0 0-659	177 0-1876	574 0-3167	8346 873-22086	18184 4456-43857	---	---
	Number of bus stop within buffer	0 0-4	0 0-4	2 0-7	19 3-50	44 6-96	---	---
Rail network	length of rail line within buffer (m)	0 0-121	0 0-401	0 0-800	0 0-9135	0 0-17044	---	---
Airport	One Km or less from the NEF25[2] contour	---	---	---	---	---	---	N = 7
Port of Montréal	Distance to nearest terminal access (m)	---	---	---	---	---	---	7646 244-26378
Division of Island	Sector East	---	---	---	---	---	---	N =18
	Sector Center							N = 49
	Sector Downtown							N = 2
	Sector West							N = 18

[1] The 87 sites included the 62 winter sites.

[2] Noise Exposure Forecast 25

Table 3 presents the results of the summer and winter multivariate GAM models. A higher R^2 was obtained for the summer (64%) compared to the winter (40%) model, with kept predictors differing between seasons. Regression coefficients for each variable are not interpreted as we did not take the collinearity between variables into account, and because we aimed to predict noise levels and not to assess the impact of each determinant on noise levels.

According to the validation results (LOOCV), the RMSE for the summer period (3.3 dBA) was better than for the winter (4.5 dBA).

Table 3. Results of the GAM model to predict LAeq24h on the island of Montreal

	Summer		*Winter*	
R-sq (adj)	0.64		0.4	
RMSE (cross-validation)	3.3 dBA		4.5 dBA	
Number of variables	11		12	
Linear relations with predictor variables	**Regression coefficient**	**Standard error**	**Regression coefficient**	**Standard error**
Intercept	62.290	2.596	7.610	5.836
Normalized Difference Vegetation Index (NDVI)	-9.226	3.908	-13.310	6.279
Area of residential low density within 50 m buffer	-0.001	0.002		
Residential medium-low density land use	---		-1.667e-5	9.60 e-6
Residential medium density land use	---		-9.598e-4	9.79 e-4
Area of residential high density within 500 m buffer	1.24 e-5	1.24 e-5	-2.042e-4	1.49 e-4
Area of commercial land use within a 750 m buffer	6.55 e-6	3.89 e-6	---	
Log(Area of industrial land use within a 750 m buffer)	0.181	9.00 e-2	0.356	0.149
Area of office building land use within a 750 m buffer	3.18 e-5	1.76 e-5	0.308	0.162
Number of residential units within 50 m buffer	-0.016	0.008	---	
Number of residential units within 150 m buffer	---		0.012	0.008
Length of arteries and highways within 50 m buffer	0.005	0.006	0.006	0.009
Log(Length of bus line within 50 m buffer)	0.971	0.194	0.800	0.316
One km or less from NEF25[1] curve	3.834	1.199	1.504	2.014
Distance to nearest terminal access			-2.438e-4	2.02 e-4
Division of Island (Reference : West)				
Downtown 2	-5.328	3.613	-9.175	5.145
Center – around downtown 4	-0.518	1.006	-3.350	2.864
East 3	2.668	1.202	-2.260	3.438

[1] Noise Exposure Forecast 25

LA$_{eq24h}$ were then estimated with the summer model across the island of Montreal at a 20 m resolution. Overall, noise levels were higher along major roads, and in proximity to industrial sectors (in the East-End of Montreal Island and close to the airport in the center-west of the Island) (see Appendix).

4. Discussion

We developed statistical models to predict summer and winter noise levels (LA$_{eq24h}$) on the Island of Montreal. Our summer model outperformed our winter model (in terms of R^2 and RMSE). Common determinants for both winter and summer LA$_{eq24h}$ models were NDVI, land use variables, the length of highways and arteries, the length of bus lines in various buffers around sampling sites, the proximity to the airport noise contours (NEF25) and the sector of the Island.

It is likely that the summer models outperformed the winter models due to the fact that a higher number of noise level measurements were available. Nonetheless, recent studies in air pollution suggest that a larger number of sites often lead to a lower R^2 because the larger number of monitoring sites captures more complexity in the combinations of predictors (Basagaña et al., 2012). Other factors may also explain differences between the models for the two seasons. For instance, in Montreal in the winter, there are snowplows and snow cover. In the summer, additional sources of noise include the gathering of people outside given the clement temperature.

Our results can be compared to similar models developed in other countries. Xie et al. 2011 developed a LUR to estimate noise levels (L_{day} for the period from 6h00 to 22h00), using land use and road categories. The R^2 of their model was not reported but the authors calculated the percentage of noise measurements predicted with an error of less than 10% (referred as PAS). Their PAS was 78% and similar figures, calculated with our LA_{eq24h} measured and predicted values were 99 % for summer and 92 % for winter.

Our results can also be compared to those of Foraster et al. 2011 in Girona (Spain). Similar to our results, these authors reported that building and road characteristics, daily traffic flows, bus lines and stops, open areas, and the proximity to the river explained a large part (73%) of the variability in noise levels estimated with a numerical model.

Finally, our results can also be compared to those of Seto et al. 2007. Seto et al. 2007 reported that in San Francisco (USA), road categories, hourly traffic flows by vehicle types and neighborhoods explained 37% of the variability in equivalent hourly noise level. As with Seto, we noted that the length of arteries and highways alone explained 26% of the variability in noise levels.

There are a number of limits inherent to our model. First, summer noise sampling was performed at the end of the holiday season. At that time, traffic flows and associated noise levels might have been lower due to reduced local traffic, which may reduce the external validity of our model and its applicability in other circumstances.

Another limit of our model is that the samplers for the measurements were attached closely to poles (10 cm). Reflection of sound waves on the surface of the poles to which the samplers were attached to is likely (ISO-1996-2). As such, actual LA_{eq24h} levels are likely to be three to six dBA lower than the ones reported in this paper.

Finally, our LA_{eq24h} Land Use Regression models predict noise levels with limited validity as the RMSEs were large in comparison to the range of measured and estimated noise values. Efforts to improve our models would include increased sampling at a number of sites, attempting to better capture the influence of airport, railways and other noise sources poorly represented in our model. Improvements would also be expected with the addition of traffic flows as predictors. Furthermore, the addition of proximity to noisy commercial zones (e.g. clubs and bars) may also be an improvement to the use of the proportion of industrial, commercial or residential land use types in various buffers.

While noise mapping is not required by an authority in Canada, improvements in methods to map noise levels are nonetheless necessary to orient interventions aiming at reducing annoyance and health risks.

Acknowledgements

We thank Louis Drouin for his support for this project. We also thank Louis-François Tétreault, Sabrina Cardin, Antoine Lewin and the Mark Goldberg's team for their contribution during the sampling campaign.

References

Alberola, J., Flindell, I. H., & Bullmore, A. J. (2005). Variability in road traffic noise levels. *Applied Acoustics, 66*(10), 1180-1195. http://dx.doi.org/10.1016/j.apacoust.2005.03.001

Babisch, W. (2006). Transportation noise and cardiovascular risk: Updated Review and synthesis of epidemiological studies indicate that the evidence has increased. *Noise Health [serial online], 8*, 1-29. http://dx.doi.org/10.4103/1463-1741.32464

Basagaña, X., Rivera, M., Aguilera, I., Agis, D., Bouso, L., Elosua, R., ..., & Künzli, N. (2012). Effect of the number of measurement sites on land use regression models in estimating local air pollution. *Atmospheric Environment, 54*, 634-642. http://dx.doi.org/10.1016/j.atmosenv.2012.01.064

Basner, M., Babisch, W., Davis, A., Brink, M., Clark, C., Janssen, S., & Stansfeld, S. (2014). Auditory and non-auditory effects of noise on health. *The Lancet, 383*(9925), 1325-1332. http://dx.doi.org/10.1016/S0140-6736(13)61613-X

Basner, M., Samel, A., & Isermann, U. (2006). Aircraft noise effects on sleep: application of the results of a large polysomnographic field study. *J Acoust Soc Am., 119*(5), 2772-2784. http://dx.doi.org/10.1121/1.2184247

Berglund, B., Lindvall, T., & Schwela, D. H. (1999). *Guidelines for community noise.* World Health Organization, Geneva.

Crouse, D. L., Goldberg, M. S., & Ross, N. A. (2009). A prediction-based approach to modelling temporal and spatial variability of traffic-related air pollution in Montreal, Canada. *Atmosphéric Environment, 43*(32), 5075-5084. http://dx.doi.org/10.1016/j.atmosenv.2009.06.040

Fahy, F., & Walker, J. (1998). *Fundamentals of Noise and Vibration.* CRC Press, p. 536.

Foraster, M., Deltell, A., Basagaña, X., Medina-Ramón, M., Aguilera, I., Bouso, R., ... & Künzli, N. (2011). Local determinants of road traffic noise levels versus determinants of air pollution levels in a Mediterranean city. *Environmental Research, 111*, 177-183. http://dx.doi.org/10.1016/j.envres.2010.10.013

Gourvil, L., & Joubert, F. (2004). *Évaluation de la congestion routière dans la région de Montréal.* Bibliothèque nationale du Québec.

Guedes, I. C. M., Bertoli, S. R., & Zannin, P. H. T. (2011). Influence of urban shapes on environmental noise: A case studiy in Aracaju – Brazil. *Science of the Total Environment,* 412-413.

Hastie, T. J., & Tibshirani, R. J. (1990). *Generalized Additive Models.* Chapman & Hall/CRC Monographs on Statistics & Applied Probability.

Huss, A., Spoerri, A., Egger, M., & Roosli, M. (2010). Aircraft noise, air pollution, and mortality from myocardial infarction. *Epidemiology, 21*(6), 829-836. http://dx.doi.org/10.1097/EDE.0b013e3181f4e634

Jackson, R. D., & Huete, A. R. (1991). Interpreting vegetation indices. *Preventive Veterinary Medicine, 11*, 185-200. http://dx.doi.org/10.1016/S0167-5877(05)80004-2

Jarup, L., Dudley, M. L., Babisch, W., Houthuijs, D., Swart, W., Pershagen, G., ... & Vigna-Taglianti, F. (2005). Hypertension and Exposure to Noise near Airports (HYENA): Study design and noise exposure assessment. *Environmental Health Perspectives, 113*(11), 1473-1478. http://dx.doi.org/10.1289/ehp.8037

Kanaroglou, P. S., Jerrett, M., Morrison, J., Beckerman, B., Arain, M. A., Gilbert, N. L., & Brook, J. R. (2005). Establishing an air pollution monitoring network for intra-urban population exposure assessment: A location-allocation approach. *Atmospheric Environment, 39*(13), 2399-2409. http://dx.doi.org/10.1016/j.atmosenv.2004.06.049

Makarewicz, R. (1993). Traffic noise in a built-up area influenced buy the ground effect. *Journal of the Acoustical Society of Japan, 14*, 5. http://dx.doi.org/10.1250/ast.14.301

Paillard, B. (2010). *Noise Sentry User's Manual.* Retrieved from http://www.convergenceinstruments.com/pdf/noise_sentry_manual.pdf.

Piercy, J. E., Embleton, T. K. W., & Sutherland, L. C. (1977). Review of noise propagation in the atmosphere. *J. Acoust. Soc. Am, 61*, 1403. http://dx.doi.org/10.1121/1.381455

Reethof, G. (1973). Effect of plantings on radiation of highway noise. *Journal of the Air Pollution Control Association, 23*(3), 185-189. http://dx.doi.org/10.1080/00022470.1973.10469763

Seto, E., Holt, A., Rivard, T., & Bhatia, R. (2007). Spatial distribution of traffic induced noise exposures in a US city: an analytic toll for assessing the health impacts of urban planning decisions. *International Journal of Health Geographics, 6*, 24. http://dx.doi.org/10.1186/1476-072X-6-24

Sørensen, M., Andersen, Z. J., Nordsborg, R. B., Jensen, S. S., Lillelund, K. G., Beelen, R., ... & Raaschou-Nielsen, O. (2012). Road traffic noise and incident myocardial infarction: a prospective cohort study. *PLoS One, 7*(6), e39283. http://dx.doi.org/10.1371/journal.pone.0039283

Stansfeld, S. A., Berglund, B., Clark, C., Lopez-Barrio, I., Fischer, P., Öhrström, E., ... & Berry, B. F. (2005). Aircraft and road traffic noise and children's cognition and health: a cross-national study. *The Lancet, 365*(9475), 1942-1949. http://dx.doi.org/10.1016/S0140-6736(05)66660-3

Statistique Canada. (2006). *Recensement de la population.* Statistique Canada.

Transport Canada. (2010). http://www.tc.gc.ca/fra/aviationcivile/publications/tp1247-partie4-partie4-2-1471.htm.

Xie, D., Liu, Y., & Chen, J. (2011). Mapping urban environmental noise: a land use regression method. *Environ Sci Technol, 45*(17), 7358-7364. http://dx.doi.org/10.1021/es200785x

Zou, B., Wilson, J. G., Zhanbd, F. B., & Zeng, Y. (2009). Air pollution exposure assessment methods utilized in epidemiological studies. *Journal of Environmental Monitoring, 11,* 475-490. http://dx.doi.org/10.1039/b813889c

Zuo, F., Li, Y., Johnson, S., Johnsona, J., Varughesea, S., Copesa, R., Liuc, F., ... & Chen, H. (2014). Temporal and spatial variability of traffic-related noise in the City of Toronto, Canada. *Science of the Total Environment, 15*(472), 1100-1107. http://dx.doi.org/10.1016/j.scitotenv.2013.11.138

Appendix A

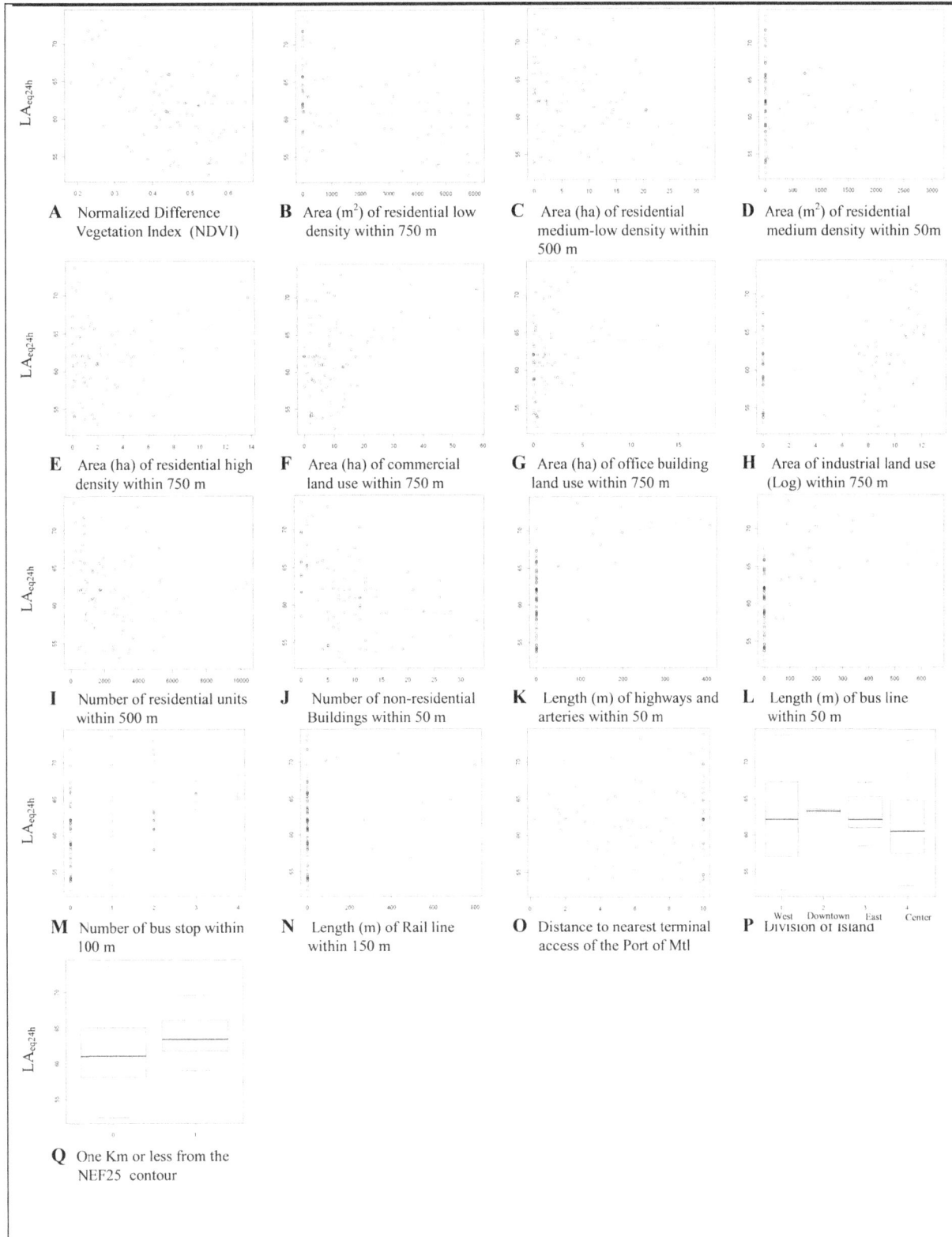

Figure A. Relation with LA_{eq24h} determinant of built environment

Appendix B

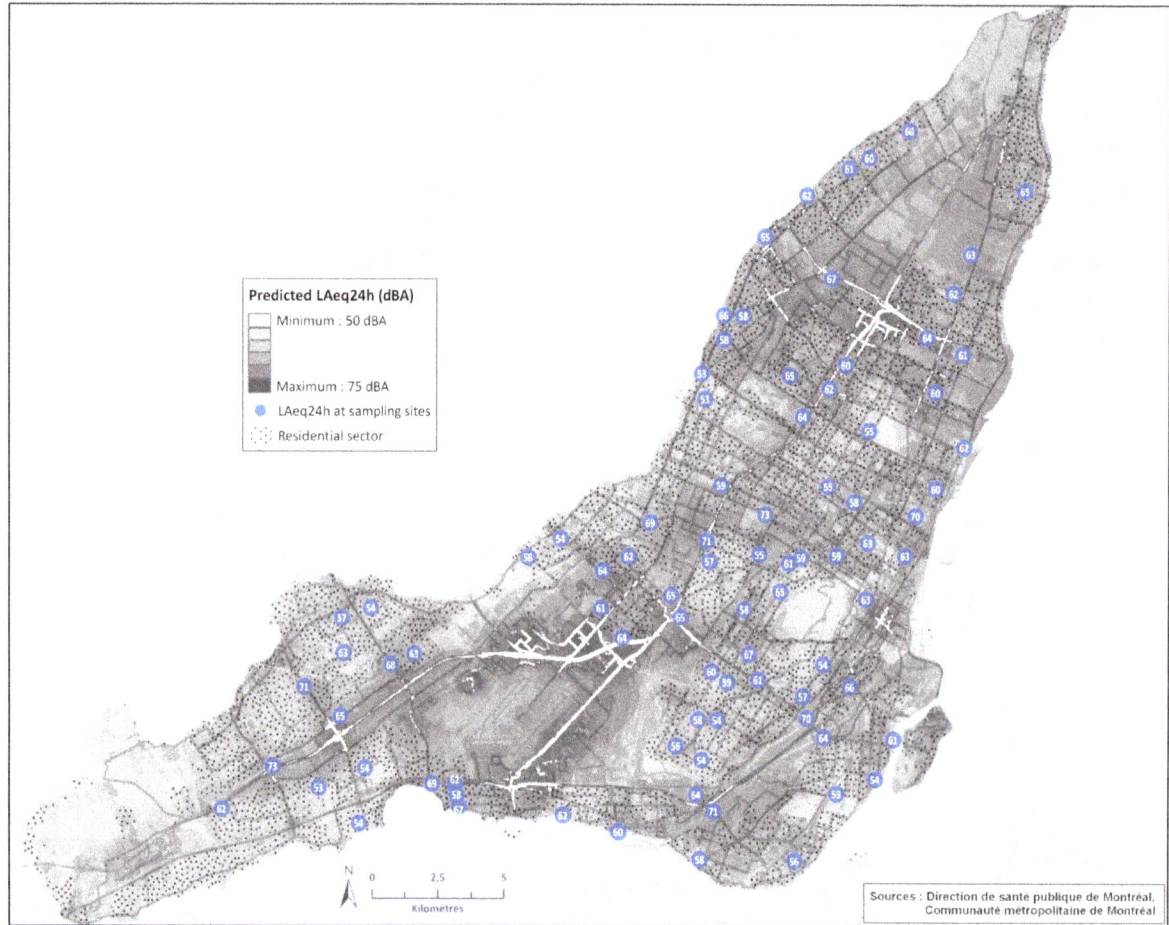

Figure B. Predicted summer noise levels (LA_{eq24h}) on the Island of Montreal (Quebec, Canada)

17

Impact of the Benefit Function Slope on the Advantage of Spatially Discriminating the Pollution Abatement Effort

François Destandau[1]

[1] Laboratory of Territorial Water and Environment Management (GESTE) UMR ENGEES-IRSTEA, Strasbourg Cedex, France

Correspondence: François Destandau, Laboratory of Territorial Water and Environment Management (GESTE) UMR ENGEES-IRSTEA, Associate researcher at BETA, UMR 7522 CNRS, Strasbourg Cedex, France. E-mail: francois.destandau@engees.unistra.fr

Abstract

Many articles have dealt with the advantages of spatially discriminating the abatement effort when the pollution is non-uniformly mixed. However, few authors have attempted to identify the impact of certain model parameters on the advantage of this discrimination. These parameters are the transfer coefficients, the parameters of the abatement cost function or the level of the quality standard. In this article, we studied the role of the slope of the environmental benefit function on the advantages of spatially discriminating pollution abatement efforts. It appears that the relative increase in the net social benefit between a uniform and a perfectly discriminated control is not affected by this slope. However, in the case of partial discrimination, a steeper benefit slope increases the optimal number of zones.

Keywords: pollution control, discrimination of abatement effort, discrimination advantage, environmental benefit function

1. Introduction

The American Clean Air Act of 1970 and, more precisely, the introduction of tradable permits to reduce air pollution in the USA, has encouraged economists to take a closer look at the spatial discrimination of pollution control instruments: prices differentiated according to the buyer and the seller or different exchange markets for tradable permits, and different tax rates depending on location for emission taxes. In theory, a uniform instrument makes it possible to minimize the cost of total emission reduction by equalizing the marginal abatement costs, but not to reduce the ambient quality at minimal cost if the impact of emissions is spatially heterogeneous (Montgomery, 1972; Tietenberg, 1974).

Two types of optimization programs were developed to demonstrate the effectiveness of a spatially discriminated policy (different marginal costs, or abatement efforts, depending on the location of the emitters). The first type aimed at reaching quality standards at a minimum cost (Atkinson & Lewis, 1974; Destandau & Nafi, 2009, 2010; Montgomery, 1972; O'Ryan, 2006), and the second type maximized the net social benefit, including total abatement costs and environmental damage, or the benefit of the pollution removed (Cabe & Herriges, 1990; Henderson, 1977; Howe & Lee, 1983; Kolstad, 1987; Krysiak & Schweitzer, 2010; Mendelsohn, 1986; Tietenberg, 1974; Wu & Babcock, 2001).

In the first type of program, the heterogeneity of emission impact is revealed by the heterogeneity of the transfer coefficients (the part of the emissions that reaches the receptor). In the second type of program, the heterogeneity of the impact is revealed by the heterogeneity of marginal contributions to damage, namely the product of transfer coefficients and the vulnerability of receptor sites (variation of the damage, or benefit, for a unit variation of pollution at the receptor site).

Tietenberg (1995, 2006) took stock of empirical studies that estimated the reduction of the total abatement cost between uniform and perfectly discriminated policies for the same quality standards. Given the disparities in the results, they deduced that the size of the savings is linked to the context. For example, Atkinson and Lewis (1974) compared the costs of reducing air pollution (particles) in Saint Louis (MO, USA) for uniform and for perfectly discriminated policies, with different quality standards. For standards of 40 to 60 $\mu g/m^3$, the uniform policy was at least twice as costly. Nevertheless, Tietenberg (2006) reported that costs seem to converge when the standard is more demanding, but provided no explanation.

Integrating regulation costs can reduce the advantage of perfectly discriminated policies, making it more costly to implement. A "partially" discriminated policy may seem to be a good compromise, with polluters grouped together in permit exchange zones or tax zones.

Mendelsohn (1986), for example, calculated the abatement cost for several possible numbers of zones, within the framework of air pollution by sulfur dioxides in the state of Connecticut. However, these breakdowns are arbitrary and he even admitted that other breakdowns of polluters would undoubtedly be more appropriate.

Destandau and Point (2000) developed a methodology that makes it possible to obtain endogenous effort zones. A simulation of a river shows that gains in efficiency rapidly decrease when the number of zones is increased. In reality, a partial discrimination with few effort zones seems to be sufficient. This can be explained by the fact that the initial breakdown from one to two zones takes place where the savings potential is the highest; the following breakdowns are done in decreasing order of "potential savings". These potential savings depend on the functions of costs and marginal contributions to damage (when the objective is to maximize the net social benefit), or transfer coefficients (when the objective is to reach a quality standard at the lowest possible cost).

Few articles have focused on the factors that have an impact on the advantage of discriminating. Two categories can be distinguished: those that compare uniform control (identical marginal abatement costs or effort) and perfectly discriminated control, and those that study the factors that have an impact on the optimal number of partially discriminated zones.

Concerning the comparison of perfect discrimination and uniform control, the heterogeneity of transfer coefficients is a factor that logically increases the advantage of discriminating (Destandau & Nafi, 2009; O'Ryan, 2006) since the equalization of marginal pollution costs is optimal with identical transfer coefficients.

Moreover, Tietenberg's (2006) intuition that costs seem to converge when the standard is more demanding was confirmed by O'Ryan (2006) who demonstrated that the advantage of discriminating increases when the standard is not very demanding. More specifically, Destandau and Nafi (2009) found that the discrimination advantage increased for moderately demanding standards. On the one hand, for a relatively undemanding standard, the low level of costs leads to small absolute efficiency gains. On the other hand, when the marginal cost functions are convex and the standard is very limiting, abatement is at a level where the marginal costs are very steep, reducing the gap between the total costs with discriminated and uniform efforts. With linear and unlimited marginal cost functions, the level of the standard has no influence on the discrimination advantage.

The impact of the shape of the cost functions was studied by Kolstad (1987). He used a simple model with two polluters with linear marginal contributions to damage and linear marginal cost functions. He concluded that the discrimination advantage increases when the slope of the marginal costs of one of the two polluters increases (the functions of the other polluter remaining the same).

O'Ryan (2006), and Destandau and Nafi (2009) combined cost function (efficiency) and transfer coefficients (harmfulness). For O'Ryan (2006), there are several polluters divided into two categories, each one with its own linear cost function and its own transfer coefficient. More specifically, some polluters-type 1-are inefficient (steep cost curve) and harmful (high transfer coefficient), and the others-type 2-are efficient and not very harmful. According to him, the discrimination advantage increases when the share of type 2 polluters increases, and when the relationship between the slopes of the cost curves is moderately high. Destandau and Nafi (2009) consider that the discrimination advantage increases when the most efficient polluters (lower abatement costs) are also the least harmful (lowest transfer coefficients), the only case considered by O'Ryan (2006).

As for partial discrimination, the factors that influence the degree of spatial discrimination (optimum number of zones) have been especially studied in the particular case of 'hot spot' risk, following the establishment of tradable permits (Krysiak & Schweitzer, 2010). The damage depends on the location of buyers and sellers. The spatial segmentation of the market therefore makes it possible to avoid the "hot spots", whereas, on the other hand, it limits potential gains linked to exchanges. The aim of these works was to study this trade-off by determining the optimal number of zones.

For Williams III (2008), polluters have identical cost functions, meaning that the grouping into zones only depends on the polluters' marginal contribution to damage. He found that the number of optimal zones decreases with the marginal cost slope. This result may seem to contradict the results of Kolstad since an increasing optimal number of zones can be linked to an increasing discrimination advantage.

The impact of the benefit slope (or environmental damage) on the advantage of discriminating the abatement effort was studied in three articles.

Kolstad (1987) found that the discrimination advantage increased when the ratio of marginal contributions to damage increased. He did not, however, distinguish the impact of the transfer coefficients that, as a result, increases the discrimination advantage and the impact of the vulnerability or the slope of the damage function.

For Williams III (2008), the optimal number of zones increases with the slope of the marginal contribution to damage. In keeping with Kolstad, he does not distinguish the transfer coefficient or the vulnerability.

Finally, Krysiak and Schweitzer (2010) were inspired by the model of Williams III, but added an uncertainty as to the spatial location of two types of polluters: "dirty" or "clean". Since both of these categories of polluters have parallel linear marginal abatement costs, only the original ordinate changes. Without knowledge of the location of pollution emitters, the breakdown into exchange zones is done exogenously, in contrast with Williams III, based on a homogeneous breakdown of the space. Krysiak and Schweitzer (2010) concluded that the breakdown of space is more interesting when the ratio between the slope of the marginal cost and the marginal damage is small.

The aim of this article is to understand how the slope of the environmental benefit function, generated by a reduction of the pollution emitted, affects the advantage of discriminating the abatement effort.

Compare to the existing literature, the innovation of this study is triple: (1) first, the impact of the slope of the environmental benefit function is studied independently of the transfer coefficients, contrary to Kolstad (1987) and Williams III (2008); (2) second, the originality of this paper is to simultaneously study the discrimination by comparing perfect discrimination and uniform control, and by calculating the optimal number of zones in partial discrimination; and (3) third, in partial discrimination, in contrast to Krysiak and Schweitzer (2010), the combination {zone, effort per zone} will be determined endogenously using the methodology of Destandau and Point (2000).

After having laid out the hypotheses of the model in Section 2, Section 3 is devoted to a comparison between uniform control and perfectly discriminated control. Finally, in Section 3, we address the role of the slope of the benefit function on the optimal number of partially discriminated zones.

2. Model Hypotheses

2.1 Static Model and Perfect Information

Like most articles that address spatial issues, we used a static model with perfect information.

Henderson (1977), Fleming and Adams (1997) and Howe and Lee (1983) used a dynamic model, varying the damage function to adapt to the movements of the victims, integrating a cumulative pollution effect, and varying transfer coefficients over time, respectively. The integration of time, however, would not provide an added value to the study of the impact of the benefit function slope on the advantage of discriminating. Likewise, Goetz and Zilberman (2000) described a relatively complex spatio-temporal model, but used a superficial approach to address the spatial issue.

Helfand and House (1995), Wu and Babcock (2001) and Cabe and Herriges (1990) assumed an uncertainty in relation to the transfer coefficients. Since the advantage itself of discrimination is based on the heterogeneity of the transfer coefficients, it is clear that discrimination would have no meaning in a model in which the transfer coefficients are unknown, and that its advantage would decrease with the degree of uncertainty of the coefficients. Once this factor was identified, we assumed that it was known, allowing us to more effectively study the role of the other factors.

2.2 The Model

We assumed a continuum of polluters: $x \in [0, X]$, classified in the order of harmfulness (increasing transfer coefficient), so that: $\theta(x) = \theta x$, where θ is a positive constant.

The polluters have the same cost function: $C(x) = \alpha_1 a(x)^{\alpha_2}$, where $a(x)$ is the pollution abatement per x.

α_1 and α_2 are positive parameters that illustrate the efficiency of polluters (efficiency increases when the value of the parameter α_1 decreases) and the convexity of the function, respectively. α_2 is a whole number greater than or equal to 2.

$\alpha_1 \alpha_2 a(x)^{\alpha_2 - 1}$ is the marginal abatement cost, which we refer to as the abatement effort. In fact, a uniform policy will lead to an equalization of the marginal abatement costs.

The environmental benefit, $B(Q)$, resulting from an improvement of the ambient quality (reduction Q of the pollution concentration), can be written as follows:

$$B(Q) = \beta Q = \beta \int_0^X \theta(x)\, a(x)\, dx$$

We thus assumed that the benefit was linearly dependent on the pollution removed. This marginal benefit, β, can be assimilated to the vulnerability of the receiving environment.

The optimization program consists of maximizing the net social benefit: Π, the difference between the environmental benefit, B, and the sum of the abatement costs, C.

3. Advantage Linked to Perfect Discrimination According to the Slope of the Environmental Benefit Function

3.1 Uniform Abatement Effort and Perfect Discrimination

Perfect discrimination of the abatement effort makes it possible to obtain a net social benefit, $\Pi_{PD} = B_{PD} - C_{PD}$, greater than the one obtained with a uniformization of the effort, $\Pi_U = B_U - C_U$.

Under perfect discrimination, we look for abatements, a(x), that maximize Π_{PD}, resulting in the following equations:

$$\frac{\partial \Pi_{PD}}{\partial a(x)} = \frac{\partial(\beta \int_0^X \theta(x).a(x)dx - \int_0^X \alpha_1 a(x)^{\alpha_2}\, dx)}{\partial a(x)} = \beta\theta.x - \alpha_1\alpha_2 a(x)^{\alpha_2-1} = 0$$

$$\frac{\partial^2 \Pi_{PD}}{\partial a(x)^2} = -(\alpha_2 - 1)\alpha_1\alpha_2 a(x)^{\alpha_2-2} < 0$$

Under perfect discrimination, the abatement is:

$$a(x) = \left(\frac{\beta\theta.x}{\alpha_1\alpha_2}\right)^{\frac{1}{\alpha_2-1}}, \quad \forall x \in [0, X] \tag{1}$$

Under uniform control, it is necessary to find the unique abatement, a_U, that maximizes Π_U, which can be written as:

$$\frac{\partial \Pi_{PD}}{\partial a_U} = \frac{\partial(\beta \int_0^X \theta(x).a_U\, dx - \int_0^X \alpha_1 a_U^{\alpha_2}\, dx)}{\partial a(x)} = \beta\theta\frac{X^2}{2} - \alpha_1\alpha_2 a_U^{\alpha_2-1} X = 0$$

$$\frac{\partial^2 \Pi_{PD}}{\partial a(x)^2} = -(\alpha_2 - 1)\alpha_1\alpha_2 a_U^{\alpha_2-2} X < 0$$

Under uniform control, the abatement is:

$$a_U = \left(\frac{\beta\theta X}{2\alpha_1\alpha_2}\right)^{\frac{1}{\alpha_2-1}} \tag{2}$$

3.2 Maximum Efficiency Gains (Meg)

In models where the aim was to reach a standard at the lowest cost, Destandau and Nafi (2010) and O'Ryan (2006) built an indicator to measure the advantage of perfect discrimination on uniform regulation. The O'Ryan ratio "R2" consists in dividing the total cost under uniform regulation by the total cost under perfect discrimination. The Destandau and Nafi indicator "Maximum Efficiency Gains", is equal to one minus the division of the total cost under perfect discrimination by the total cost under uniform regulation.

In this study, we measure the relative advantage of perfect discrimination by means of the indicator, MEG, described below:

$$MEG = \frac{\Pi_{PD} - \Pi_U}{\Pi_U}$$

With results (1) and (2), the net social benefits can be written as follows:

$$\Pi_{PD} = \beta \int_0^X \theta \, x \, a(x) dx - \int_0^X \alpha_1 a(x)^{\alpha_2} dx \;\; = \;\; \frac{(\alpha_2 - 1)^2 (\beta \theta X)^{\frac{\alpha_2}{\alpha_2 - 1}} X}{(2\alpha_2 - 1)\alpha_2 (\alpha_2 \alpha_1)^{\frac{1}{\alpha_2 - 1}}}$$

$$\Pi_U = \beta \int_0^X \theta \, x \, a_U \, dx - \int_0^X \alpha_1 a_U^{\alpha_2} dx \;\; = \;\; \frac{(\alpha_2 - 1)(\beta \theta X)^{\frac{\alpha_2}{\alpha_2 - 1}} X}{2\alpha_2 (2\alpha_2 \alpha_1)^{\frac{1}{\alpha_2 - 1}}}$$

Therefore:

$$MEG = \frac{2^{\frac{\alpha_2}{\alpha_2 - 1}}(\alpha_2 - 1) - 2\alpha_2 + 1}{2\alpha_2 - 1} \tag{3}$$

The relative gain of perfect discrimination is independent of the slope of the environmental benefit function. The impact of marginal contribution to damage reported by some authors is therefore due to the heterogeneity of the transfer coefficients. The impact of the heterogeneity of the transfer coefficients cannot be observed here because of the assumption of linearity of these coefficients, but has already been demonstrated by Destandau and Nafi (2010).

The only factor of influence here is the degree of convexity of the cost functions. The greater the convexity is, the more the discrimination appears to serve no purpose. For linear marginal costs ($\alpha_2 = 2$), the value of MEG is 1/3. This indicator is less than 5% for a value of α_2 greater than 6 (Figure 1).

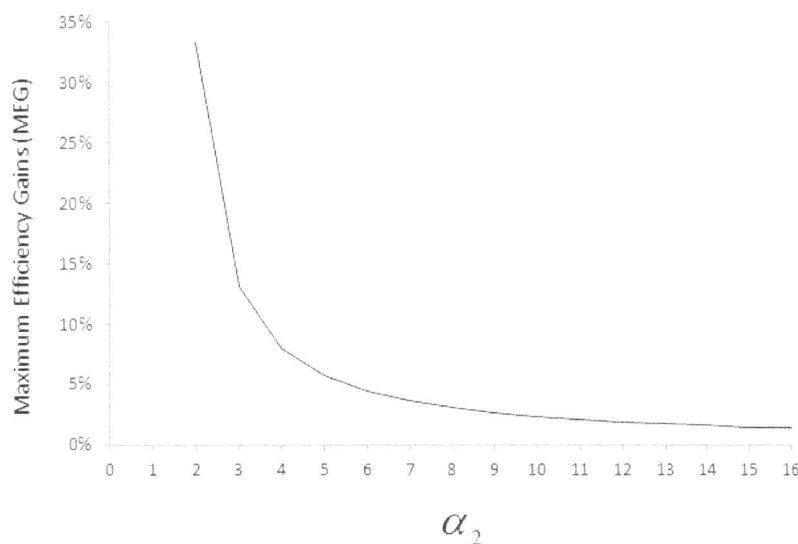

Figure 1. Decrease of the MEG value as a function of the convexity of the cost functions

4. Optimal Number of Zones According to the Slope of the Environmental Benefit Curve

In this section, we have simplified the cost function with $\alpha_2 = 2$, so that the marginal costs are linear, as in Krysiak and Schweitzer (2010).

As explained in the introduction, Mendelsohn (1986) calculated the abatement cost for several possible numbers of zones. However, these breakdowns are arbitrary. He only took the marginal contributions to damage of each pollution source into account, calculating the average for the zone and breaking down the zone into two parts according to whether the contribution was below or above the average. Finally, the most effective breakdown is the one that separates highly populated zones (since its damage function only depends on that) and rural zones. Destandau and Point (2000), on the other hand, developed a methodology that makes it possible to obtain endogenous effort zones. Their methodology consists of four "problems": (1) the "representation problem", making it possible to allocate a social benefit to all types of combinations {zone, effort per zone}; (2) the "allocation problem", to determine zones; (3) the "community problem", to assign an effort to each zone; and, finally (4) the "complexity problem" whose aim is to determine the number of zones.

In this section, we used the methodology of Destandau and Point (2000) to determine partial discrimination.

4.1 Dividing Up the Space

Dividing up the space into endogenous zones depends on the cost functions and the transfer coefficients of the polluters. With the hypotheses described in Section 1, according to which the cost functions are identical and the transfer coefficients increase linearly, the optimal breakdown of $[0, X]$ into N effort zones ('allocation problem') is done by homogeneous division according to size:

$$[0, X/N], \quad [X/N, 2X/N], ..., \quad [(n-1)X/N, nX/N], ..., \quad [(N-1)X/N, X].$$

The effort will be uniform in each effort zone:

$$a(x) = a_n, \quad \forall x \in n, \quad \forall n \in [1, N]$$

Krysiak and Schweitzer also divided up the space into homogeneous zones, but exogenously. They allocated the same quantity of permits to each zone, whereas we determined (see Section 4.3 below) the pollution that should be removed from each zone endogenously, what Destandau and Point (2000) refer to as the "community problem".

4.2 Net Social Benefit as A Function of a_n

We integrated a unit cost for zone z that was proportional to the number of zones.

The net social benefit is:
$$\Pi_N = B(Q) - \int_0^X C(x)dx - zN$$

Then:

$$\Pi_N = \beta \sum_{n=1}^{N} \int_{\frac{(n-1)X}{N}}^{\frac{nX}{N}} \theta x a_n dx - \sum_{n=1}^{N} \int_{\frac{(n-1)X}{N}}^{\frac{nX}{N}} \alpha_1 a_n^2 dx - zN$$

$$\Leftrightarrow \quad \Pi_N = \frac{\beta\theta X^2}{2N^2} \sum_{n=1}^{N} a_n(2n-1) - \frac{\alpha_1 X}{N} \sum_{n=1}^{N} a_n^2 - zN \quad (4)$$

The net social benefit makes it possible to characterize each combination {zone, effort per zone}. This is considered to be a "representation problem".

4.3 Effort a_n for Each Zone

On the basis of Equation (4), we can deduce the abatement cost (marginal abatement cost) that will be required in each of the zones ("community problem"), by deriving the net social benefit per a_n.

$$\frac{\partial \Pi_N}{\partial a_n} = \frac{\beta \theta X^2 (2n-1)}{2N^2} - \frac{2\alpha_1 X}{N} a_n = 0$$

$$\frac{\partial^2 \Pi_N}{\partial a_n^2} = -\frac{2\alpha_1 X}{N} < 0$$

Therefore:

$$a_n = \frac{\beta \theta X (2n-1)}{4\alpha_1 N}, \quad \forall n \in [1, N] \tag{5}$$

4.4 Net Social Benefit as a Function of N

On the basis of Equations (4) and (5), the net social benefit becomes:

$$\Pi_N = \frac{\beta \theta X^2}{2N^2} \sum_{n=1}^{N} \frac{\beta \theta X}{4\alpha_1 N} (2n-1)^2 - \frac{\alpha_1 X}{N} \sum_{n=1}^{N} \frac{\beta^2 \theta^2 X^2 (2n-1)^2}{16\alpha_1^2 N^2} - zN$$

$$= \frac{\beta^2 \theta^2 X^3}{16\alpha_1 N^3} \sum_{n=1}^{N} (2n-1)^2 - zN = \frac{\beta^2 \theta^2 X^3}{16\alpha_1 N^3} [4N^3 - N] - zN$$

$$\Leftrightarrow \Pi_N = \frac{\beta^2 \theta^2 X^3}{16\alpha_1} \left[\frac{4N^2 - 1}{N^2} \right] - zN \tag{6}$$

4.5 Optimal Number of Zones

The optimal number of zones ('complexity problem') can be deduced from the derivative of the net social benefit (6) in relation to N:

$$\frac{\partial \Pi_N}{\partial N} = \frac{\beta^2 \theta^2 X^3}{8\alpha_1 N^3} - z = 0$$

$$\frac{\partial^2 \Pi_N}{\partial N^2} = -\frac{3\beta^2 \theta^2 X^3}{8\alpha_1 N^4} < 0$$

The optimal number of zones is:

$$N = \left(\frac{\beta^2 \theta^2 X^3}{8\alpha_1 z} \right)^{1/3} \tag{7}$$

The optimal number of zones increases with the slope of the benefit function, with the transfer coefficients and with the number of polluters. It decreases with the unit cost of discrimination and the slope of the marginal costs.

This result casts a new light on the results obtained in Section 5. The slope of the environmental benefit function has an impact on discrimination, not in terms of the perfect discrimination advantage in relation to uniform control, but by increasing the optimal number of partially discriminated zones.

5. Discussion and Conclusion

Since the beginning of the 1970s, many articles have dealt with the advantages of discriminating the abatement effort when pollution has a heterogeneous impact according to the location of the emitter. However, it is only recently that some authors have attempted to identify the impact of certain model parameters on the advantage of this type of discrimination. We know that the advantage of discriminating the abatement effort increases when the transfer coefficients are the most heterogeneous, when the most efficient polluters are the least harmful ones, and when the quality standard to be reached is not too demanding.

The aim of this article was to fill in the gaps concerning the impact of the slope of the environmental benefit function that had not been adequately studied in the past.

In this paper, the impact of the slope of the environmental benefit function is studied independently of the transfer coefficients; at the same time, in comparing perfect discrimination and uniform control, and in calculating the optimal number of zones in partial discrimination. In partial discrimination the number of zones was determined endogenously using the methodology of Destandau and Point (2000).

The results show that the slope of this function does not have an impact on the relative gain when we change from uniform control to perfectly discriminated control. This result was obtained by isolating this parameter of the transfer coefficients, in contrast with some authors who hold that the marginal contribution to damage (product of the transfer coefficients and the benefit function) has an impact on the discrimination advantage. We now know that this result was only due to the role of the transfer coefficients.

It therefore appears here that the degree of vulnerability of the receptor site has no influence on the advantage of perfectly discriminating the abatement effort of the emitters. However, in the models whose objective was to reach a standard at a minimal cost, it appears that a greater demand on these standards made discrimination less interesting. The difference is due to the fact that in a model whose objective is to maximize the net social benefit, the environmental quality obtained is not the same under uniform and discriminated control. Therefore, the two methods may seem to be comparable in terms of costs, but discrimination may generate a greater benefit by improving the quality.

Our second result, however, casts a new light on the preceding one because by considering a partial discrimination of the abatement effort, the number of zones must be greater if the slope of the benefit function increases. This parameter therefore increases the discrimination advantage.

The majority of the policies concerning water pollution fix standards to be reached at the lowest cost. However, the European Water Framework Directive (European Commission, 2000) should be mentioned here since a cost-benefit analysis can allow the member states to adjust the standard to their local characteristics. In this case, our results show that the advantage of discriminating does not depend on the local sensibility. Nevertheless, in partial discrimination, the higher the local sensibility is, the greater the number of regulation zones must be (for example, zones where an effluent tax rate will be different).

It could be interesting in the future to explore which complementary results could be obtained with other shapes of cost or benefit functions.

References

Atkinson, S. E., & Lewis, D. H. (1974). A Cost-Effectiveness Analysis of Alternative Air Quality Control Strategies. *Journal of Environmental Economics and Management, 1,* 237-250. http://dx.doi.org/10.1016/0095-0696(74)90005-9

Cabe, R., & Herriges, J. H. (1990). *The Regulation of Heterogeneous Non-Point Sources of Pollution Under Imperfect Information* (Unpublished results). Center for Agricultural and Rural Development. Iowa State University.

Destandau, F., & Nafi, A. (2009). Programmes de mesures de la Directive Cadre sur l'eau : doit-on discriminer l'effort de dépollution? *Revue d'Economie Politique, 119*(1), 96-118.

Destandau, F., & Nafi, A. (2010). What is the best distribution for pollution abatement efforts? Information for optimizing the WFD Programs of measures. *Environmental and Resource Economics, 46*(3), 337-358. http://dx.doi.org/10.1007/s10640-010-9344-y

Destandau, F., & Point, P. (2000). Cheminement d'impact et partition efficace de l'espace. *Revue Economique 51*(3), 609-620. http://dx.doi.org/10.2307/3503149

European Commission. (2000). *Directive 2000/60/EC of the European Parliament and of the Council establishing a framework for Community action in the field of water policy.* OJL327, 22.12.2000.

Fleming, R. A., & Adams, R. M. (1997). The Importance of Site-Specific Information in the Design of Policies to Control Pollution. *Journal of Environmental Economics and Management, 33,* 347-358. http://dx.doi.org/10.1006/jeem.1997.0990

Goetz, R. U., & Zilberman, D. (2000). The dynamics of spatial pollution: The case of phosphorus runoff from agricultural land. *Journal of Economic Dynamics & Control, 24,* 143-163. http://dx.doi.org/10.1016/S0165-1889(98)00067-0

Helfand, G. E., & House, B. (1995). Regulating Nonpoint Source Pollution under Heterogeneous Conditions. *American Journal of Agricultural Economics, 77*(4), 1024-1032. http://dx.doi.org/10.2307/1243825

Henderson, J. V. (1977). Externalities in a spatial context. *Journal of Public Economics, 7,* 89-110. http://dx.doi.org/10.1016/0047-2727(77)90038-X

Howe, C. W., & Lee, D. R. (1983). Priority Pollution Rights: Adapting Pollution Control to a Variable Environment. *Land Economics, 59*(2), 141-149. http://dx.doi.org/10.2307/3146044

Kolstad, C. D. (1987). Uniformity versus Differentiation in Regulating Externalities. *Journal of Environmental Economics and Management, 14,* 386-399. http://dx.doi.org/10.1016/0095-0696(87)90029-5

Krysiak, F. C., & Schweitzer, P. (2010). The optimal size of a permit market. *Journal of Environmental Economics and Management, 60,* 133-143. http://dx.doi.org/10.1016/j.jeem.2010.05.001

Mendelsohn, R. (1986). Regulating Heterogeneous Emissions. *Journal of Environmental Economics and Management, 13,* 301-312. http://dx.doi.org/10.1016/0095-0696(86)90001-X

Montgomery, W. D. (1972). Markets in Licences and Efficient Pollution Control Programs. *Journal of Economic Theory, 5*(3), 395-418. http://dx.doi.org/10.1016/0022-0531(72)90049-X

O'Ryan, R. E. (2006). Factors that determine the cost-effectiveness ranking of second-best instruments for environmental regulation. *Journal of Regulatory Economics, 30*(2), 179-198. http://dx.doi.org/10.1007/s11149-006-0014-5

Tietenberg, T. H. (1974). Derived Decision Rules for Pollution Control in a General Equilibrium Space Economy. *Journal of Environmental Economics and Management, 1,* 3-16. http://dx.doi.org/10.1016/0095-0696(74)90014-X

Tietenberg, T. H. (1995). Tradeable Permits for Pollution Control when Emission Location Matters: What have we learned? *Environmental and Resource Economics, 5,* 95-113. http://dx.doi.org/10.1007/BF00693018

Tietenberg, T. H. (2006). *Emissions Trading: Principles and Practice* (2nd ed.). Resources for the Future, Washington DC, USA.

Williams III, R. C. (2008). *Cost-Effectiveness vs. Hot spots: Determining the Optimal size of Emissions Permits Trading Zones* (Unpublished results). Department of Economic, University of Texas, Austin.

Wu, J., & Babcock, B. A. (2001). Spatial Heterogeneity and the Choice of Instruments to Control Nonpoint Pollution. *Environmental and Resource Economics, 18,* 173-192. http://dx.doi.org/10.1023/A:1011164102052

18

An Exploration of Transportation Source Contribution to Noise Levels Near an Airport

Laura Margaret Dale[1], Maximilien Debia[2], Olivier Christian Mudaheranwa[2], Céline Plante[3] & Audrey Smargiassi[2,4]

[1] McGill School of Environment, McGill University, Montréal, Canada

[2] Département de santé environnementale et de santé au travail, Université de Montréal, Montréal, Canada

[3] Direction de la santé publique de l'Agence de la santé et des services sociaux de Montréal, Montréal, Canada

[4] Institut National de Santé Publique du Québec, Montréal, Canada

Correspondance: Audrey Smargiassi, Département de santé environnementale et de santé au travail, Université de Montréal, Montréal, QC., H3C 3J7, Canada.

Abstract

Our study aimed to explore the contribution of transportation activities to environmental noise levels near an international airport. A-weighted equivalent and maximum noise levels (LAeq, 6hr; LAmax, 6hr; LAeq, 1hr) were monitored at six different locations characterized by their varying proximity to transportation sources. The values for LAeq, 6hr were in the range of 55.3 to 75.6 dBA, and LAmax, 6hr were 4.1 to 9.1 dBA higher than their respective LAeq, 6hr values. Standard deviations were low across all sites and indicators (0.20-1.83 dBA). We found that at each site measured, the WHO's noise exposure guidelines of 55 dBA were exceeded, including sites located in residential areas or near a school. In one residential area near the airport (but away from other transportation sources), noise levels were 63.1 dBA. In another residential area closer to the airport, the contribution of airport noise to environmental noise was estimated to be 72.3 dBA, which is roughly as high as the contribution of two other transportation sources (highway and railway) in this area. In proximity to the Montreal International Airport, noise levels may have been elevated by airport operations and by noise from flights, the latter of which likely has a weaker effect on, and the former of which is unlikely to contribute to, noise levels at locations farther from the airport. At farther distances however, aircraft passages raised noise levels to a greater extent above those levels when no flight was passing, reflecting a sporadic quality of aircraft noise.

Keywords: environmental noise, urban, residential, traffic, airport, highway, railway

1. Introduction

An exposure quantification of urban pollutants is an important step in understanding the risks associated with their sources. One such pollutant, environmental noise, has been linked with several non-auditory malaises in both children and adults, including cognitive impairment (Stansfeld, Hygge, Clark, & Alfred, 2010), cardiovascular effects (Stansfeld & Crombie, 2011; Haralabidis et al., 2008), sleep disturbance (Perron, Tétreault, King, Plant, & Smargiassi, 2012) and annoyance (Passchier-Vermeer & Passchier, 2000).

Environmental noise refers to noise emitted from all sources except for that at industrial workplaces (Berglund, Lindvall, & Schwela, 1999). The noise sources contributing to the most widespread health effects in people tend to be related to transportation activities, namely road, rail and air traffic. The impacts of various sources, even at equal levels, however, are not uniform. Studies suggest that features of air traffic noise, such as its intensity, variability and unpredictability may contribute to greater health outcomes compared to the relatively more constant noise emitted by road traffic. Increased cognitive impairment in children, for example, has been associated with aircraft, but not road traffic noise exposure (Stansfeld et al., 2005). Similarly, exposure to rail traffic noise has been shown to result in a lower annoyance response than that of both air and road traffic (Fields & Walker, 1982). A greater understanding of each source's contribution to exposure levels is needed to better assess the health response in people.

The present study measured and compared noise levels and variability at 6 sample sites characterized by their varying proximity to highway, airport and railway activities. The primary objective of this study was to explore the contribution of these transportation activities to environmental noise in the region of the Montreal airport.

2. Method

2.1 Sampling Sites

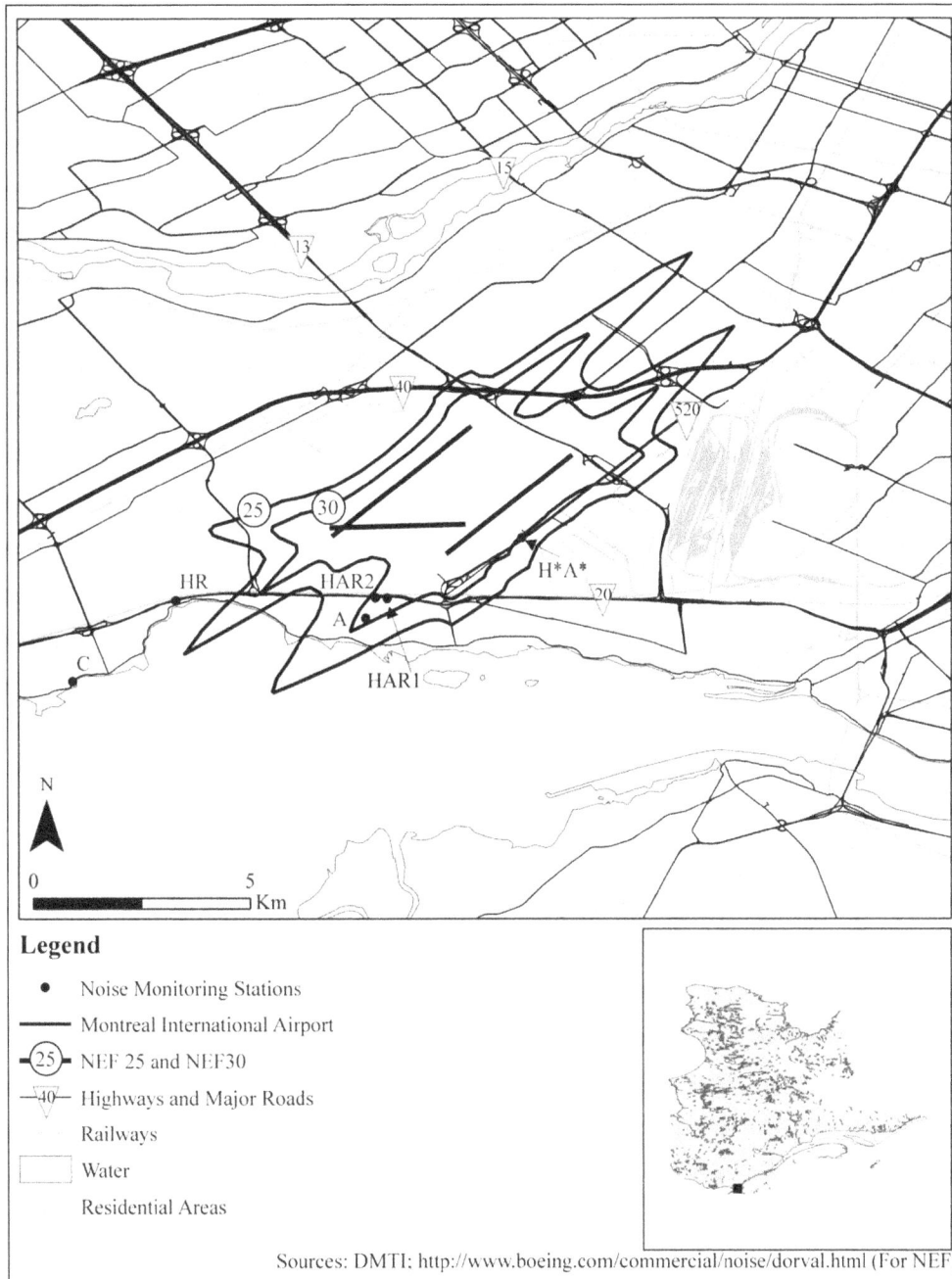

Figure 1. Map of sampling sites in Dorval City, Montreal Island

Sampling sites within Dorval, a city on the island of Montreal, were chosen based on NEF 30 and NEF 25 Noise Exposure Forecasts around the Montreal International Airport for the year 2009 (Boeing, 2011) (Figure 1). The NEF 30 zone refers to the area surrounding the airport in which there is a noise exposure forecast (NEF) value of 30. Values are calculated based on effective perceived noise levels for aircraft operations in order to encourage compatible land use in surrounding areas. The NEF 25 zone represents a lower level of noise exposure and thus corresponds to a wider contour, which surrounds that of the NEF 30 zone. The sampling sites are defined as follows:

- **HAR1:** the Highway + Airport NEF 30 + Railway site was located about 1400 m southwest of the Montreal International Airport, along the axis of its runways. It was within the NEF 30 zone and along the A20 highway.

- **HAR2:** the Highway + Airport NEF 30 + Railway site was located 200 m west of HAR1.

- **HR:** the Highway + Railway site was located along the A20 highway, outside the NEF 30 and NEF 25 zones.

- **H*A*:** the Highway* + Airport NEF 25* site was located along the Côte-de-Liesse Expressway, just outside of the NEF 30 zone but within the contours of the NEF 25 zone. It was about 650 m southeast of the airport's runways.

- **A:** the Airport NEF 30 site was situated about 2000 m southwest of the Montreal International Airport, along the axis of its runways. It was in a residential area within the NEF 30 zone and close to a school.

- **C:** The Control site was located in a residential area outside of both the NEF 30 and 25 zones.

Sites A and C were located in residential areas. Sites HAR1, HAR2 and HR were slightly outside their closest respective residential zones, but less than 50 m away at each location. Each of sites HAR1, HAR2, HR and H*A* were situated just off the curb of a highway access road. It was not possible to use the same highway for each of these four sites since the railway was always situated parallel to the A20 highway. The Highway* + Airport* site is thus labelled with two asterisks to draw attention to two considerations. The first is that the Côte-de-Liesse Expressway was selected for this site rather than the A20 highway, and the second is that this site was within the NEF 25 contours, but just outside the NEF 30 zone. Highway traffic flow data for the year 2010 suggests that the traffic flow at site H*A* (about 32 000 vehicles) was roughly half of that at sites HR, HAR1 and HAR2 (in the range of 63 000 to 75 000 vehicles) (Transports Québec, 2010), thus the contribution of road traffic noise was assumed to be equal at sites HR, HAR1 and HAR2, but not at H*A*.

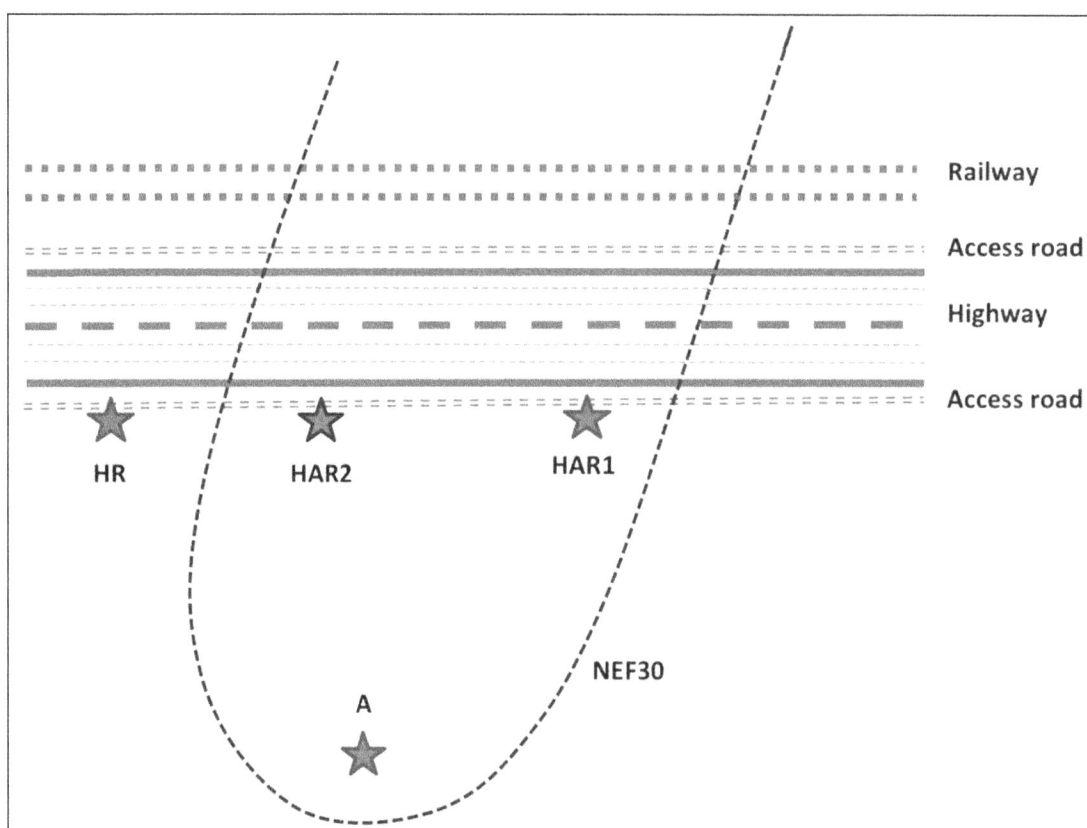

Figure 2. Schematic representation of sampling sites HR, HAR1, HAR2 and A

At sites HR, HAR1 and HAR2, the railway track, freight trains and passenger trains were situated parallel to the A20 highway, north of the sampling locations (Figure 2). The distance from the highway to the railway was roughly 60 m at each of these sites and so the contribution of railway noise was assumed to be equal. At sites A, HAR1, HAR2, and H*A*, samples were taken within visible range of passing aircrafts, within the NEF 25 or 30 zone.

2.2 Sampling Strategy

In the summer of 2012, sites HAR1, H*A*, and A were sampled; in the fall of 2012, sampling was done at sites HAR2, HR and C. The two HAR sites were in the same area, but were termed HAR1 and HAR2 to differentiate between the times of their sampling campaigns, in the summer and fall, respectively. The replication of this measurement acted to ensure the representativity of the results for this zone, and that the seasonal variation in the data was negligible between the summer and fall sampling events.

Table 1. Days sampled at each site in 2012

	Monday	Tuesday	Wednesday	Thursday	Friday
HAR1	July 30	Aug 7 Aug 21		Aug 2	
H*A*	Aug 6	Aug 14	Aug 22	Aug 9	
A	Aug 13	July 31	Aug 8	Aug 16	
HAR2	Nov 5	Nov 6	Nov 7		Nov 9
HR	Nov 12		Nov 14	Nov 15	Nov 16
C				Nov 22	Nov 23

Sites HAR1, HAR2, HR, H*A* and A were sampled on four days from Monday to Friday over 6 hours (noon to 6pm), and site C was sampled on two days over 6 hours (11am to 5pm). Data was collected every minute to give a total of 360 measurements for each of 22 days of sampling. The distribution of sampling sites by day is shown in Table 1.

Samples were collected such that wind speed did not equal or exceed 5 m/s, humidity was much less than 90%, there was no precipitation, and ambient temperatures were within the tolerable limits for the instruments used. No sampling was done on days with atypical noise events, such as construction and public work. At sites HAR1 and A, a fieldworker recorded the times of aircraft passage over the sites in a notebook for consideration in the analysis. It was assumed that each flight lasted one minute over the site.

2.3 Equipment

(1) **The Sound Level Meter:** A Larson Davis SoundTrack LxT2 (Class 2) sound level meter was used to determine noise levels and the distribution of sound frequencies at the sites. The sound level meter was equipped with an anti-wind shield, and mounted on a tripod at a height of 1.5 m from the ground. The data was processed first by the firmware version 2.112 included in the sound level meter and then by the software Slm Utlity-G3 installed on a computer.

(2) **The Calibrator:** The sound level meter was calibrated at the beginning and end of each measurement with the Larson Davis CAL150 Sound Level Calibrator.

(3) **The Air Velocity Meter:** The TSI Model 8360-M-GB VelociCalc Plus Air Velocity Meter was used to monitor the wind speed, humidity and ambient temperature at each site.

2.4 Analysis of Results

Three sound level indicators were calculated: 1) LAeq, 6 hr, representing the A-weighted equivalent sound energy during 6 hours of sampling; 2) LAeq, 1 hr, representing the A-weighted equivalent sound energy over one hour of sampling; and 3) LAmax, 6hr, which expresses the A-weighted peak sound events within the 6-hour sampling interval, usually generated by a particular source such as the passing of an aircraft or a large truck.

Student's t-tests were used to assess the differences in the means for both LAeq, 6hr and LAmax, 6hr between sites HAR1 and HAR2. We used an analysis of variance (ANOVA) to test the differences in means for these same indicators among all sites. This was followed by a post hoc pairwise comparison with a Bonferroni adjustment. There was not enough data in the samples to test for the normality of the distributions. However, normality was assumed based on previous data obtained on Laeq, 24h from another sampling campaign (84 sites including similar locations in Montréal) which were close to normal (S. Goudreau, personal communication).

LAeq, 6hr and LAmax, 6hr values were also calculated for periods when aircrafts were (LAeq, P & LAmax, P) or were not passing (LAeq, NP & LAmax, NP) for sites HAR1 and A. These calculations were made based on averages of unlogged minute LAeq data obtained from the sound level meter. Data on equivalent sound levels that corresponded to different frequency components was also obtained from the sound level meter. The unlogged form of this data was expressed as a percentage of the overall equivalent sound level at each site to illustrate the relative frequency distribution.

Lastly, the contribution of airport noise to site HAR (using compiled data from HAR1 and HAR2) was calculated using the following equation, as described by Barron (2003):

$$L_A = 10 \log (10^{\frac{L_{AB}}{10}} - 10^{\frac{L_B}{10}}) \, dBA \tag{1}$$

Where L_A = LAeq, 6hr calculated for A, L_{AB} = LAeq, 6hr measured at HAR, and, L_B = LAeq, 6hr measured at HR.

3. Results

Table 2. LAeq, 6hr, LAmax, 6hr, and standard deviation at each site

	HAR1	HAR2	HR	H*A*	A	C
LAeq, 6hr (dBA)	75.6 ± 0.33	75.6 ± 0.36	72.9 ± 0.35	70.9 ± 0.20	63.1 ± 1.52	55.3 ± 1.58
(number of days)	(N=4)	(N=4)	(N=4)	(N=4)	(N=4)	(N=2)
LAmax, 6hr (dBA)	82.5 ± 0.97	80.0 ± 0.23	77.0 ± 0.58	79.2 ± 0.40	72.2 ± 1.83	62.5 ± 1.80
(number of days)	(N=4)	(N=4)	(N=4)	(N=4)	(N=4)	(N=2)

Table 2 shows that the LAeq, 6hr and LAmax, 6hr were highest at the HAR1 (75.6 dBA and 82.5 dBA, respectively) and HAR2 site (75.6 dBA and 80.0 dBA, respectively) where there were 3 principal sources contributing to the noise levels: air, road and train traffic. A similar pattern in LAeq, 6hr and LAmax, 6hr across the sites was observed. The HR site however exhibited a lower LAmax, 6hr (77.0 dBA) than the H*A* site (79.2 dBA), a trend that was not apparent in the LAeq, 6hr results. The elevation of LAmax, 6hr values above LAeq, 6hr values ranged from 4.1 to 9.1 dBA.

Noise levels were quite similar between days, exhibiting consistently low standard deviations. This was particularly true at sites H*A* (0.20 dBA and 0.40 dBA), HR (0.35 dBA and 0.58 dBA), HAR1 (0.33 dBA and 0.97 dBA) and HAR2 (0.36 and 0.23) for LAeq, 6hr and LAmax, 6hr, respectively. Standard deviations were higher at sites A (1.52 dBA and 1.83 dBA) and C (1.58 dBA and 1.80 dBA), though still small.

A t-test confirmed that the difference between the means of site HAR1, which was sampled in the summer, and HAR2, which was sampled in the fall, was negligible for LAeq, 6hr (t=0.0024; p=0.9981). However, for the LAmax, 6hr indicator, the difference between HAR1 and HAR2 was not equal to zero (t=4.845; p=0.01296). We thus combined the LAeq, 6hr from site HAR1 with that of HAR2 so as to obtain an overall LAeq, 6hr value for site HAR to be used in the remainder of the analysis. The combined LAeq, 6hr value for HAR based on values at HAR1 and HAR2 was 75.6 ± 0.32 (N=8). For analyses with the LAmax, 6hr indicator, the separated HAR1 and HAR2 values were used.

An ANOVA revealed that there was significant variability among the means of LAeq, 6hr at each of the sites HAR, HR, H*A*, A and C (F=369.75, p<0.01). Furthermore, post hoc pairwise t-tests also showed a significant difference between each pair of sites for this indicator (p<0.05). For the LAmax, 6hr values, the ANOVA suggested that the means at the sites were different (F = 127.21, p<0.01). A subsequent pairwise comparison confirmed the results of the ANOVA for most comparisons (p<0.05), however, it did suggest that the means of H*A* and HR, as well as of H*A* and HAR2 were not significantly different (p=0.1330 and p=1.0000, respectively). In addition, with the Bonferroni adjustment, the results of the pairwise comparison between HAR1 and HAR2 differed from the t-test, suggesting now that the difference between these two sites was not significant (p=0.0754).

The LAeq, 6hr value calculated using Equation 1 for the airport contribution to site HAR was 72.3 dBA. Comparatively, the LAeq, 6hr level measured at site A was 63.1 dBA. There was therefore at least a 9.2 dBA discrepancy between the estimated contribution of airport noise to noise levels in these regions (given that the value measured at site A also reflects background noise levels).

Table 3. LAeq, LAmax, and standard deviation from four days of sampling at sites HAR1 and A with (P) or without (NP) an aircraft passing

Site	HAR1	A
LAeq, P (dBA)	76.5 ± 1.10	67.0 ± 2.06
LAeq, NP (dBA)	75.4 ± 0.42	61.8 ± 2.62
LAmax, P (dBA)	85.0 ± 3.16	75.9 ± 1.81
LAmax, NP (dBA)	82.0 ± 0.57	71.0 ± 2.87

The passage of aircrafts over site HAR1 raised LAeq, P by 1.1 dBA and LAmax, P by 3.0 dBA above the noise levels when no aircraft was passing. Aircraft passage over site A raised LAeq, P and LAmax, P by 5.2 dBA and 4.9 dBA, respectively (Table 3).

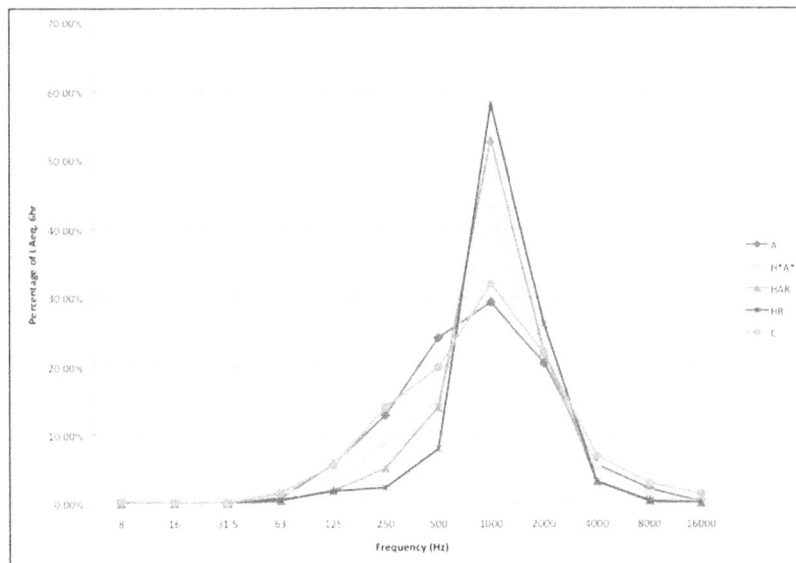

Figure 3. Relative frequency distribution of LAeq, 6hr at each site

Figure 3 shows that the frequency distribution is most broad for sites A and C, followed by H*A*, and lastly HAR and HR, both of which are concentrated quite narrowly at a frequency of 1000 Hz. This frequency represents a peak for each of the other sites sampled as well. At 1000 Hz, the HR peak exceeds that of HAR, and the C peak exceeds that of A. At 500 Hz, the peak at site A exceeds all other sites, and the peaks at sites H*A* and HAR exceed that at site HR.

4. Discussion

Our study aimed to explore the contribution of transportation sources to environmental noise in the Dorval region of the international airport of Montreal. Our main findings were that noise levels were above optimal levels at each of the sites measured and that airport noise may contribute to a large portion of environmental noise in residential areas, roughly as much as two other transportation sources combined.

Where the NEF exceeds 30, as it does at sites HAR and A, Transport Canada (2012) suggests that residential development should not proceed. Residential housing however already exists within the NEF 30 zone of the Montreal International Airport; in fact, 2,150 persons were residing within this zone in 2009 (Boeing, 2011). Outside the NEF 30 contours at the Control site (C), equivalent noise levels were the lowest measured (55.3

dBA). The World Health Organization Guidelines for Community Noise, suggest that daytime outdoor sound levels on balconies, terraces and outdoor living areas should not exceed either 55 dB LAeq or 50 dB LAeq in order to protect the majority of the population from being seriously or moderately annoyed, respectively (Berglund et al., 1999). Thus, even in terms of serious annoyance, these guidelines are exceeded at every site measured. Furthermore, these guidelines were developed based on responses from the average or "normal" person, and certain groups such as children, the elderly, the ill or the disabled may be at risk at even lower levels of exposure; they may also be less able to cope with environmental noise in general. However, it is important to note that our measurements were not conducted directly on residential properties and do not act explicitly to quantify noise levels to assess population exposure. Additionally, only two sites could be considered as residential sites (A and C). Finally, the WHO guideline for outdoor sound levels was developed based on an LAeq, 16hr indicator. Our results on the other hand were based on an LAeq, 6hr indicator. It is possible that the contribution of evening measurements would have resulted in lower LAeq values than those reported here.

During playtime, in the outdoor playgrounds of schools, it is suggested that sound levels from external sources should not exceed 55 dB LAeq (Berglund et al., 1999). The A site was located in proximity to a school and the equivalent noise level monitored was 63.1 dBA. When an aircraft passed, the LAeq, P reached on average 67.0 dBA, and when no aircraft was passing, the LAeq, NP was much smaller (61.8 dBA), but still well above the WHO's recommendation of 55 dBA. The passage of individual flights were also able to raise noise levels in the summer at site HAR1, though not to the same extent as at site A, where noise levels when no aircraft was passing were not already elevated appreciably by highway and railway traffic. These results may be important in the context of evidence suggesting that intermittent patterns of noise output may foster greater disturbance than more constant sources. The RANCH study, for example, showed a direct link between aircraft noise and impaired reading comprehension in children that could not be paralleled by road traffic noise (Clark et al., 2005). Overall, when evaluated against public health standards, each of the measured sites appeared to be problematic, including sites A and C, which are in residential areas.

The analysis of the LAmax, 6hr indicator suggested that for site HAR, there existed variability between summer (HAR1) and fall (HAR2) sampling events; this may suggest that aircraft noise is more predominant in the summer. The pairwise comparison that followed our ANOVA indicated, contrary to our expectations, that there was a negligible difference between the sites HAR2 and H*A*, as well as sites HR and H*A*. The LAmax, 6hr values indicate also that sites A, H*A*, and C experienced the greatest elevations in noise (9.1 dBA, 8.3 dBA and 7.2 dBA, respectively) compared to their LAeq, 6hr values. This may be a reflection of the relatively lower LAeq, 6hr values used as a starting point at these sites, which would allow specific noise events to elevate noise levels considerably. Overall, LAmax, 6hr values were mostly consistent with the trend observed for LAeq, 6hr values, but one difference was that the trend for LAmax, 6hr was reversed at sites HR (77.0 dBA) and H*A* (79.2 dBA) (Table 2). This result may be explained by the non-significant difference found in the pairwise comparison between the means of H*A* and HR for LAmax, 6hr. However, it may also reflect a difference in the characteristics of aircraft noise compared to railway noise, another testament to aircraft noise's intermittency. Indeed, this feature is further emphasized when looking at the elevation of LAmax, 6hr above LAeq, 6hr at comparable sites HAR [6.9 dBA (HAR1); 4.4 dBA (HAR2)] and HR (4.1 dBA), as well as A (9.1 dBA) and C (7.2 dBA), the larger elevation of which contains an aircraft component in each case. It is apparent that aircraft noise represents an important contributor to the LAmax, 6hr indicator, which reflects peak noise events.

Unlike equivalent noise levels, the frequency distribution of environmental noise at the sites does not appear to be particularly problematic from a public health standpoint. Our results show a tendency for frequencies to be concentrated largely in the mid-range values at each of the sites. On the other hand, health effects tend to manifest in the extreme frequency ranges, particularly at low frequencies in the range of 10 Hz to 200 Hz (Leventhall, 2004). Importantly however, our results are A-weighted, meaning that low frequency noise is under-represented compared to its un-weighted form (Leventhall, 2004). Given that the WHO recommends that noise with low-frequency components should have lower exposure guidelines (Berglund et al., 1999), these results should be interpreted prudently.

At 1000 Hz, a peak in noise levels exists for each of the sites measured. The HR peak here exceeds that of HAR and H*A*, and the C peak exceeds that of A. This result may suggest that airport noise has a greater influence at other frequencies while other traffic sources, particularly rail, contribute largely to 1000 Hz. For example, at 500 Hz, the A site exceeds all other sites, and the H*A* and HAR sites exceed the HR site. Thus, aircraft noise may be an important contributor to this frequency.

In residential areas, our results suggest that the contribution of airport noise to environmental noise levels is highly variable. Noise levels measured at site A were 63.1 dBA. This value reflects an airport contribution as well as background noise, but no other major transportation sources. Comparatively, the estimated contribution of airport noise 600 m closer the airport at site HAR was 72.3 dBA (Equation 1). Thus, a comparison of these two values suggests that the contribution of airport noise at site HAR is at least 9.2 dBA higher than that at site A. This large discrepancy may not be adequately explained by differences in highway or railway traffic at the two sites compared. We suggest that it may also be accounted for by airport activities and operations other than flights, as well as the greater elevation of aircrafts on their flight paths at site A. Since the HAR site is located 600 m closer to the airport than the A site, it is in a better position to be exposed to noise from airport operations such as parking lot activities, access road traffic, public transportation stations, and other factors not accounted for in the NEF 30 calculations. In addition, the contribution of flights alone at site HAR may be somewhat greater due to site A's location, which is closer to the boundaries of the NEF 30 zone and 600 m farther from the airport. This greater distance from the airport allows flights to reach a higher elevation at site A and thereby increase ambient noise levels to a lesser extent. To put this calculated value for the airport noise contribution at site HAR (72.3 dBA) into context, it can be compared to the value measured at site HR (72.9 dBA). The noise levels measured at the HR site roughly equate the level of background noise (compared to airport noise) at the HAR site; thus, these results suggest that the contribution of airport noise is roughly as large as the contribution of two other major noise sources: highway and railway traffic.

The strategy of aircraft noise quantification above background levels presented here represents a less fastidious and time consuming alternative to the use of microphone arrays, as suggested by Genescà et al. (2009). Our study, however, was not free of limitations. Our measurements were taken exclusively during the daytime and so potentially important data on nighttime noise is missing. Another possible drawback was the combination of summer and fall data. Previous studies have suggested that seasonal variation may exist in environmental noise levels, perhaps due to variations in flight numbers (Berglund et al., 1999). In the present study, such a difference was observed for the LAmax, 6hr indicator. However, an important secondary finding of our study is the representativity of our results for equivalent noise levels. The agreeability between sites HAR1 and HAR2 for the LAeq, 6hr indicator suggested a lack of fall versus summer variability in equivalent environmental noise levels for the year studied. Similarly, a lack of day-to-day variability in noise measurements was observed, indicated by the consistently low standard deviations for both indicators at all sites. Although this low variability may indicate that this limitation is minor, it should be noted that days with rain or snow were not considered in this study, which restricts the generalizability of this result.

Another consideration is that our results were based on the assumption that the highway and railway contributions to sites HR and HAR were equal. Due to a variety of site factors, this may not necessarily have been the case. Our study would have benefitted from an additional Highway (H) and/or Railway (R) site that lacked contribution from other transportation activities. Of course, this was not possible given the setting. Prospective studies around different airports should look for such sites where possible to allow for a more practicable comparison.

We recommend that future studies assess both indoor and nighttime noise levels in proximity to traffic sources, and also that socioeconomic inequalities within the exposed groups be considered. Finally, as just one of many factors in the impact of transportation sources on the environment, noise pollution should be studied in the context of other pollutants (e.g. air) to gain a more holistic conception of the influence of these sources.

5. Conclusion

Our study suggests that near the Montreal International Airport, noise sources other than aircrafts, such as railways and highways, may be important contributors to environmental noise levels. However, close to the airport, the estimated contribution of airport noise was roughly equal to that of these two other transportation sources combined. In this area, noise levels may be elevated by airport operations in addition to noise from flights, the former of which is unlikely to be a contributing factor at farther distances from the airport. At farther distances however, aircraft passages can be observed to raise noise levels to a greater extent above levels when no flight is passing, reflecting a sporadic quality of aircraft noise that may be more annoying than noise from other sources. Expectedly, noise levels were found to be the highest where there were three contributing transportation sources. However, at all sites measured, the WHO's noise exposure guidelines of 55 dBA were exceeded, including those sites located in residential areas or near a school.

Acknowledgments

We would like to acknowledge Allan Brand for his contribution of the map of Montreal.

References

Barron, R. F. (2003). *Industrial noise control and acoustics.* New York, NY: Marcel Dekker, Inc.

Berglund, B., Lindvall, T., & Schwela, D. H. (1999). *Guidelines for Community Noise.* Geneva: World Health Organization.

Boeing. (2011). *Airport noise and emissions regulations.* Retrieved July 17, 2013, from http://www.boeing.com/commercial/noise/dorval.html

Clark, C., Martin, R., van Kempen, E., Alfred, T., Head, J., Davies, H. W., ... Stansfeld, S. A. (2005). Exposure-effect relations between aircraft and road traffic noise exposure at school and reading comprehension. *Am J Epidemiol, 163,* 27-37. http://dx.doi.org/10.1093/aje/kwj001

Fields, J. M., & Walker, J. G. (1982). Comparing the relationships between noise level and annoyance in different surveys: A railway noise vs. aircraft and road traffic comparison. *J Sound Vib, 81,* 51-80. http://dx.doi.org/10.1016/0022-460X(82)90177-8

Genescà, M., Romeu, J., Arcos, R., & Martín, S. (2013). Measurement of aircraft noise in a high background noise environment using a microphone array. *Transportation Research Part D, 18,* 70-77. http://dx.doi.org/10.1016/j.trd.2012.09.002

Genescà, M., Romeu, J., Pàmies, T., & Sánchz, A. (2009). Real time aircraft fly-over noise discrimination. *J Sound Vib, 323,* 112-129. http://dx.doi.org/10.1016/j.jsv.2008.12.030

Genescà, M., Romeu, J., Pàmies, T., & Sánchz, A. (2010). Aircraft noise monitoring with linear microphone arrays. *IEEE Aerospace Electron Syst Mag, 25,* 14-18. http://dx.doi.org/10.1109/MAES.2010.5442149

Haralabidis, A. S., Dimakopoulou, K., Vigna, T. F., Giampaolo, M., Borgini, A., Dudley, M. L., ... Jarup, L. (2008). Acute effects of night-time noise exposure on blood pressure in populations living near airports. *European Heart Journal, 29,* 658-664. http://dx.doi.org/10.1093/eurheartj/ehn013

Leventhall, H. G. (2004). Low frequency noise and annoyance. *Noise & Health, 6,* 59-72.

Passchier-Vermeer, W., & Passchier, W. F. (2000). Noise Exposure and Public Health. *Environ Health Perspect, 108,* 123-131.

Perron, S., Tétreault, L. F., King, N., Plante, C., & Smargiassi, A. (2012). Review of the effect of aircraft noise on sleep disturbance in adults. *Noise & Health, 14,* 58-67. http://dx.doi.org/10.4103/1463-1741.95133

Stansfeld, S., & Crombie, R. (2011). Cardiovascular effects of environmental noise: Research in the United Kingdom. *Noise & Health, 13,* 229-233. http://dx.doi.org/10.4103/1463-1741.80159

Stansfeld, S., Hygge, S., Clark, C., & Alfred, T. (2010). Night time aircraft noise exposure and children's cognitive performance. *Noise & Health, 12,* 255-262. http://dx.doi.org/10.4103/1463-1741.70504

Stansfeld, S. A., Berglund, B., Clark, C., Lopez-Barrio, I., Fischer, P., Öhrström, E., ... Barry, B. F. (2005). Aircraft and road traffic noise and children's cognition and health: a cross-national study. *Lancet, 365,* 1942-49. http://dx.doi.org/10.1016/S0140-6736(05)66660-3

Transport Canada. (2012). *Noise exposure forecast and related programs.* Retrieved July 18, 2013, from http://www.tc.gc.ca/eng/civilaviation/standards/aerodromeairnav-standards-noise-nef-924.htm

Transports Québec. (2010). *Atlas des transports: Débits de circulation.* Retrieved July 17, 2013, from http://transports.atlas.gouv.qc.ca/NavFlash/SWFNavFlash.asp?input=SWFDebitCirculation_2010

19

Follow-up on High Lead Concentrations in New Decorative Enamel Paints Available in Egypt

Scott Clark[1], William Menrath[1], Yehia Zakaria[2], Amal El-Safty[2], Sandy M Roda[1], Caroline Lind[1], Essam Elsayed[2] & Hongying Peng[1]

[1] Department of Environmental Health, University of Cincinnati, USA

[2] Department of Occupational and Environmental Health, Cairo University, Cairo, Egypt

Correspondence: Scott Clark, Department of Environmental Health, University of Cincinnati, USA. E-mail: clarkcs@ucmail.uc.edu

Abstract

The average total lead concentration of new enamel household paints in Egypt was previously reported to be the second highest among the seven countries from Africa, Asia and South America that were included in a 2009 publication. The follow up study reported in this paper includes more than twice as many brands (11 versus 4) and samples (45 versus 20) as the initial study. Paints from three of the four brands included in the initial study were sampled again to examine possible changes. Paint from the eight brands not examined in the initial study had lower lead concentrations (4,150 ppm average) compared to brands in the initial study, 26,200 ppm resulting in an average concentration of 11,900 ppm in the follow up study. These two averages are 291- and 132-times higher, respectively, than the current U.S. limit of 90 ppm in new paints for consumer use.

Paint lead concentrations in brands/colors manufactured at different times did not exhibit any overall pattern of increase or decrease. The data from the follow up and initial studies were combined using the sample collected more recently for those brand/colors collected twice, resulting in a total of fifty-two (52) samples with an average lead concentration of 14,300 ppm. The presence of lead in new paints continues to represent a threat to children and efforts are needed to cease the use of lead compounds in making paints by using readily available substitutes.

In a 1997 report of the analyses of fifteen (15) new paints intended for use on the interior of houses, the median concentration, 370 ppm, and the maximum, 19,200 ppm, were much lower than those presented in this report. This is consistent with a statement in the 1997 report that anecdotal evidence that some paint companies may be starting or increasing the production of lead-based paint.

1. Background and Introduction

In recent years there has been growing world-wide concern as awareness of the continued use of lead (Pb) in new household paint becomes more widespread. Data from paint analyses of new enamel decorative paints in twelve countries (Adebamowo et al., 2007; Clark et al., 2009) show average concentrations ranging between 7,000 ppm and 32,000 ppm total Pb. Most of the brands tested had concentrations exceeding 10,000 ppm in at least one sample, 111 times the current limit in the United States (CPSIA, 2008). Other studies have also reported very high Pb concentrations (e.g. Van Alphen, 1999; Clark et al., 2006, Lin, Peng, Chen, Wu & Du, 2008; Kumar & Gottesfeld, 2008; Toxics Link/IPEN, 2009; Ewers et al., 2011; Nganga, Clark & Weinberg, 2012; Gottesfeld, Kuepouo, Tetsopgang & Durand, 2013).

Lead poisoning from the former use of leaded paints is a hazard that continues for decades after use of Pb in the manufacturing of paint has ended. For example, in the United States, after the use of Pb in paints was banned thirty-five years earlier, significant Pb -based paint hazards still exist in over 20 million housing units as reported by Jacobs et al. (2002). Several studies of existing housing in other countries have found paints that have high Pb concentrations: in India by Kuruvilla et al. (2004) and in South Africa by Montgomery and Mathee (2005), In China, Lin et al. (2008) reported high concentrations of Pb in paints on walls, furniture and toys of kindergarten and primary schools. For children, the major Pb exposure pathway in housing involves ingestion of Pb-contaminated settled dust on floors and window sells (Jacobs et al., 2002). More than one-half of the housing units in New Delhi that were examined by Kumar and Clark (2009) had at least one dust Pb sample that exceeded the USEPA standard for dust lead loading in floor or window sill dust.

Concern from the findings of high Pb concentrations in new paints resulted in the creation by the United Nations of the Global Alliance for the Elimination of Lead in Paints (UNEP, 2013) involving the United Nations Environment Programme (UNEP) and the World Health Organization (WHO) (UNEP, 2013). With a goal of fostering the elimination of the use of Pb in new paints the European Union has provided support for the creation of the multi-year Asian Lead Paint Elimination Project directed by IPEN, an international network of non-governmental organizations involving its member organizations in seven Asian countries (IPEN, 2013).

As global awareness of Pb paint hazards increase it will also become important to conduct follow up studies to determine whether the awareness has resulted in reductions in the Pb concentrations in new paints. One such country where a follow-up study has occurred is Egypt. The average paint Pb concentration reported in the initial paint Pb sampling in Egypt of 26,200 ppm total Pb, second highest of that found in seven countries in Africa as reported from several studies as shown in Table 1. The average concentration of Pb in the paints from Egypt was also the second highest of those reported for eleven countries in Africa, Asia and South America as reported by Clark et al. (2009).

Table 1. Total Pb concentrations in new decorative enamel paints from countries in Africa

Country	Date of Study/ Report	# Samples	Average ppm Pb	% Samples greater than 90 ppm lead	% Samples greater than 600 ppm lead	% Samples greater than 10,000 ppm lead
Cameroon	2011[a]	60	22,800	67	65	25
Egypt	2006[b]	20	26,200	65	65	45
Kenya	2012[c]	31	14,900	87	81	39
Nigeria	2009[b, e]	25	15,800	96	96	44
Nigeria	2009[d, e]	23	37,000	100	100	65
Senegal	2009[d]	21	5,870	86	76	19
South Africa	2009[d]	29	19,900	65	62	28
Tanzania	2009[d]	20	14,500	100	95	25

[a] Gottesfeld et al. (2013). [b] Clark et al. (2009), [c] Nganga et al. (2012), [d] Toxics Link (2009).

[e] Samples with average of 15,800 ppm purchased in 2006; those with 37,000 ppm avg. purchased in 2008. For the four (4) paints of same brand/color included in each survey, the ratio of the average concentrations (50,000/23,000 = 2.2) was about the same as that for all samples in each survey as shown above (37,000/15,800 = 2.2).

The purpose of this project is to determine if leaded house paints are still available in Egypt and if there is any evidence of changes in the concentration of Pb in these paints.

2. Materials and Methods

2.1 Sample Collection

Enamel household paints from retail shops in the Greater Cairo area were analyzed for total Pb on two occasions, one reported by Clark et al. (2009) and the other presented in this manuscript. Results from the initial analysis of five colors from each of four major brands were included in with data from a total of eleven (11) countries in three continents. The current follow-up study of forty-five (45) new paint samples includes (1): collection of paints from brands not included in the initial study, (2) paints from some colors of these additional brands that were manufactured at different times and (3) repeat sampling of some paints from brands previously analyzed in initial study (Table 2).

Table 2. Paint samples collected in follow-up study

BRANDS	# Brands	# Samples
1.Brands included in Follow-Up Study but not in Initial Study	8	34
2. Brands included in Follow-up Study but not in Initial Study and for which some paints manufactured at two different times were included	2[a]	6[a]
3. Brands included in Initial study that were re-sampled in Follow-up	3	11
4. Brand/colors included in Initial study that were re-sampled in Follow-up	2[b]	7[b]
5. Total number of brands and samples in Follow-up Study (sum of Lines 1 and 3)	11	45
6. Total number of samples of specific brand/colors included in Follow-up Study (Total Sample number in line 5 minus Total Sample number in Line 2)	11	39

[a] these samples are included in line 1 above and are thus not part of the Total in line 5.

[b] these samples are included in line 3 above and thus are not part of the Total in line 5.

A combined data set was also developed using data from both studies but only using a single result for each brand/color combination. If more than one sample was collected of a particular brand/color, the result from the most recent sample was included in the combined data set. The combined data set has a total of fifty-two (52) samples with unique brand/color combinations.

2.2 Sample Preparation and Analyses

Each paint sample was thoroughly mixed with a clean single-use stirrer. For each sample a single coat of paint was carefully applied to a clean piece of polycarbonate with an unused clean paint brush that was discarded after use with one sample. In the initial Clark et al. (2009) survey paint samples were applied to unused clean wood. After drying overnight at room temperature, samples were placed in an unused clean plastic bag for delivery to the H & E Laboratory at the University of Cincinnati. In the H & E Laboratory, paint samples were carefully removed from the painted surface by means of a clean sharp paint scraper, using care not to remove portions of the substrate. The paint scrapings were extracted with nitric acid and hydrogen peroxide according to the method: Standard Operating Procedures for Lead in Paint by Hotplate or Microwave-based Acid Digestions and Atomic Absorption or Inductively Coupled Plasma Emission Spectroscopy, EPA, PB92-114172, September 1991 (US EPA, 2001). Extracts were analyzed by flame atomic absorption spectroscopy using a Perkin-Elmer 5100 spectrometer.

The H & E Laboratory was accredited by the American Industrial Hygiene Association (AIHA) as an environmental Pb laboratory under the National Lead Laboratory Accreditation Program. The laboratory participated in the Environmental Lead Proficiency Analytical Testing (ELPAT) program. Strict quality control procedures were maintained according to the accreditation guidelines. The accreditation program operated by AIHA meets all international program requirements of ISO/IEC 17025 and subsequently ISO/IEC 17011. AIHA is a full member of the International Laboratory Accreditation Cooperation (ILAC).

3. Results

Total Pb concentration data in parts per million (ppm) dry weight were compared by color and paint brand. Results are presented as average, minimum and maximum and percentages exceeding 90 ppm, 600 ppm and 10,000 ppm. Ninety (90) ppm is the current limit for Pb in decorative paints in the US (CPSIA, 2008), 600 ppm is the former limit in the US (CPSC, 1977) and the current limit in several countries including Brazil (Brazilian Federal Law, 2008) and Singapore (Singapore NEA, 2004) and 10,000 ppm represents a very high concentration. Analysis of results was also by paint brand and date of paint manufacture or purchase.

3.1 Pb Concentrations of Paints in Follow-up (Current) Survey

Forty-five (45) paint samples were collected in the follow up survey. For six brand/color combinations two samples from each were collected with different dates of manufacture. Thus there was a resulting thirty-nine (39) brand/colors included in the current survey.

3.1.1 Comparison of Results of Samples From Paint Manufactured at Different Dates

Results from the duplicate samples for these six brand/color combinations (Table 3) are similar. The four samples with the lowest concentrations in the samples manufactured earlier decreased in concentration from the earlier to the later date of manufacture and the two with higher concentrations increased from the earlier to the later date. The average concentration increased from 6,540 ppm to 9,320 ppm. The two-sided p-value of Wilcoxon two-sample test is 0.69, which indicates no statistically significant difference in Pb concentration between samples collected on different dates.

Table 3. Results of Pb analyses (ppm) of samples of paints of the same brand/color in follow-up study that were manufactured on different dates

Brand	Color	Pb Concentration (ppm) (date of manufacture)	Pb Concentration (ppm) (date of manufacture)
A	Black	146 (7/08)	30.4 (10/09)
A	Blue	39.8 (5/08)	26.1 (1/09)
A	Red	6,080 (5/08)	10,600 (10/09)
A	White	13.9 (5/08)	8.6 (9/09)
A	Yellow	32,800 (5/08)	45,200 (9/09)
B	White	132 (9/09)	63.6 (10/09)
Average		6,540	9,320

3.1.2 Comparisons of Pb Concentrations in Paint Brands/Colors in Initial Study that Are Also Included in Follow-up

Seven (7) of the forty-five (45) samples analyzed in the current study were the same brand/color as included in the initial study. The average Pb concentrations in the samples of the same brand/color that were collected in the two time periods (Table 4) are similar but one sample from one brand decreased substantially while two from the other brand had large increases. The average of these brand/colors in the initial study was 35,100 ppm and 36,900 ppm in the follow-up study. The two-sided p-value of Wilcoxon two-sample test is 1, which indicates no statistically significant difference between the Pb concentration in these samples collected in the initial and follow-up studies.

Table 4. Comparison of total Pb concentrations (ppm) in new enamel decorative paints included in both initial and follow-up studies

Brand	Color	Concentration (ppm) Initial Study	Concentration (ppm) Follow-up Study
C	Green	39,000	17,000
C	Yellow	140,000	122,000
D	Blue	106	70.9
D	Green	14,300	16,600
D	Red	5,620	9,040
D	White	46.6	60.5
D	Yellow	46,600	110,000
Overall Average		35,100	36,900

3.1.3 Distribution of Pb Concentrations of Samples in Follow-up Study

For paint samples of the same brand/color manufactured on different dates that were analyzed, results from the most recent sample, as shown in Table 3, are included in the distribution of Pb concentrations presented in Table 5. Data from a total of thirty-nine (39) samples are presented in Table 5 by brand and Pb concentration. Three or more samples were analyzed for seven of the eleven brands.

Table 5. Distribution of total Pb concentration (ppm) in the thirty-nine brand/color combinations of new enamel decorative paints purchased in Egypt in current follow-up survey

Brand	Number of Samples	Average Parts Per Million (ppm) Lead	Percent of Samples Above 90 ppm Lead	Percent of Samples Above 600 ppm Lead	Percent of Samples Above 10,000 ppm Lead	Minimum, ppm	Maximum ppm
E	1 (gold)	35.3	0	0	0	35.3	35.3
C	4	51,600	100	100	75	5,260	122,000
F	4	2,590	100	50	0	92.5	6,790
G	1 (blue)	3,110	100	100	0	3,110	3,110
H	1(white)	93.3	100	0	0	93.3	93.3
I	3	134	100	0	0	101	193
A	8	8,750	50	38	38	8.6	45,200
D	6	22,600	67	50	33	60.9	110,000
B	7	5,690	86	86	14	63,6	16,500
J	3	49.2	0	0	0	44.1	54.3
K	1 (red)	272	100	0	0	272	272
Overall	39	11,900	72	49	23	8.6	122,000

Overall 72% of samples exceeded 90 ppm Pb concentration and 49% exceeded 600 ppm. The average concentration was 11,900 ppm. Six of the eleven brands have an average Pb concentration above 1,000 ppm; four of eleven brands had at least one sample with a very high lead concentration > 10,000 ppm. The two brands with the highest average concentrations, 51,600 ppm (Brand C) and 22,600 ppm (Brand D) had maximum lead concentrations: 122,000 ppm and 110,000 ppm, respectively. These two brands were among the four examined in the initial study.

3.2 Analysis of Combined Results From the Follow up and Initial Studies

For the thirteen (13) paints shown in Tables 3 and 4 samples that were manufactured on two different dates were analyzed. Statistical analysis presented earlier revealed that results did not vary by date. Therefore, for these paints only the results from the samples that were manufactured more recently were included in the combined sample analysis. The resulting combined data set contains a total of fifty-two (52) samples. The number of 52 paints was determining by adding the 39 samples of specific brands/colors in follow-up study (Table 3 line 5) to the number of samples in initial study, 20, (as indicated in Abstract) minus the seven (7) paints in the intial study that were re-sampled in the follow-up study (Table 3 line 4). These 52 paints are from twelve (12) brands and have an average Pb concentration of 14,600 ppm (Table 6). Seventy-three percent (73%) had Pb concentrations > 90 ppm, fifty-three percent (53%) were > 600 ppm, twenty-seven (27)% had concentrations > 10,000 ppm and the highest concentration was 122,000 ppm.

Table 6. Distribution by Brand of Total Pb Concentration (ppm) in Combined Data Set of New Enamel Decorative Paints Purchased in Egypt in Initial and Follow-up Studies

Brand	Number of Samples	Average ppm Lead	Percent of Samples Above 90 ppm Pb[a]	Percent of Samples Above 600 ppm Pb[b]	Percent of Samples Above 10,000 ppm Pb	Minimum, ppm	Maximum ppm
E	1	35.3	0	0	0	35.3	35.3
C	7	37,100	86	86	57	`4.5	122,000
F	4	2,590	100	50	0	92,5	6,790
G	6	23,000	100	100	50	2,170	68,200
H	1	93.3	100	0	0	93.3	93.3
I	3	134	100	0	0	101	193
A	8	8,750	50	38	38	8.6	45,200
D	6	22,600	67	50	33	60.9	110,000
B	7	4,990	86	86	14	63,6	16,500
J	3	49.2	0	0	0	44.1	54.3
L	5	18,100	20	20	20	4.5	90,600
K	1 (red)	272	100	0	0	272	272
Overall	52	14,300	69	48	28	4.5	122,000

[a] 90 ppm total Pb is the current standard in the US for new domestic paints (CPSIA, 2008)

[b] 600 ppm total lead is the current standard in several countries including Brazil (Brazilian Federal Law 2008) and Singapore (Singapore NEA, 2004).

The four brands included in the 2009 study (Clark et al., 2009) (C, D, L, and G, as identified in Table 6) had the highest average Pb concentration in the combined data set, with a range from 18,100 ppm to 37,100 ppm and an average of 26,000 ppm. The average concentrations of the eight (8) brands that were included only in the follow up study were lower, ranging from 35.3 ppm and 8,750 ppm with an average of 4,150 ppm. The median two-sample test statistic for these data equals 13, and its standardized Z value is 1.74. The two-sided Chi-Square p-value Pr > Z equals 0.08. Although the difference in Pb concentration between the two groups of paint brands is not significant at the p level of 0.05, the difference observed may be considered to be of borderline significance.

4. Discussion

The lower average Pb concentration in samples in the follow up study,11,900 ppm, than found in the earlier study, 26,200 ppm, resulted from the lower average concentrations of samples from brands in the follow up that were not included in the earlier study.

For the thirteen paints of various colors of four brands of paint that were collected on two different dates (Tables 3 and 4) the Pb concentrations were not significantly different between the two sets of samples.

Pb concentrations in fifteen (15) new interior household paints from Cairo analyzed by Chappell et al. (1997) were much lower than those found in this follow-up study. The concentrations reported by Chappell et al. (1977) ranged from 0.02 to 19,200 ppm and with median 370 ppm.The maximum concentration was only about one-sixth of that in this follow up study and one-seventh of that in the earlier study. Five samples in the follow-up study had maximum concentrations higher than that reported by Chappell et al. (1977), ranging from 1.7 times higher to more than six times higher. The median concentration in the follow-up study (1,690 ppm) was 2.8 times higher than that reported by Chappell et al. (1997). These differences may be due to different brands and colors being sampled or to actual increases in the amount of Pb used in producing paints. Chappell et al. (1997) reported that there was anecdotal evidence, that could not be substantiated, that some private paint manufacturers may be starting or increasing the production of Pb-based paints.

5. Conclusions

Seventy-two (72) percent of the paints in the follow-up study had Pb concentrations that exceeded the allowable limit in the United States. Forty-nine (49) percent exceeded 600 ppm, the former limit in the US and the current limit in some other countries. The average Pb concentration of 11,900 ppm in the new enamel decorative paints in this follow up study was lower, than the average of 26,200 ppm found in the earlier study. This difference is consistent with the lower average concentrations in samples from brands included in the follow up that were not included in the earlier study

For the thirteen (13) paints of the same brand and color for which samples manufactured on two different dates were analyzed there was no significant difference in the Pb concentrations over time. This suggests that Pb concentrations in new paints in Egypt do not appear to be increasing or decreasing with time. Therefore, health authorities and the general public need to be aware of the continuing Pb threat to children and others from lead-based paints.

Concentrations of lead in new paints in the follow-up and initial studies were much higher those in paints manufactured earlier as reported by Chappell et al. (1997).

References

Adebamowo, E. O., Clark, C. S., Roda, S., Agbede, O. A., Sridhar, M. K. C., & Adebamowo, C. A. (2007). Lead Content of Dried Films of Domestic Paint Currently Sold in Nigeria. *Science Total Environ, 388*, 116-120. http://dx.doi.org/10.1016/j.scitotenv.2007.07.061

Brazilian Federal Law 11.762 is dated August 1, 2008.

Chappell, R., Billig, P., Brantly, E., Ault, S., & Ezzeldin, H. S. (1997). Lead Exposure Abatement Plan for Egypt: Results of Environmental Sampling for Lead, US. Agency for International Development, USAID Mission to Egypt, Environmental Health Project No. 936-5994, Wash. DC 20523.

Clark, C. S., Rampal, K. G., Thuppil, V., Chen, C. K., Clark, R., & Roda, S. (2006). The lead content of currently available new residential paint in several Asian countries. *Environmental Research, 102*, 9-12. http://dx.doi.org/10.1016/j.envres.2005.11.002

Clark, C. S., Rampal, K. G., Thuppil, V., Roda, S. M., Succop, P., Menrath, W., ... Yu, J. (2009). Lead levels in new enamel household paints from Asia. *Africa and South America Environmental Research, 109*, 930-936. http://dx.doi.org/10.1016/j.envres.2009.07.002 PMid:19656507

Consumer Product Safety Commission (CPSC). 42 Federal Register 44193 (September 1, 1977).

Consumer Product Safety Improvement Act (CPSIA) of 2008, Public Law 110-314, August 11. Washington, DC.

Ewers, L., Clark, C. S., Peng, H., Roda, S. M., Menrath, B, Lind, C., & Succop, P. (2011). Lead Levels in New Residential Enamel Paints in Taipei, Taiwan and Comparison with Those in Mainland China. *Environmental Research, 111*(6), 757-760.

Gottesfeld, P., Kuepouo, G., Tetsopgang, S., & Durand, K. (2013). Lead Concentrations and Labeling of New Paint in Cameroon. *Journal of Occupational and Environmental Hygiene, 10*, 243-249. http://dx.doi.org/10.1080/15459624.2013.768934 PMid:23472856

IPEN, Asian Lead Paint Elimination Project. Retrieved September 25, 2013, from http://www.ipen.org/projects/asia-project-2012-2015

Jacobs, D. E., Clickner, R. P., Zhou, J. Y., Viet, S. M., Marker, D. A., Rogers, J. W., ... Friedman, W. (2002). The Prevalence of Lead-Based Paint Hazards in US Housing (2002). *Environmental Health Perspectives, 110*, 599-607. http://dx.doi.org/10.1289/ehp.021100599

Kumar, A., & Clark, C. S. (2009). Lead Loadings in Household Dust in Delhi, India. *Indoor Air, 19*, 414-420. http://dx.doi.org/10.1111/j.1600-0668.2009.00605.x PMid:19659889

Kumar, A., & Gottesfeld. P. (2008). Lead Content in Household Paint in India. *Science of the Total Environment, 407*, 333-337. http://dx.doi.org/10.1016/j.scitotenv.2008.08.038

Kuruvilla, A., Pillay, V. V., Venkatesh, T., Adhikari, P., Chakrapani, M., Clark, C. S., ... Sinha,S. (2004). Portable Lead Analyzer to Locate Sources of Lead. *Indian Journal of Pediatrics, 41*, 495-499. http://dx.doi.org/10.1007/BF02724287

Lin, G. Z., Peng, R. F., Chen, Q., Wu, Z. G., & Du, L. (2008). Lead in housing paints: an existing source still not taken seriously for children lead poisoning in China. *Environmental Research, 109*, 1-5. http://dx.doi.org/10.1016/j.envres.2008.09.003

Montgomery, M., & Mathee, A. (2005). A Preliminary Study of Residential Paint lead Concentrations in Johannesburg. *Environmental Research, 98*, 279-283. http://dx.doi.org/10.1016/j.envres.2004.10.006
Nganga, C., Clark, S., & Weinberg, J. (2012). Lead in Kenyan Household Paint, September, 2012, iLima, Nairobi, Kenya, IPEN, University of Cincinnati.

Singapore National Environmental Agency (NEA). (2004). List of controlled hazardous substances. Retrieved June 16, 2004 from http://app.nea.gov.sg/cms/htdocs/article.asp?pid=1428S

Toxics Link-IPEN Global Study: Lead in New Decorative Paints. (2009). Retrieved from http://www.ipen.org/ipenweb/documents/work%20documents/global_paintstudy.pdf

United Nations Environmental Program (UNEP). (2013). Retrieved September 25, 2013, from http://www.unep.org/hazardoussubstances/hazardoussubstances/LeadCadmium/PrioritiesforAction/GAELP/tabid/6176/Default.aspx

U.S. Environmental Protection Agency (USEPA). (2001). Standard Operating Procedures for Lead in Paint by Hotplate or Microwave-based Acid Digestions and Atomic Absorption or Inductively Coupled Plasma Emission Spectroscopy, EPA, PB92-114172, Sept. 1991

Van Alphen, M. (1999). Lead in Paints and Water in India. In A. M. George (Ed.), *Lead Poisoning Prevention & Treatment: Implementing a National Program in Developing Countries* (pp. 265-272). The George Foundation, Bangalore, India.

20

Phthalates and Other Plastic Additives in Surface Sediments of the Cross River System, S.E. Niger Delta, Nigeria: Environmental Implication

Orok E. Oyo-Ita[1], Bassey O. Ekpo[1], Inyang O. Oyo-Ita[1] & John O. Offem[1]

[1] Environmental and Petroleum Geochemistry Research Group, Department of Pure and Applied Chemistry, University of Calabar, C.R. State, Nigeria

Correspondence: Orok E. Oyo-Ita, Environmental and Petroleum Geochemistry Research Group, Department of Pure and Applied Chemistry, University of Calabar, C.R. State, Nigeria. E-mail: orokoyoita@yahoo.com

Abstract

Quantitative determination of phthalates and other plastic additives was carried out using GC-MS in order to understand the distribution and fate of these compounds in surface sediments of the Cross River System. Results show the concentration ranges (mean±standard deviation) of the three phthalates as: di(ethylhexyl)phthalate (DEHP; 1.97– 86.76 [24.06 ± 29.88 mg/kg dry weight dw]); di(n-butyl)phthalate (DnBP; 0.16 – 17.41 [3.25 ± 5.03 mg/kg dw]); di(isobutyl)phthalate (DiBP; 1.14 – 29.64 [9.82 ± 10.23 mg/kg dw]). However, examination of n-hexane procedural blank (used also in the study for clean-up protocol) GC trace revealed the presence of certain amounts of DEHP and tris-2,6-di(t-butyl)phenylphosphite (PhP) which interfere with the targeted analytes. Therefore, extract/blank (E/B) ratios were calculated and were in the range 1.05 – 12.54, indicating that the concentrations of DEHP and PhP previously assigned to the sediments were partly derived from laboratory artifacts. Differences in grain size distribution, partitioning behavior, volatility and solubility in the aqueous phase as well as localized influx may account for the observed spatial variation of phthalates and other plastic additives in the river system. The primary sources of these phthalates and tris-2,6-di(t-butyl)phenylphosphate (PhP') were considered to be the result of direct discharge of untreated effluent/solid waste and emission arising from burning of refuse containing plastic materials, respectively. The occurrence of certain anti-oxidant degradation products in sediments (not in the blank) such as 2,6-di(t-butyl)-4-hydroxybenzaldehyde (HBA) and 2,6-di(t-butyl)quinine (Qn) and PhP' (a compound proposed here as a possible marker for plastic combustion for the region/air basin), suggests that phthalates contamination had occurred before sample contact with laboratory artifacts.

Keywords: Phthalate, tris-2,6-di(t-butyl)phenylphosphate, source, distribution, environmental implication, cross river system

1. Introduction

Over the past few decades, phthalate compounds have become one of the most abundant pollutants in the environment from a variety of industrial discharges, particularly in tropical developing economies (such as Nigeria) with their attendant poor disposal systems. In 1999, the top U.S 100 companies reported total di(n-butyl)phthalate (DnBP) discharges of more than 300, 000 pounds. The same year, the top U.S 100 companies emitting DEHP reported nearly 1.2 million pound of discharges. These reports however, exclude phthalate discharges to the environment from the use or disposal of finished consumer products (Staples, Charle, Dennis, & Peterson., 1997). Similar estimate has not yet been documented for the sub-saharan tropical region of Africa. Therefore, the present study provides baseline data for future inventory estimate of phthalate load in sedimentary environment for the region.

Phthalates are used as plasticizers to soften plastics and as solvents in a wide range of products such as perfumes, nail polish, hair spray, detergent, toys and food packages. Most of the mid- to high-molecular weight phthalates are used in the manufacture of PVC while the low molecular weight counterparts such as DnBP are used in epoxy resin, cellulose ester bound plastics and specialized formulation. Thus for this reason, phthalates have been found in virtually in all compartments of the environment including fresh water, lake sediments, marine environments, sewage treatment plants and sewage sludge (Aranda, O'Connor, & Ekeman, 1989) as well as in buried PVC in landfill leachates and in atmospheric aerosols (Simoneit, Standley, & Cox, 1988). Phthalates have also been detected in rain water of Los Angeles, USA with DEHP as the dominant homolog (Simoneit & Mazurek, 1989).

According to Simoneit et al. (2005), the major components of plastic extracts are n-alkanes, DEHP (plasticizer) and anti-oxidants (AO) as well as lubricant/anti-adhesve (IRGANOX 1075, IRGAFOX 168). We have also detected phthalates, non-specific n-alkanes (mainly of even carbon numbered predominance), 4-hydroxybenzoic acid and biphenyl as well as minor amounts of polycyclic aromatic hydrocarbons in organic solvents believed to have been kept in plastic containers prior to dispensing into bottles by commercial chemical vendors in Nigeria (Ekpo, Oyo-Ita, Oros, & Simoneit, 2011).

The general populace is exposed to phthalates through food, water, air and the use of phthalate-containing consumer products that may be eaten, inhaled or applied directly to the skin and absorbed. Phthalates and their by-products have been detected in urine and blood of individuals exposed to phthalates (National Institute of Public Health and Environmental Protection [NIPHEP], 1991; Meeker, Sheela, & Shama, 2009). DEHP is a compound of special concern among the plasticizers because it has been described as a probable human carcinogen by the United State Environmental Protection Agency (USEPA, 1992). Toxic and carcinogenic effects of phthalates are well established in experimental animals (Ashby, 1994; Jim-Carlisle, Henkel, Ling-Hongly, & Pagepainter, 2009), even though such ability to produce similar effects in human is yet to be documented. Evidence that humans are not at serious risk from exposure to phthalates has also been proven (NIPHEP, 1991; USEPA, 1992).

Phthalates may come in contact with sediments during sample collection, storage and analysis. One of the challenges in the present study was to establish whether the phthalates and other plastic additives found in sediment samples from the Cross River System were indigenous or that they were introduced into the samples during laboratory protocols. To achieve this, the composition and concentrations of phthalates and other plastic additives in the sediment samples were determined in relation to the laboratory blank (n-hexane). Since these compounds may be globally distributed (Simoneit, Medeiros, & Didyk, 2005), the additional thrust in the present study was to present data that could add to knowledge of the fate and factors that control the distribution of these compounds in the environment. To achieve this, we compared these compositions and concentrations in samples (which were obtained about 10 years ago) with those samples recently obtained from the same location for analysis of lipids.

Compounds such as tris-2,4-di(t-butyl)phenylphosphate and 1,3,5-triphenylbenzene have been identified as markers in smoke particles from burning of plastic materials in the USA air-basin (Simoneit et al., 2005). For purposes of improving emission inventories and air quality management for policy making strategy, this is the first elucidation and measurement of a compound (proposed here as a marker) in sediments of the Niger Delta river network. This proposed marker compound could be utilized as an assessment tool for the contribution of refuse combustion to Nigerian air-basin and other environments worldwide where plastics with similar chemical composition may be in use. The main objectives of the study therefore were to report on the levels, fate and distribution of phthalates and certain plastic additives in sediments of the Cross River System, as well as evaluate the environmental implication of their occurrence.

1.1 Study Area

The study area (Figure 1) extends from Mbo River to 1 km beyond Itu bridge and included part of the upper Calabar River, a distance of approximately 195 km covering remote areas. The Cross River System is one of the river systems of the southeastern Niger Delta of Nigeria, which lies between longitudes 2°03'E and 10°00'E and latitudes 4°00'N and 8°00'N and covers an area of 54,000 km^2, of which 14,000 km^2 lies in Cameroon and 39,000 km^2 is in Nigeria. The river is formed from numerous tributaries arising from the western slopes of the Cameroon mountains and flows south-westwards into the Atlantic Ocean with a discharge of between 879 and 2533 m^3/sec (Asuquo, Ogri, & Bassey, 1999). The system is exposed to temporal flooding depending on the tides and season, and has large fluctuation in hydrographical conditions (Lwenberg & Knzel, 1992; Pisani., Oros, Oyo-Ita, Ekpo, Jaffe, & Simoneit, 2013). Calabar and Great Kwa Rivers are the major tributaries of the Cross River System.

1.2 Sample Collection, Preservation and Preparation

Cross River System was divided into four zones, each depicting a peculiar type of regional impact: Zone I consists of samples collected from the estuary-near Oron beach. This is a highly populated area that receives a significant amount of untreated domestic sewage, agricultural run-off, garbage/refuse discharge, possible oil spills from offshore operations and ship/boat traffic pollution. Zone II consists of samples obtained from area near Oku Iboku beach and thick mangrove swamp forest that receives effluents from a paper mill and garbage/refuse discharge; Zone III consists of samples collected from area near Itu beach that receives domestic sewage discharge, agricultural run-off as well as boat traffic pollution and garbage/refuse discharge; Zone IV consists of samples collected from the upper Calabar River that receives mainly untreated waste water/solid waste from a rubber processing/plastic industry as well as emission arising from burning of garbage/refuse containing plastic materials at a nearby waste dump.

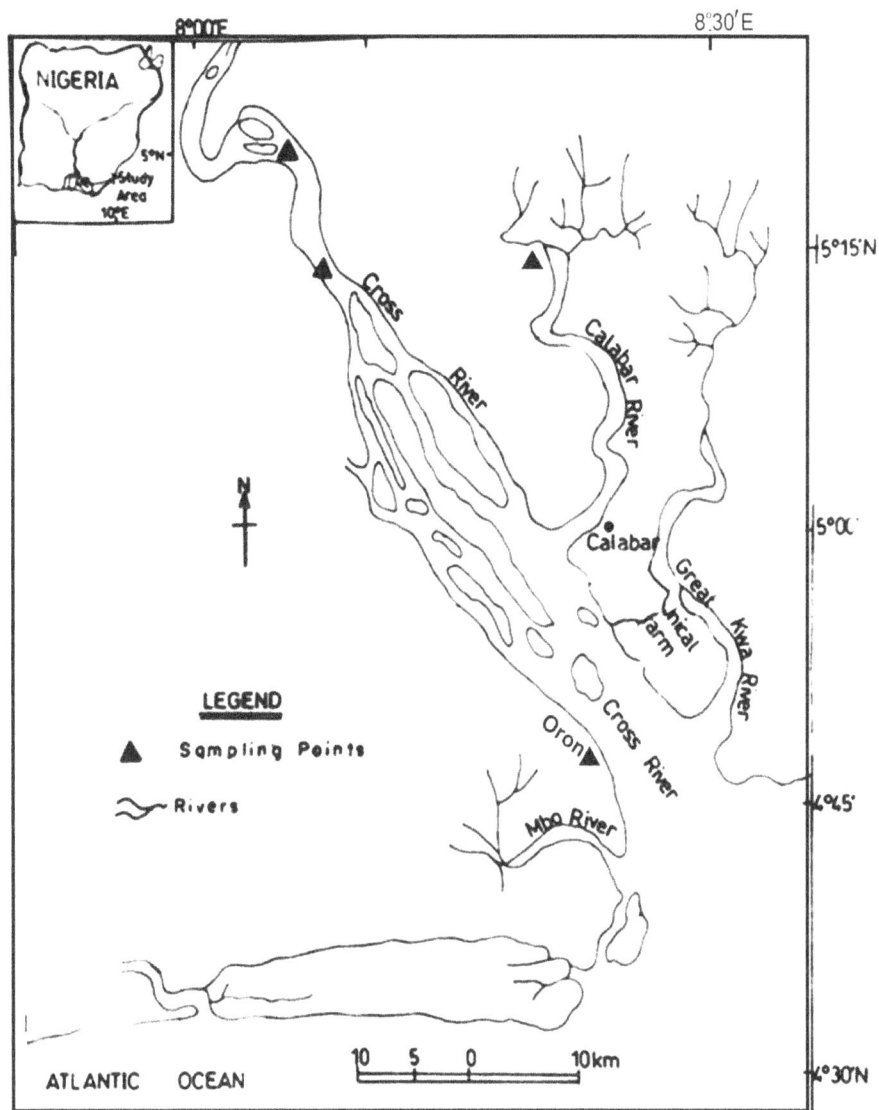

Figure 1. Map of the cross river system showing the sampling locations

Sediments were collected about 10 years ago (April, 2003) during raining season with a Van-Veen grab sampler (0.1 m^2), wrapped in aluminum foil and stored frozen at -4 °C until further analysis is carried out. Freeze-dried sediments were ground using a *mortar/pestle* and subsequently sieved through 230 mesh sieve to obtain < 63 m fraction. Grain size analysis was performed on the sediment according to the method described by Folk (1974). Similar treatments were performed on sediments collected recently from the same locations which were being analyzed for lipids in another study.

2. Materials and Methods

2.1 TOC Determination, Extraction and Clean-up

The following solvents-dichloromethane (DCM), methanol (MeOH), hexane, ethyl acetate and acetone-were used for the analysis. The vessels used for analysis were pre-cleaned with ethyl acetate followed by acetone and heated overnight at 300 °C. Prior to the extraction protocol, freeze-dried sediment samples were decarbonated in 37% hydrochloric acid in several stages until bubbling stopped and rinsed in deionised water to neutral pH. The total organic carbon (TOC) content of the sediments was determined by flash combustion at 1024 °C, followed by thermal conductivity detection in triplicate in a CHNS Elemental Analyser, Carlo Erbar 1108.

Extraction of 2 g sediment sample in a test tube was performed sequentially by sonication with DCM/MeOH (2:1) and DCM in triplicate (Grimalt, Vam-Drogge, Ribes, Vilanova, Fernandez, & Appleby, 2004). Clean-up was performed by adsorption chromatography in an open glass column packed with 1 g anhydrous sodium sulfate (top), 2 g neutral alumina (middle; activated at 400 °C, 5% water deactivated) and glass wool (bottom). Elution with hexane yielded fractions enriched in aromatic compounds. The collected fractions were concentrated in a stream of N_2 gas to almost dryness.

2.2 Instrumental Analysis

Individual phthalates and associated plastic additives were identified and quantified using gas chromatography-FID and gas chromatography-mass spectrometer (GC-MS) on a Hewlett-Packard model 6890 GC coupled to a Hewlett-Packard model 5973 quadrupole MSD. Separation was achieved on a fused silica capillary column coated with DB5 (30 m x 0.25 mm i.d., 0.25 um film thickness). The GC operating conditions were as follows: temperature hold at 65 °C for 2 min, increased from 65 to 300 °C at a rate of 6 °C min⁻¹, with final isotherm hold at 300 °C for 20 min. Helium was used as carrier gas. The sample was injected in "splitless" mode with the injector temperature at 300 °C. The mass spectrometer was operated in the electron impact mode (EI) at 70 eV ionization energy and scanned from 50 to 650 dalton or using selected ion monitoring (SIM) mode. Data were acquired and processed with the Chemstation software. Individual compounds were identified by comparison of mass spectra resolution with authentic standards and interpretation of mass spectrometric fragmentation pattern, as well as comparison of their GC retention indices with those of reference standards.

One analytical blank (hexane) was run with every batch of 2 - 3 samples to check background contamination during the clean-up step. To remove background contamination, we corrected phthalates and plastic additives concentrations by subtracting mean concentration of the analytical blank from the concentrations of individual target compounds. A calibration curve (detector response versus amounts injected) *was* performed for each compound to be quantified. The linear range of the detector was estimated from the curve generated by plotting detector signal versus amount injected. All measurements were performed in the linear ranges for each target compound. In few cases, the samples were re-diluted and re-injected to fit within the linear range of the instrument. The limit of detection (LOD) in the SIM mode ranged from 0.02 to 0.10 ng/g dw. The lower limit of determination of these target compounds was estimated from the smallest peak (signal-to-noise ratio > 3:1 in the chromatograms) that could be integrated. The triplicate sample relative percent difference (i.e. relative standard deviation) for all the samples were less than 20% for all compounds Validation of the accuracy of data arising from the quantitative measurements of these compounds was done by analyzing with the certified reference material "CRM-104" (Resource Technology Corporation, USA), obtaining accuracy from 79% to 109% compared with the certified values.

3. Results and Discussion

3.1 Sediment Bulk Geochemical Properties

The results of bulk geochemical parameters for the sediments including the characteristic features of the environment are presented in Table 1. The grain size distribution data revealed predominance of clay fraction at the upper Calabar River stations while the estuary stations were predominated by silt/sand fraction. The TOC contents for the sediments ranged between 1.27 and 4.56%, with a maximum value recorded at station CR4 and a minimum value recorded at station CR8, while soluble organic matter (SOM) varied between 1,140 and 4,140 mg/kg dw with maximum and minimum values found at stations CR6 and CR9, respectively. There was no significant relationship between grain size distribution (silt/clay fraction) and TOC contents of sediments (correlation coefficient $r^2 = 0.2314$: p > 0.01). This result is at variance with documented data of other sedimentary environments which (in most cases) show strong relationship between these two variables (indicating greater ability of fine-grained particles to adsorb organic matter than coarse-grained particles; Unlu & Alpar, 2006; Oyo-Ita & Oyo-Ita, 2013).

3.2 Composition and Concentrations of Phthalates and Other Plastic Additives in Sediments and Blank

Three phthalate compounds were detected in surface sediments of the Cross River System with DEHP exhibiting the highest abundance in almost all the sampling stations. The concentrations of DEHP were in the range 1.97–86.76 mg/kg dw with an average of 24.06 ± 29.88, while those for DnBP ranged from 0.16 to 17.41 mg/kg dw with an average of 3.25 ± 5.03 and $1.14 - 29.64$ mg/kg dw with an average of 9.82 ± 10.23 were found for DiBP. Other associated plastic additives identified in sediments were 2,6-di(t-butyl)-4-hydroxybenzaldehyde (HBA), 2,6-di(t-butyl)quinine (Qn), tris-2,6-di(t-butyl)phenylphosphite (PhP) and tris-2,6-di(t-butyl)phenylphosphate (PhP'). On the other hand, variable levels of DEHP, di-octyladipate (DOA), PhP and trace amounts of n-alkanes and polycyclic aromatic hydrocarbons were found in the blank. The mass chromatogram for the phthalate compounds in sediments is shown in Figure 2a while the total ion current GC trace for n-hexane blank is presented in Figure 2b.

Blanks are required to determine the fraction of individual contaminants in sediment extracts that are derived from laboratory background. To determine whether the DEHP and PhP found in the sediments were indigenous or exclusively/partially introduced during laboratory protocol (clean-up stage), we computed extract/blank (E/B) ratios. E/B is the concentrations of individual target compounds in sediment extract relative to laboratory blank. E/B values < 20 were previously reported to be regarded conservatively as indicators for background contamination (Brocks, Buick, Logan, & Summons, 2003a; Brocks, Grosjean, Logan, & Graham, 2008). In our case study, we estimated E/B ratios for DEHP and PhP as these were the only target compounds detected in both the sediments and blank, and values < 20 were recorded (Table 1). The low E/B_{DEHP} and E/B_{PhP} values indicate that DEHP and PhP detected in the sediments were not exclusively derived from analytical procedure and that some levels of these compounds were indigenous. As DEHP is one of the most commonly used plasticizers, its relative abundances in the blank and sediments are expectedly high. Since DEHP and PhP in the blank may interfere with those in the sediments, we subtracted their concentrations in the sediments from those in the blank. The corrected DEHP and PhP concentrations in the sediments varied as presented above.

Comparing data, higher concentrations of DEHP were found in levels exceeding 90 mg/kg dw in sediments from lake Mead, a national park in US whereas lower levels of DEHP (25 mg/kg dw) were recorded in Ocean sediments at sewage outfall point, New Jersey, USA (Staples et al., 1997).

3.3 Identification and Sources of Phthalates and Other Plastic Additives

The three phthalate compounds were identified on the basis of their retention times window compared with authentic standards, monitored using key fragment ion m/z 149 (Figure 2a), while identification of the plastic additives was done by interpretation of mass spectra fragmentation patterns. A compound detected in some sediment samples (not in the blank) with molecular ion m/z 234 and a major fragment ion at m/z 219 was identified as 2,6-di(t-butyl)-4-hydroxybenzalaldehyde (HBA). This compound appears as underivatized parent compound in the mass spectrum (Figure 3a) as silylation is inhibited by the steric hindrance of tert-butyl groups in the ortho positions to the aldehyde or hydroxyl group. The compound and its oxidized form (Figure 3b) having a molecular ion at m/z 220 and a major fragment ion at m/z 177, identified as 2,6-di(t-butyl)quinine are considered here as diagenetic products of butylatedhydroxylbenzene (BHB) based anti-oxidants used as additives in a wide range of products including petroleum based lubricants and plastics (Grosjean & Logan, 2007).

Table 1. Characteristic features of the environment, bulk geochemical parameters and extract/blank ratios of sediments from the Cross River System

ZONE	I				II			III			IV		
Sample code	CR1	CR2	CR3	CR4	CR5	CR6	CR7	CR8	C-9	CR10	CR11	CR12	CR13
Coordinate	N4°	N4°	N4°	N4°	N4°	N5°	N5°	N5°	N5°	N5°	N5°	N5°	N5°
	43.96'	46.53'	49.92'	52.67	56.87'	00.43'	04.31'	04.31'	12.72'	12.25'	18.05'	18.15'	18.22'
	E8°	E8°	E8°	E8°'	E8°	E8°	E8°	E8°	E8°	E8°	E8°	E8°	E8°
	21.32'	18.90'	15.50'	12.74'	09.33'	07.06'	06.25'	06.25'	03.49'	00.22'	23.16'	23.18'	23.21'
Location name	Oron beach area				Oku Iboku beach area			Itu beach area			Upper Calabar/Great Kwa Rivers area		
Characteristic features of the environment.	Untreated sewage, agricultural waste, oil spills, ship/boat pollution, garbage/refuse discharge.				Paper mill untreated waste effluents and garbage/refuse discharge.			Boat traffic Pollution, domestic sewage, agricultural waste, garbage/refuse discharge			Rubber processing/plastic industry discharge and agricultural waste, garbage/refuse discharge.		
Sediment texture	Silty	silty	Sandy	Sandy	Sandy	Silty	Silty	Sandy	Sandy	Silty	Clayey	clayey	Clayey
TOC(%)	4.03	3.64	4.35	4.56	4.2	4.38	2.77	1.27	2.66	4.26	3.78	3.95	4.08
SOM (mg/kg dry wt.)	3,000	3,680	1,920	2,950	3,710	4,140	2,650	1,510	1,140	1,850	2,376	2,124	3,468
E/B$_{(DEHP)}$	9.95	4.21	4.94	1.65	3.16	1.05	3.17	1.61	1.65	1.42	10.23	0.89	12.54
E/B(PhP)	5.76	3.23	4.98	2.86	3.98	1.82	2.95	1.56	2.05	2.74	5.23	1.21	1.75

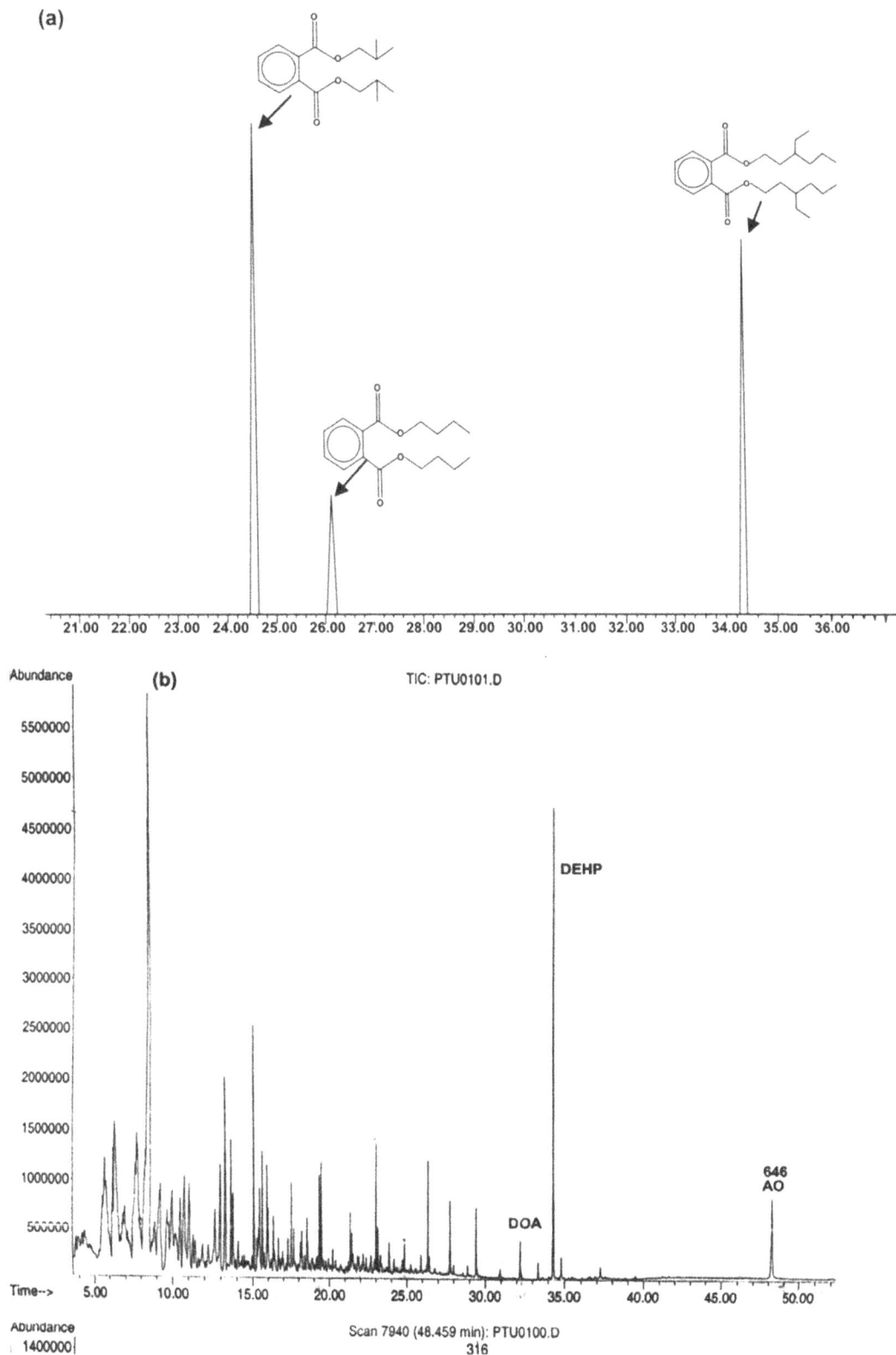

Figure 2. (a) Mass chromatogram of phthalate compounds monitored with m/z 149 key fragment ion detected in sediments of the cross river system. and (b) total ion current GC trace for the hexane blank

(a)

(b)

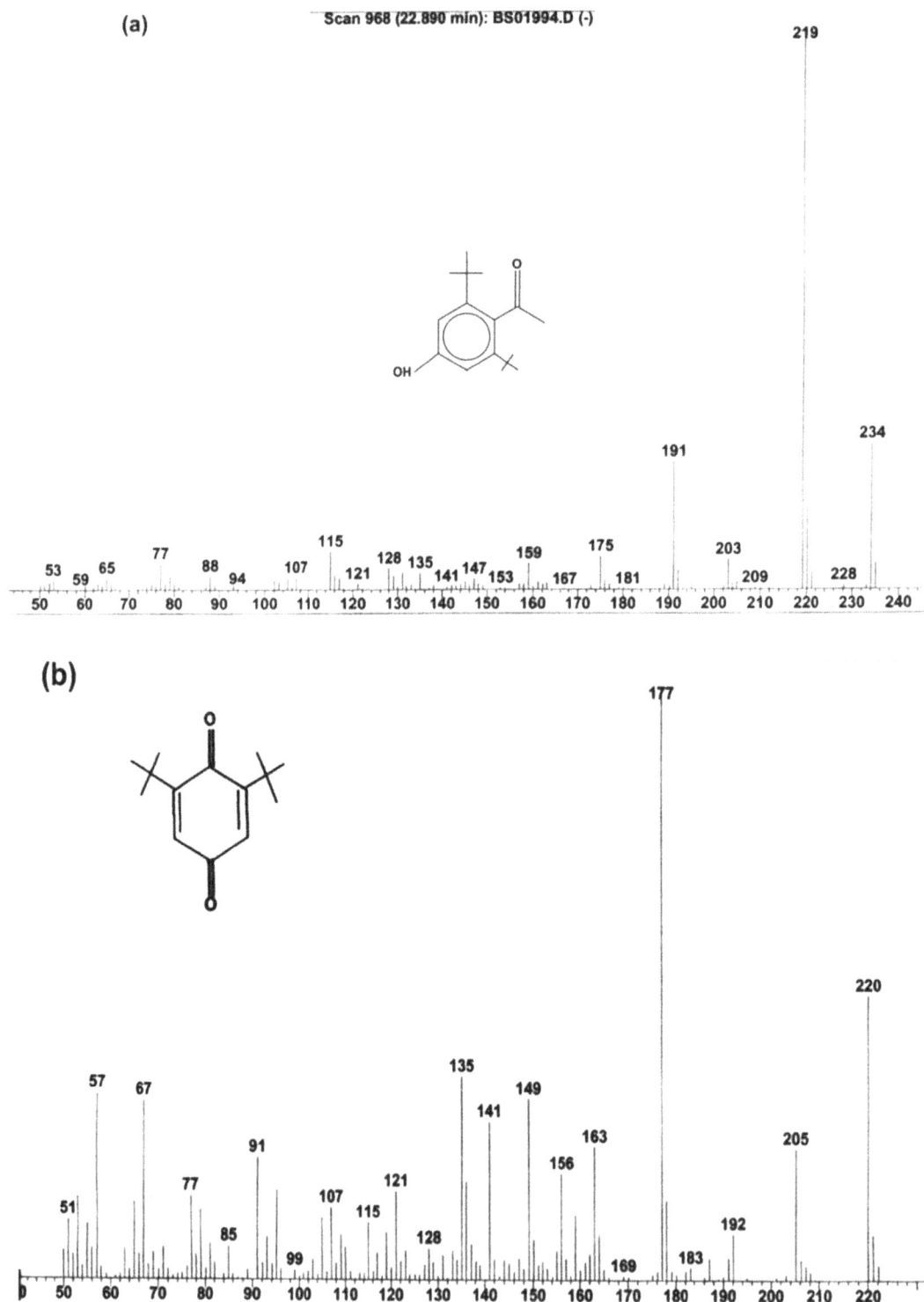

Figure 3. Mass spectra of (a) 2,6-di(t-butyl)-4-hydroxybenzaldehyde and (b) 2,6-di(t-butyl)quinone detected in sediments of the cross river system

An anti-oxidant compound (detected in both sediments and blank), with a weak molecular ion at m/z 646, eliminates a di(t-butyl)phenoxy moiety on fragmentation in the ion chamber to give the base peak at m/z 316 and minor fragment ions at m/z, 367, 291, 191, 147 and 57, was identified as tris-2,6-di(t-butyl)phenylphosphite (PhP; Figure 4a). A related compound with a base peak at m/z 647, key fragment ion at m/z 316 and a molecular ion at m/z 662, identified as tris-2,6-di(t-butyl)phenylphosphate (PhP'; Figure 4b), is analogous to tris(2,4-di-butyl)phenylphosphate (a compound previously identified in plastic combustion products (Simoneit et al., 2005). In this study, PhP' is proposed as a marker for plastic combustion for the region/air basin. Therefore, future measurement of the proposed plastic combustion marker in the Nigerian air-basin should be a good assessment tool for the contribution of refuse combustion to air-borne contamination and should provide further information to improve emission inventories and air quality management.

The detection of PhP in the blank suggests contact of the blank with plastic material probably during clean-up procedure while its presence in sediments at higher levels most likely indicates additional input arising from leeching of plastic materials to the sedimentary environment. However, the detection of PhP' in the sediments (not in the blank) implies inpit from atmospheric deposition during or after burning of refuse containing plastic materials. It appears that during this process, the volatilized PhP became oxidized in the atmosphere at high temperature to PhP'. Once in the sediments, the marker compound underwent degradation and may be responsible for the occurrence of HBA and Qn in some sediment samples of the study area. This scenario indicates that sediments of the Cross River System were influenced by phthalates and plastic additives contamination before sample collection, storage and contact with laboratory artifacts.

Primarily, phthalate compounds in the river system was considered to originate from direct discharge of untreated effluent/solid waste produced by the plastic and rubber processing industry located about 5 km north-east of the study area. This deduction is based on the detection of higher phthalate levels in sediments from the Upper Calabar River (zone IV).

3.4 Spatial Distributions of Phthalates and PhP' in Sediments

The low molecular weight phthalate components (DnBP and DiBP) occurring at relatively reduced levels in all the sampling stations (except CR6) may be linked to their relatively high volatility, and being more soluble, might have been retained in the aqueous phase more than being adsorbed in the solid matrix (Figure 5). Experimental data on physicochemical properties of phthalates such as partitioning behavior and vapor pressure show an eight order of magnitude in octanol-water partition coefficient, and a four order of magnitude in vapor pressure decreases as alkyl chain length increases from 1 to 13 (Staples et al., 1997). These data support the observed low levels of low molecular weight phthalates in the sediments. The implication here is that the low molecular weight phthalates (DnBP and DiBP) might not have been strongly partitioned on the solid matrices compared to the high molecular weight homolog (DEHP), and could become easily solubilized and volatilized.

In addition, the spatial variation in the concentrations of these phthalate compounds may be partly a reflection of the differences in sediment grain size distribution (Table 1; Figure 5), proximity of samples to contamination source and the effect of sediment re-suspension at the river mouth arising from turbulent mixing. This scenario suggests that long range regional eolian or river current transport may not be an important factor that determines the distribution of these compounds in the system and that localized contamination was more significant.

Additives are released to the environment from plastics by leaching, contact interference and direct volatilization into smoke during burning of refuse containing of plastic materials (Braun, 2004). The variation in spatial distribution of the marker compound (PhP') in the study area is shown in Figure.5. Among the sampling stations, those of Zone IV show much higher levels of PhP', supporting the fact that long range eolian transport or river water current trajectory was not important in the distribution of the marker compound. Besides the strong affinity of PhP' towards sedimentary organic carbon predominated by clay textural fraction, wet precipitation of air contaminant load following plastic combustion emission as well as differences in refuse disposal pattern among inhabitants of the catchments might have played a significant role.

3.5 Environmental Implication

Although phthalates are lipophilic and of low volatility, experimental evidence shows that these compounds are not persistent and are rapidly photo-chemically and biologically degraded in the environment (Shanker, Ramakrishna, & Seith, 1985; Schnitzer, Scheunert, & Korte, 1988; Barron, Albro, & Hayton, 1995; Adeniyi, Dayomi, & Okedeyi, 2008). The sediment samples in the present report were collected about 10 years ago, and it is worthy of note that analysis of lipids in recently obtained sediment samples from the same sampling area (reported elsewhere) shows absence or trace levels of phthalates. This observation may be linked to the rapid photochemical and biological degradation of phthalates in the sediments. In addition, the non-operation of the existing plastic industry in the vicinity of the upper Calabar River for the past five years might have also accounted for this absence.

(a)

(b)

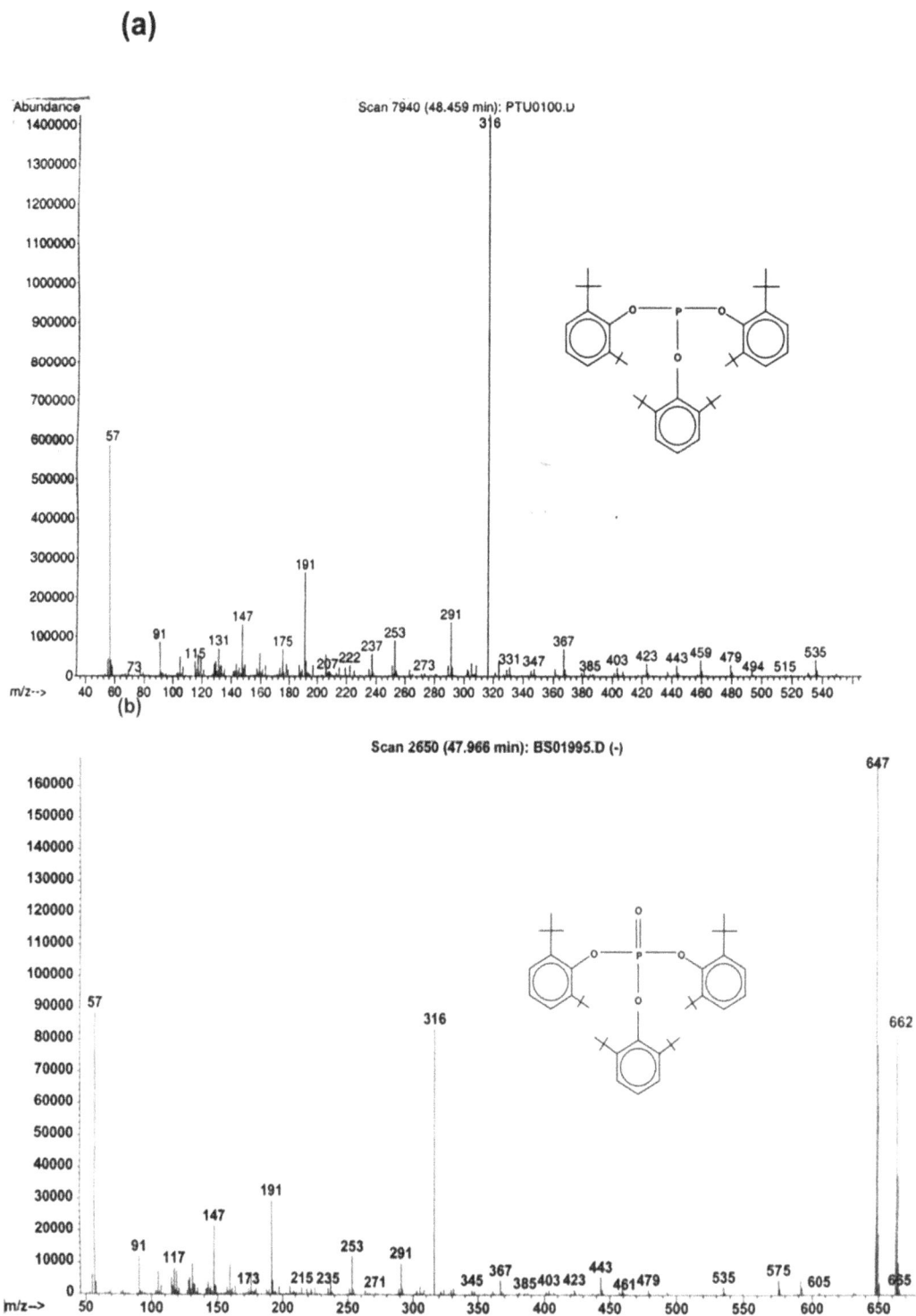

Figure 4. Mass spectra of (a) tris(2,6-di(t-butyl)phenylphosphite detected in sediments and blank and (b) tris(2,6-di(t-butyl)phenylphosphate detected in sediments of the cross river system

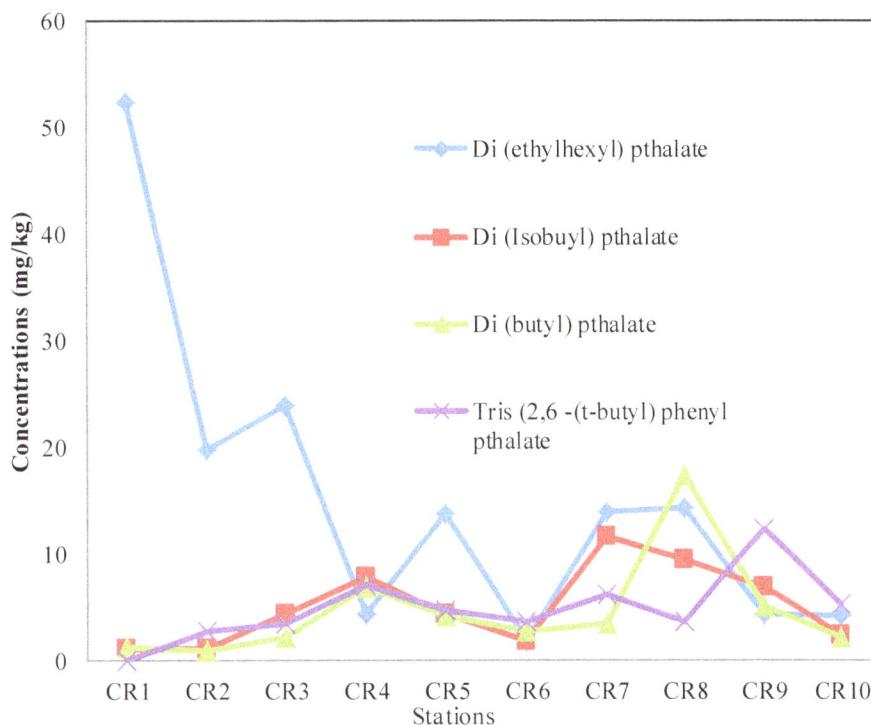

Figure 5. Spatial variation in the concentrations of Phthalate and other plastic additives amongst sampling stations of the cross river system

Staples et al. (1997) show that under aerobic and anaerobic conditions, phthalates in sewage sludge undergo > 50% ultimate biodegradation within 28 days. According to these authors, these phthalates may be used by aerobic and anaerobic microbes as a source of carbon and energy. Numerous experimental data have shown that the bioaccumulation of phthalates in the aquatic and terrestrial food chains is limited by their biotransformation which increases with the increasing trophic level (Staples et al., 1997). Therefore, the relatively high levels of phthalates found in sediments of the Cross River System might not have posed any serious health threat to both resident organisms and humans. The compounds might have been rapidly degraded in the sediments and metabolized by aquatic organisms to less harmful compounds.

4. Conclusions

The present study has demonstrated the efficacy of extract/blank (E/B) ratio as indicator of background contamination in environmental samples and indicates that DEHP and PhP found in sediments of the Cross River System were not exclusively derived from laboratory artifacts, and that some levels of these compounds in the sediments were indigenous. The detection of certain anti-oxidant degradation products indicates that the sediments were influenced by phthalates contamination before sample collection, storage and contact with laboratory artifacts.

The occurrence of phthalates and PhP' in the sediments are considered to originate primarily from direct discharge of untreated effluent/solid waste by the rubber processing/plastic industry and emissions arising from burning of refuse containing plastic materials, respectively. Long range eolian transport or river water current trajectory did not play a significant role in the distribution of these compounds in the sediments, rather wet precipitation and localized influx were more important. Therefore, future measurement of the proposed plastic combustion marker in the Nigerian air-basin should be a good assessment tool for the contribution of refuse combustion to air-borne contamination and should provide further information to improve emission inventories and air quality management for policy making strategy.

The levels of phthalates contamination found in sediments of the river system, though relatively high, might not have posed serious health threat to resident organisms and subsequently humans via food chain as these compounds are readily degraded in the sediment and metabolized in aquatic organisms to less harmful products.

Acknowledgement

This study would not have been completed without the assistance of Emeritus Professor Bernd R. T. Simoneit and the corporation of technical staff and management of the College of Oceanic and Atmospheric Science, Oregon State University, Corvalis, USA. The GC-MS of our fractions were run at no cost. The GC-FID analysis of the recently obtained sediment samples was made possible by Professor JosepBayona of the Institute of Environmental Assessment and Water Research, Spanish Council of Scientific Research (CSIC). Barcelona, Spain.

References

Adeniyi, A., Dayomi, M., & Okedeyi, O. (2008). An assessment of the levels of phthalate esters andmetals in the Muledane open dump Thohoyanou, Limpopo Province, South Africa. *Chemistry Central Journal, 2*, 1-9. http://dx.doi.org/10.1186/1752-153X-2-9

Aranda, J. N., O'Connor, G., & Ekeman, G. (1989). Effects of sewage slude on di(-2-ethylhexyl)phthalate uptake by plants. *Environmental Quality, 18*, 40-45. http://dx.doi.org/10.2134/jeq1989.00472425001800010008x

Ashby, J., Brady, A., Ficomber, G., Elliot, B., Isreal, J., Olum, J., & Purchase, I. (1994). Mechanically based human hazard assessment of previxisome proliferation induced heptocarlinogens. *Human & Environmental Toxicology, 3*, 52-61.

Asuquo, E. A., Ogri, O. R., & Bassey, E. S. (1999). Distribution of heavy metals and total hydrocarbonsin coastal waters and sediments of Cross River System, S. E. Nigeria. *International Journal of Tropical Environment, 2*, 229-242.

Barron, M. G., Albro, P. W., & Hayton, W. L. (1995). Bioitransformation of di(2-ethylhexyl)phthalate byRainbow Trout. *Environmental Toxicology & Chemistry, 14*(5), 873-876. http://dx.doi.org/10.1002/etc.5620140519

Braun, D. (2004). Poly(vinyl chloride) on the way from 19th century to the 21st century. *Polymer Science, 42*, 578-586.

Brocks, J. J., Buick, R., Logan, G. A., & Summons, R. E. (2003a). Composition of Syngeneity of molecularfossils from the 2.78-2.45 billion year old Mount Bruce Supergroup, Pilbara Cration, Western Australia. *Geochimicaet Cosmochimica Acta, 67*, 4289-4319. http://dx.doi.org/10.1016/S0016-7037(03)00208-4

Brocks, J. J., Grosjean, E., Logan, R., & Graham, A. (2008). Assessing biomarker syngeneity usingbranchrd alkanes with quaternary carbon (BAQCs) and other plastic contaminants. *Geochimicaet Cosmochimica Acta, 72*, 871-888. http://dx.doi:10.1016/j.gca.2007.11.028

Ekpo, B. O., Oyo-Ita, O. E., Oros, D. H., & Simoneit, B. R. T. (2011). Organic Contaminants in solvents and implication for geochemistry and environmental forensics: an example from local vendors in Ngeria. *Environmental Forensics, 13*, 1-6. http://dx.doi.org/10.1080/15275922.2011.643339

Folk, R. L. (1974). *Petrology of sedimentary rock*. Texas, Hemphill publishing company.

Grimalt, J. O., Van-Drooge, B. L., Ribes, A., Vilanova, R. M., Fernandez, P., & Appleby, P. (2004). Persistent organochlorine compounds in soils and sediments of European High Altitude Mountain lakes. *Chemosphere, 54*, 1549-1561. http://dx.doi:10.1016/j.chemosphere.2003.09.047

Grosjean, E., & Logan, G. A. (2007). Incorporation of organic contaminants into geochemical samples and an assessment of potential sources: examples from Geosciences Australia marinesurveyS282. *Organic Geochimistry, 38*, 583-591. http://dx.doi.org/10.1016/j.orggeochem.2006.12.013

Jim-Carlisle, D. V., Henkel, S., Ling-Hongly, E., & Pagepainter, D. (2009). Toxicological profile for di 2-ethylhexyl)phthalate (DEHP). *Environmental Health Hazard Assesssment, 23*, 41-50.

Lwenberg, U. H., & Knzel, T. H. (1992). Investigations on the hydrology of the lower Cross River, Nigeria. *Anim. Research and Development, 35*, 72-85.

Meeker, J. D., Sheela, S. Y., & Shama, H. (2009). Phthalates and other additives in plastics-human exposure and associated health outcomes. *Phylosophical Transformation of the Royal Society, 364*, 2097-2113. http://dx.doi.org/10.1098/rstb.2008.0268

National Institute of Public Health and Environmental Protection (NIPHEP). (1991). *Update of Exploratory Report: Phthalates Report, 7*, 1008-1040.

Oyo-Ita, O. E., & Oyo-Ita, I. O. (2013). PAHs depositional history and sources in recent sediment core from the Ukwalbom Lake, SE Nigeria. *Environmental Geochemistry & Health, 35*, 180-199. http://dx.doi.org/10.1007/S10653-012-9475-X

Pisani, O., Oros, D. H., Oyo-Ita, O. E., Ekpo, B. O., Jaffe, R., & Simoneit, B. R. T. (2013). Biomarkers in surface sediments of Cross river and estuarine system, SE Nigeria: assessment of organic matter sources of natural and anthropogenic origins. *Applied Geochemistry, 31*, 239-250. http://dx.doi.org/10.1016/j.apgeochem.2013.01.010.

Schnitzer, J., Scheunert, L., & Korte, F. (1988). Fate of bis(2-ethylhexyl)phthalate in laboratory and out-door soil-plant system. *Agriculture & Food Chemistry, 36*, 210-215. http://dx.doi.org/10.1021/jf00079a053

Shanker, R., Ramakrishna, C., & Seith, P. (1985). Degradation of some phthalic acid esters in soils. *Environmental Pollution, 39*, 1-7. http://dx.doi.org/10.1016/0143-1471(85)90057-1

Simoneit, B. R. T., Standley, L. J., & Cox, R. E. (1988). Organic matter in the troposphere IV: lipids in harmattan aerosols of Nigeria. *Atmospheric Environment, 22*, 983-1004. http://dx.doi.org/10.1016/0004-6981(88)90276-4

Simoneit, B. R. T., & Mazurek, M. A. (1989). Organic tracers in ambient aerosols and rainwater. *Aerosol Science & Technology, 10*, 267-291. http://dx.doi.org/10.1080/02786828908959264

Simoneit, B. R. T., Medeiros, P. M., & Didyk, B. N. (2005). Combustion products of plastics asindicators of refuse burning in atmosphere. *Environmental Science & Technology, 10*, 267-291.

Staples, A., Charle S, Dennis, R., Peterson, B., Thompson, F., Parketon, A., & William, J. (1997). The environmental fate of phthalate esters: a literature review. *Chemosphere, 35*(4), 667-749. http://dx.doi.org/10.1016/S0045-6535(97)00195-1

United State Environmental Protection Agency (USEPA). (1992). *Characterization of municipal solid waste in the united states: Report No. EPA/530-s92.* Washington, DC, United State Environmental Protection Agency.

Unlu, S., & Alpar, B. (2006). Distribution and sources of hydrocarbons in surface sediments of Gemlik Bay (Marmara Sea, Turkey). *Chemosphere, 64*, 764-777. http://dx.doi:10.1016/j.chemosphere.2005.10.064

Permissions

List of Contributors

B. B. Amos
Department of Geography, Gombe State University, Gombe, Nigeria

I. Musa
Department of Geography, Gombe State University, Gombe, Nigeria

M. Abashiya
Department of Geography, Gombe State University, Gombe, Nigeria

I. B. Abaje
Department of Geography and Regional Planning, Federal University Dutsin-Ma, Katsina State, Nigeria

Haruna Adamu
Department of Chemistry, University of Aberdeen, UK
Department of Environmental Management Technology, Abubakar Tafawa Balewa University, Nigeria

Leke Luter
Department of Chemistry, University of Aberdeen, UK
Department of Chemistry, Benue State University, Nigeria

Mohammed Musa Lawan
Department of Chemistry, University of Aberdeen, UK
Department of Chemistry, Yobe State University, Damaturu, Nigeria

Bappah Adamu Umar
Department of Geology and Petroleum Geology, University of Aberdeen, UK
National Centre for Petroleum Research and Development, Energy Commission of Nigeria, Abubakar Tafawa Balewa University Research Centre, Nigeria

Yusri Yusup
Environmental Technology, School of Industrial Technology, Universiti Sains Malaysia, Penang, Malaysia

Mardiana Idayu Ahmad
Environmental Technology, School of Industrial Technology, Universiti Sains Malaysia, Penang, Malaysia

Norli Ismail
Environmental Technology, School of Industrial Technology, Universiti Sains Malaysia, Penang, Malaysia

Chun-You Zhu
College of Resources and Environment, University of Chinese Academy of Sciences, Beijing, P. R. China

Peng Bao
State Key Lab of Urban and Regional Ecology, Research Center for Eco-Environmental Sciences, Chinese Academy of Sciences, Beijing, P. R. China

Yu-Xin Ba
College of Resources and Environment, University of Chinese Academy of Sciences, Beijing, P. R. China

Jing Hua
College of Resources and Environment, University of Chinese Academy of Sciences, Beijing, P. R. China

Xiao-Ning Liu
College of Resources and Environment, University of Chinese Academy of Sciences, Beijing, P. R. China

Guo-Hua Hou
State Key Lab of Urban and Regional Ecology, Research Center for Eco-Environmental Sciences, Chinese Academy of Sciences, Beijing, P. R. China

Chun-Zao Liu
State Key Lab of Urban and Regional Ecology, Research Center for Eco-Environmental Sciences, Chinese Academy of Sciences, Beijing, P. R. China

Zheng-Yi Hu
College of Resources and Environment, University of Chinese Academy of Sciences, Beijing, P. R. China

Ikama E. Uwah
Department of Pure & Applied Chemistry, University of Calabar, Calabar, Nigeria

Solomon F. Dan
Department of Marine Biology, Faculty of ocean Science and Technology, Akwa Ibom State University Ikot Akpaden, Nigeria

Rebecca A. Etiuma
Department of Pure & Applied Chemistry, University of Calabar, Calabar, Nigeria

Unyime E. Umoh
Department of Pure & Applied Chemistry, University of Calabar, Calabar, Nigeria

Carlos Montalvo
Universidad Autónoma del Carmen, Dependencia de Ciencias Químicas y Petrolera, México

Claudia A. Aguilar
Universidad Autónoma del Carmen, Dependencia de Ciencias Químicas y Petrolera, México

Luis E. Amador
Universidad Autónoma del Carmen, Centro de Investigación de Ciencias Ambientales, México

Julia G. Cerón
Universidad Autónoma del Carmen, Dependencia de Ciencias Químicas y Petrolera, México

Rosa M. Cerón
Universidad Autónoma del Carmen, Dependencia de Ciencias Químicas y Petrolera, México

Francisco Anguebes
Universidad Autónoma del Carmen, Dependencia de Ciencias Químicas y Petrolera, Méxi

Atl V. Cordova
Universidad Autónoma del Carmen, Dependencia de Ciencias Químicas y Petrolera, México

Bernadette Quémerais
School of Medicine, University of Alberta, Edmonton, Canada

Haruki Shimazu
School of Science and Engineering, Kinki University, Osaka, Japan

Rodrick Hamvumba
Department of Plant Sciences, School of Agricultural Sciences, University of Zambia, Zambia

Mebelo Mataa
Department of Plant Sciences, School of Agricultural Sciences, University of Zambia, Zambia

Alice Mutiti Mweetwa
Department of Soil Sciences, School of Agricultural Sciences, University of Zambia, Zambia

Rémi Labelle
Université de Montréal. Département de Santé Environnementale et de Santé au Travail, Montréal, Québec, Canada

Allan Brand
Institut National de Santé Publique du Québec (INSPQ), Canada

Stéphane Buteau
Institut National de Santé Publique du Québec (INSPQ), Canada
Department of Medecine, McGill University, Montréal, Québec, Canada

Audrey Smargiassi
Institut National de Santé Publique du Québec (INSPQ), Canada
Chaire sur la pollution de l'air, les changements climatiques et la santé, en partenariat avec la Direction de Santé publique de l'Agence de la santé et des services sociaux de Montréal et l'INSPQ, Université de Montréal, Montréal, Québec, Canada

Emad A. Almuhanna
College of Agricultural Scneces and Food, King Faisal University, Hofuf, Saudi Arabia

Inkyu Han
Division of Epidemiology, Human Genetics, and Environmental Sciences, University of Texas Health Science Center at Houston School of Public Health, Houston, TX, USA
Southwest Center for Occupational and Environmental Health, University of Texas Health Science Center at Houston, Houston, TX, USA

Je-Seung Lee
Seoul Metropolitan Government Research Institute of Public Health and Environment, Seoul, Korea

Soo-Mi Eo
Seoul Metropolitan Government Research Institute of Public Health and Environment, Seoul, Korea

Guangwei Huang
Graduate School of Global Environmental Studies, Sophia University, Tokyo, Japan

Deshar Bashu Dev
Environmental Economic System, Rissho University, Tokyo, Japan

Rupak Aryal
Centre for Water Management and Reuse, School of Natural and Built Environments, University of South Australia, Mawson Lakes, SA, Australia

Aeri Kim
Department of Civil and Environmental Engineering, University of Ulsan, Ulsan, Republic of Korea

Byeong-Kyu Lee
Department of Civil and Environmental Engineering, University of Ulsan, Ulsan, Republic of Korea

Mohammad Kamruzzaman
Centre for Water Management and Reuse, School of Natural and Built Environments, University of South Australia, Mawson Lakes, SA, Australia

Simon Beecham
Centre for Water Management and Reuse, School of Natural and Built Environments, University of South Australia, Mawson Lakes, SA, Australia

Sophie Goudreau
Direction de santé publique de Montréal, Montréal, Canada

Céline Plante
Direction de santé publique de Montréal, Montréal, Canada

Michel Fournier
Direction de santé publique de Montréal, Montréal, Canada

Allan Brand
Institut National de Santé Publique du Québec, Montréal, Canada

Yann Roche
Département de Géographie, Université du Québec à Montréal, Montréal, Canada

Audrey Smargiassi
Institut National de Santé Publique du Québec, Montréal, Canada
Département de santé environnementale et de santé au travail, Université de Montréal, Montréal, Canada

François Destandau
Laboratory of Territorial Water and Environment Management (GESTE) UMR ENGEES-IRSTEA, Strasbourg Cedex, France

Laura Margaret Dale
McGill School of Environment, McGill University, Montréal, Canada

Maximilien Debia
Département de santé environnementale et de santé au travail, Université de Montréal, Montréal, Canada

Olivier Christian Mudaheranwa
Département de santé environnementale et de santé au travail, Université de Montréal, Montréal, Canada

Céline Plante
Direction de la santé publique de l'Agence de la santé et des services sociaux de Montréal, Montréal, Canada

Audrey Smargiassi
Département de santé environnementale et de santé au travail, Université de Montréal, Montréal, Canada
Institut National de Santé Publique du Québec, Montréal, Canada

Scott Clark
Department of Environmental Health, University of Cincinnati, USA

William Menrath
Department of Environmental Health, University of Cincinnati, USA

Yehia Zakaria
Department of Occupational and Environmental Health, Cairo University, Cairo, Egypt

Amal El-Safty
Department of Occupational and Environmental Health, Cairo University, Cairo, Egypt

Sandy M Roda
Department of Environmental Health, University of Cincinnati, USA

Caroline Lind
Department of Environmental Health, University of Cincinnati, USA

Essam Elsayed
Department of Occupational and Environmental Health, Cairo University, Cairo, Egypt

Hongying Peng
Department of Environmental Health, University of Cincinnati, USA

Orok E. Oyo-Ita
Environmental and Petroleum Geochemistry Research Group, Department of Pure and Applied Chemistry, University of Calabar, C.R. State, Nigeria

Bassey O. Ekpo
Environmental and Petroleum Geochemistry Research Group, Department of Pure and Applied Chemistry, University of Calabar, C.R. State, Nigeria

Inyang O. Oyo-Ita
Environmental and Petroleum Geochemistry Research Group, Department of Pure and Applied Chemistry, University of Calabar, C.R. State, Nigeria

John O. Offem
Environmental and Petroleum Geochemistry Research Group, Department of Pure and Applied Chemistry, University of Calabar, C.R. State, Nigeria